T0262253

Deciphering Spacecraft Systems and Orbital Calculations

Deciphering Spacecraft Systems and Orbital Calculations

Edited by **Jean Tabor**

LANRYE
INTERNATIONAL

New Jersey

Published by Clanrye International,
55 Van Reypen Street,
Jersey City, NJ 07306, USA
www.clanryeinternational.com

Deciphering Spacecraft Systems and Orbital Calculations
Edited by Jean Tabor

© 2015 Clanrye International

International Standard Book Number: 978-1-63240-135-9 (Hardback)

Contents

Preface

This book focuses on the limitations of existing technology and the creation of new technologies which are utilized for space missions. Various experts have contributed in the attempt to create a connection between current limitations and advancing technology to facilitate the field's progress. It consists of information regarding methods to calculate the orbit which is based on researches in the field of astrophysics. This book also gives a comprehensive summary of spacecraft systems inclusive of the reliability of inexpensive AOCs, sliding mode controlling, and a novel view on attitude controller design based on sliding mode, with thrusters. Technological instructions for the optimization of HVAC are also discussed. Finally, this book provides a complete analysis of the attempt to resolve problems regarding space missions with the correlation of current technologies and novel developments. Hence, this book intends to give a comprehensive synopsis of the growth of technology in spacecraft systems.

The information contained in this book is the result of intensive hard work done by researchers in this field. All due efforts have been made to make this book serve as a complete guiding source for students and researchers. The topics in this book have been comprehensively explained to help readers understand the growing trends in the field.

I would like to thank the entire group of writers who made sincere efforts in this book and my family who supported me in my efforts of working on this book. I take this opportunity to thank all those who have been a guiding force throughout my life.

Editor

Part 1

Orbit Determination and Spacecraft Systems

Physics-Based Control Methods

T. A. Sands
Columbia University,
USA

1. Introduction

Spacecraft control suffers from inter-axis coupling regardless of control methodology due to the physics that dominate their motion. Feedback control is used to robustly reject disturbances, but is complicated by this coupling. Other sources of disturbances include zero-virtual references associated with cascaded control loop topology, back-emf associate with inner loop electronics, poorly modeled or un-modeled dynamics, and external disturbances (e.g. magnetic, aerodynamic, etc.). As pointing requirements have become more stringent to accomplish missions in space, decoupling dynamic disturbance torques is an attractive solution provided by the physics-based control design methodology. Promising approaches include elimination of virtual-zero references, manipulated input decoupling, which can be augmented with disturbance input decoupling supported by sensor replacement. This chapter introduces these methods of physics-based control. Physics based control is a method that seeks to significantly incorporate the dominant physics of the problem to be controlled into the control design. Some components of the methods include elimination of zero-virtual reference, observers for sensor replacements, manipulated input decoupling, and disturbance-input estimation and decoupling. In addition, it will be shown that cross-axis coupling inherent in the governing dynamics can be eliminated by decoupling a normal part of the physics-based control. Physics-based controls methods produce a idealized feedforward control based on the system dynamics that is easily augmented with adaptive techniques to both improve performance and assist on-orbit system identification.

2. Physics-based controls

2.1 Zero-virtual references

Zero-virtual references are implicit with cascaded control loops. When inner loops reference signals are not designed otherwise, the cascaded topology results in zero-references, where the inner loop states are naturally zero-seeking. It is generally understood that if any control system demands a positive or negative rate, the inner position loop (seeking zero) would essentially be fighting the rate loop, since a positive or negative rate command with quiescent initial conditions dictates non-zero position command. Elimination of the zero-virtual reference may be accomplished by using analytic expressions for both position and rate eliminating the nested, cascaded topology. Using analytic expressions for both position and rate commands implies the utilization of commands that both correspond to achieving the same desired end state, essentially eliminating the conflict between the position and rate commands inherit in the cascaded topology.

2.2 Manipulated Input Decoupling (MID)

Manipulated input is the actual variable that can be modified by a control design. Very often in academic settings control u is the goal of a design, but in reality a voltage command is sent to a control actuator, and this voltage command should be referred to as the true manipulated input. The importance of this distinction lies in the fact that electronics may not properly replicated the desired control u, unless the control designer has accounted for internal disturbance factors like the resistive effects of back-emf (inherit in any electronic device where current is generated and modified in the presence of a magnetic field). The manipulated input signal should be designed to decouple these effects.

2.3 Sensor replacement

Due to simplicity of the approach, observer-based augmentation of motion control systems is becoming a ubiquitous method to increase system performance [4], [8], [11], [12]. The use of observers also permits (in some cases) elimination of hardware associated with sensors, or alternatively may be used as a redundant method to obtain state feedback. Velocity sensors may be eliminated using speed observers based on position measurement without. Estimation methods such as Gopinath-styled observers and Luenberger-styled Observers are robust to parameter variation and sensor noise. Both position and velocity estimates may be used for state feedback eliminating the effects of sensor noise on the state feedback controller. Luenberger-styled and Gopinath-styled observer topologies will be compared. Luenberger-styled observers (henceforth simply referred to as Luenberger observers) are a simple method to estimate velocity given position measurements that will prove superior to Gopinath-styled observers (which remain a viable candidate for sensor replacement). Additionally, the Luenberger observer may be used to provide estimates of external system disturbances, since the observer mimics the order of the actual systems dynamic equations of motion. When used the Luenberger disturbance observer bestows robustness to system parameter variations.

Often used terminology from current literature [11], [12] is maintained in here where the modification of the signal chosen as the disturbance estimate establishes a "modified" Luenberger observer. The modified Luenberger observer as referred in the cited literature is clearly superior (with respect to disturbance estimation) to the nominal Luenberger observer, so it is assumed to be the baseline Luenberger observer for disturbance estimation. Recent efforts [12], [14] seeking to improve estimation performance augments the architecture with a second, identical Luenberger observer. The two observers are tuned to estimate velocity and external disturbances respectively. The approach improves estimation accuracy and system performance, but still suffers from estimation lag, motivating these more recent improved methods eliminating estimation lag. Methods to improve estimation performance will be presented. Together with estimation improvement, motion control will be enhanced with disturbance input decoupling (which also aids estimation performance).

2.4 Disturbance Input Decoupling (DID)

Augmentation of speed observers with a command feedforward path permits near-zero lag estimation, even in a single-observer topology. Elimination of estimation lag improves estimation accuracy which subsequently improves the performance of the motion controller. Augmentation of the motion controller with disturbance input decoupling extends the bandwidth of nearly-zero lag estimation considerably again even in a single-observer topology. The estimates from the observer are frequently used for state feedback eliminating the requirements for both velocity sensors and position measurement smoothing. Adding command feedforward to the observer establishes nearly-zero lag estimation with good accuracy. Furthermore, augmenting the motion controller with disturbance input decoupling improves motion control.

2.5 Idealized feedforward control based on predominant physics

Decoupling the cross-motion motivates an idealized feedforward control. Section 7 of this chapter will introduce a feedforward control for accomplishing commanded trajectories that is designed using the predominant physics and decouples the particular solution to the differential equation of motion that results from the commanded trajectory.

2.5.1 Cross-axis motion decoupling

Newton's Law is commonly known: the sum of forces acting on a body is proportional to its resultant acceleration, and the constant of proportionality is the body's mass. This applies to all three axes of motion for 3-dimensional space, so the law can also be stated as "the summed force vector [3x1] acting on a body is proportional to its resultant acceleration vector [3x1], and the constant of proportionality is the body's mass matrix [3x3]". One crucial point is that this basic law of physics applies only in an inertial frame that is not in motion itself. A similar law may be stated for rotational motion just as we have stated Newton's Law for translational motion. The rotational motion law is often referred to as Newton-Euler, and it may be paraphrased as: "the summed torque vector [3x1] acting on a body is proportional to its resultant angular acceleration vector [3x1], and the constant of proportionality is the body's mass inertia matrix [3x3]." Newton-Euler also only applies in a non-moving, inertial frame. The equations needed to express the spacecraft's rotational motion are valid relative to the inertial frame and may be expressed in inertia. The motion measurement relative to the inertial frame is taken from onboard sensors expressed in a body fixed frame. The resulting cross product that accounts for relative motion of the body frame contains the key cross coupled terms often casually referred to as "roll-yaw coupling" for example in the case of a spacecraft whose inertia matrix produced relatively pronounced coupling between the roll and yaw axes. Decoupling the cross-product nonlinearities eliminates undesired motion.

2.6 Reference trajectory

A reference trajectory is introduced in section 9.2 to improve performance still further. The main motivation is that a controller should recognize that the plant is not (cannot) instantaneously achieve the commanded trajectory. Time is required for motion to occur, so when it is desired to maneuver more rapidly, a reference trajectory may be used.

2.7 Adaptive control and system identification

Taken together, an idealized feedforward control (designed using the dynamics of the system) together with a classical feedback controller and a reference trajectory lead to the ability to introduce adaptive control schemes that can learn a spacecraft's new physical parameters and adjust the control signal to accommodate things like fuel sloshing and spacecraft damage.

3. Equations of motion

Newton's Law is commonly known: the sum of forces acting on a body is proportional to its resultant acceleration, and the constant of proportionality is the body's mass. This applies to all three axes of motion for 3-dimensional space, so the law can also be stated as "the summed force vector [3x1] acting on a body is proportional to its resultant acceleration vector [3x1], and the constant of proportionality is the body's mass matrix [3x3]". One crucial point is that this basic law of physics applies only in an inertial frame that is not in motion itself. A similar law may be stated for rotational motion just as we have stated Newton's Law for translational motion.

The rotational motion law is often referred to as Newton-Euler, and it may be paraphrased as: "the summed *torque* vector [3x1] acting on a body is proportional to its resultant *angular* acceleration vector [3x1], and the constant of proportionality is the body's mass *inertia* matrix [3x3]." Newton-Euler also only applies in a non-moving, inertial frame. The equations needed to express the spacecraft's rotational motion are valid *relative to* the inertial frame (indicated by subscript "B/i" often assumed) and may be *expressed* in inertia. The motion measurement *relative to* the inertial frame is taken from onboard sensors *expressed in* a body fixed frame

$$\dot{\vec{H}} = \left\{\frac{d\vec{H}}{dt}\right\}_i = \left\{\frac{d\vec{H}}{dt}\right\}_B + \{\vec{\omega}\}^{B/i} \times \{\vec{H}\}_B \text{ where } \{\vec{H}\}_B = [J] \cdot \{\vec{\omega}\}^{B/i} \tag{1}$$

$$\Sigma\{\vec{T}\}_{B/i} \rightarrow \{\tau\}_B = \{\dot{H}\} = [J]\{\dot{\omega}\} + \{\omega\} \times [J]\{\omega\} \tag{2}$$

Note the cross product that accounts for relative motion of the body frame contains the key cross coupled terms often casually referred to as "roll-yaw coupling" for example in the case of a spacecraft whose inertia matrix produced relatively pronounced coupling between the roll and yaw axes. Decoupling the cross-product nonlinearities eliminates undesired motion and makes the equation more similar to the basic Newton's Law in an inertial reference frame. Decoupling the cross-product may be done in feedforward or feedback fashion, but should account for both the homogenous solution to the governing differential equation (response to initial conditions) and also the particular solution (response to the command input). Well-behaved, decoupled dynamics would behave with simple double-integrator dynamics, so the mathematical expressions of force dynamics and torque dynamics would look similar.

4. Virtual-zero references and mid

Spacecraft torque-actuators contain electronic that often contain other force or torque motors. Control moment gyroscopes for example are said to exhibit "torque magnification" since a small amount of torque applied to the gimbal motor produces a resultant large spacecraft torque. Motors associated with electronics are cascaded inner-loops, and they are often paid less attention in the control design [9], [11]. Such cascaded inner loops often reduce the overall system bandwidth due to zero-virtual references. Lacking designed references, the cascaded inner loops seek zero. Design engineers should consider eliminating zero-virtual reference and decoupling the cascaded electronics to increase overall system performance. Consider four voice-coil force actuators (as an example), and pay particular attention to the fact that force output is coupled due to back-emf and armature resistance which physically desire to seek a virtual-zero reference. In accordance with the definition of MID in section 2.4, the goal is to design the voltage signal that accounts for the predominant physics (both electrical and physical motion). The manipulated input is a voltage signal (e.g $E^*(s)$), not control signal u.

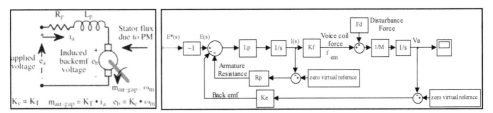

Fig. 1. LEFT: DC servo drive (cascaded current loop). RIGHT: Voice coil actuator. Note the presence of cross coupled armature resistance and back-emf.

An initial goal is to regulate i(t) to regulate f_{em}, (since i(t) and f_{em} are identical variables for this class of machines) to get well-behaved dynamics for the motion states. Since velocity-dependent back-emf complicates the electrical dynamics (it is cross-coupled state feedback), feedback decoupling was implemented. Especially since K_e and K_f are often quite high, back-emf can be quite a factor if not dealt with. Note that positive feedback for $\hat{K}_e$ approximately nulls K_e. Next, the effects of voice-coil resistance R_p were decoupled with feedback decoupling (i.e. decouple the effects of the armature resistance). Neither of these activities (decoupling back-emf and armature resistance) improves dynamics stiffness rather they yield well behaved force modulators. As a matter of fact, decoupling back-emf results in system inertia being the only remaining system disturbance rejection property.

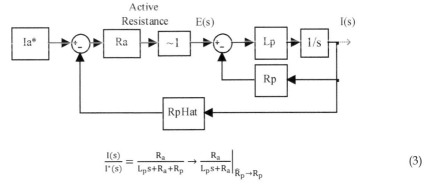

$$\frac{I(s)}{I^*(s)} = \frac{R_a}{L_p s + R_a + R_p} \rightarrow \left.\frac{R_a}{L_p s + R_a}\right|_{\hat{R}_p \rightarrow R_p} \tag{3}$$

Fig. 2. Decoupling armature resistance.

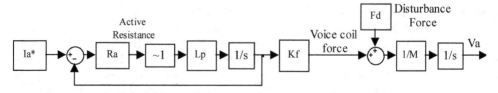

Fig. 3. Well-behaved voice coil actuator.

Figure 4 is a simplified block diagram that displays the back-emf and armature resistance decoupling (driven to near-zero). Mason's rule analysis (similar to the one done for decoupling armature resistance in Figure 2) demonstrates that decoupling yields unity gain current regulators.

Notice this remains *strictly* true as the armature resistance estimation is accurate. In reality, it is okay if it is not *strictly* true. The goal is to reduce the effects of armature resistance to allow the active resistance to dominate yielding well-behaved current regulators (i.e. within the regulators bandwidth, the behavior is nearly exactly as desired). Since these are the cascaded low energy states that feed the high energy motion states, the active resistance was tuned to a high bandwidth, 100 Hz (resulting value of $R_a=4$).

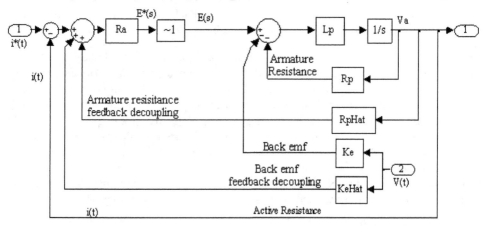

Fig. 4. Decoupling of armature resistance and back-emf.

Placing the force actuators into the equations of motion yields the following block diagram. After decoupling back-emf and armature resistance, the simplified block diagram reveals the now-dominant active resistance that may be tuned for system performance. The equivalent full-form displayed in block diagram form below.

Neglecting armature resistance and back-emf decoupling, the resultant dynamic stiffness is:

$$\frac{F_d(s)}{V(s)} = \frac{M_v L_p s^4}{(b_a s^2) + K_{rsas} + K_{risa}(L_p s - R_p)(K_f) + K_e L_p s^4} \tag{4}$$

The effects of decoupling may be observed on dynamic stiffness by setting an terms to zero to expose the individual effects of each loop on disturbance rejection.

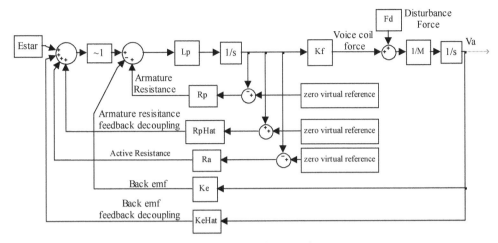

Fig. 5. Voice-coil state block diagram with virtual-zero reference.

5. Sensor replacement: Observers

This section of the paper evaluates the effect of observer types on two, observer-based, incremental motion format, state feedback motion controllers with a cascaded current loop applied to the dc servo drive in Fig. 1 with state feedback decoupling, but not disturbance input decoupling (to be performed in a later section of the paper). Luenberger and Gopinath observer topologies (Figure 6) will be compared [12], [14] (as taught in ME746 at the University of Wisconsin at Madison). Luenberger-styled observers (henceforth simply referred to as Luenberger observers) are a simple method to estimate velocity given position measurements. Additionally, the Luenberger observer may be used to provide estimates of external system disturbances, since the observer mimics order of actual systems dynamic equations of motion. When used the Luenberger disturbance observer bestows robustness to system parameter variations.

J_p	$= 0.015 \times 10^{-3} \text{ kg m}^2$	polar moment of inertia
K_T	$= 0.14 \text{ Nm/Amp}$	torque constant
K_e	$= 0.14 \text{ volts/rad/sec}$	back emf constant
R_p	$= 2.6 \text{ ohms}$	armature resistance
L_p	$= 4.3 \text{ millihenries}$	armature inductance
e_s	$= \text{applied terminal voltage in volts}$	
i_a	$= \text{armature current in amperes}$	
m_{ag}	$= \text{electromagnetic air-gap torque (moment)} = K_T i_a$	
e_b	$= \text{induced (back emf) voltage} = K_e \omega_m \text{ in volts}$	
ω_m	$= \text{load angular velocity in rad per sec}$	
θ_m	$= \text{load angular position in rad}$	

Table 1. Parameter Values and Variable Definitions.

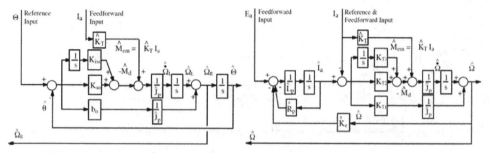

Fig. 6. LEFT: Luenberger-Styled Observer. RIGHT: Gopinath-Styled Velocity Observer.

5.1 Observer gain tuning

For desired observer eigenvalues λ_1=12.5, λ_2=50, λ_3=200, desired motion controller gains (tuned for disturbance rejection) λ_{c1}=6, λ_{c2}=25, λ_{c3}=100, and current regulator gain λ_i=800, the general form of the characteristic equation may be equated to the specific observer forms, controller form and current regulator form revealing gains. Tuning was directed in the problem statement to be identical to permit apples-to-apples comparison of effects on estimation accuracy.

5.2 Luenberger tuning (actual current)

This method uses the actual current from the actuator circuit (rather than modeled or predicted current) to provide the feedforward element of the observer. This position would normally include the actual current or control, u in typical observer designs (recalling that observer design is a *dual* process of controller design). Utilizing the reference input and actual circuit moment, you can produce an estimate of remaining disturbance (normally fed back to feedback controllers to handle).

$$\text{C.E.} = (s+\lambda_1)(s+\lambda_2)(s+\lambda_3) = \hat{J}_p s^3 + b_o s^2 s + K_{so} s + K_{iso} \tag{5}$$

$$b_o = \hat{J}_p(\lambda_1 + \lambda_2 + \lambda_3) \tag{6}$$

$$K_{so} = \hat{J}_p([\lambda_1(\lambda_2 + \lambda_3) + \lambda_2 \lambda_3] \quad K_{iso} = \hat{J}_p(\lambda_1 \lambda_2 \lambda_3) \tag{7}$$

5.3 Gopinath tuning

$$\frac{\hat{\omega}(s)}{\omega(s)} = \frac{(K_{T1}s^2 + K_{T2}s + K_{T3})\left(\dfrac{J_p s^2 (L_p - \hat{L}_p) + J_p (R_p - \hat{R}_p)s}{K_T} + \dfrac{\hat{K}_t}{K_t}J_p s(\hat{L}_p + \hat{R}_p)\right)}{\hat{J}_p \hat{L}_p s^3 + (\hat{J}_p R_p + \hat{K}_e K_1)s^2 + \hat{K}_e K_2 s + \hat{K}_e K_3} \tag{8}$$

Equating coefficient of 's' and solve for gains:

$$(s+\lambda_1)(s+\lambda_2)(s+\lambda_3) = \hat{J}_p s^3 + (\hat{J}_p \hat{R}_p + \hat{K}_e K_1)s^2 + \hat{K}_e K_2 s + \hat{K}_e K_3 \tag{9}$$

$$K_{T1} = \frac{\hat{J}_p \hat{L}_p (\lambda_1 + \lambda_2 + \lambda_3) - \hat{J}_p \hat{R}_p}{\hat{K}_e} \tag{10}$$

$$K_{T3} = \frac{\hat{J}_p \hat{L}_p}{\hat{K}_e} (\lambda_1 \lambda_2 \lambda_3) \tag{11}$$

$$K_{T2} = \frac{\hat{J}_p \hat{L}_p (\lambda_1 \lambda_2 + \lambda_3) - \hat{J}_p \hat{R}_p (\lambda_1 (\lambda_2 + \lambda_3) + \lambda_2 \lambda_3)}{\hat{K}_e} \tag{12}$$

Motion Controller:

$$(s+\lambda_{c1})(s+\lambda_{c2})(s+\lambda_{c3}) = \hat{J}_p s^3 + b_a s^2 s + K_s s + K_{is} \tag{13}$$

Current regulator: $\qquad\qquad (s+\lambda_i) = L_p s + R_a \tag{14}$

Observer *estimation* frequency response functions were calculated and plotted first for ±20% estimated-inertia error then for the case of ±20% error in estimate of $K_e = K_t$ (Figure 7). Notice first that for all cases of zero-error, both observers exactly estimate the angular velocity of motion. Overall, the Gopinath-styled observer (referred to as simply "Gopinath" for brevity) performed poorer than the Luenberger-styled observer indicating the Luenberger observer is less parameter-sensitive with respect to inertia, K_e, and K_t.

Luenberger Gains			Gopinath Gains			Motion Controller Gains			
B_o	K_{so}	K_{iso}	K_{T1}	K_{T2}	K_{T3}	B_p	K_s	K_{is}	R_a
Nm/m/s	Nm/m	Nms	rad/s Nms /A	Nm/A	Nms/A	Nm/m/s	Nm/m	Nms	V/A
24.74	7772.3	465090	0.4813	238.7	14285	12.3	1924.6	55811	21.6

Table 2. Observer Gains.

While the Luenberger observers diverge very close to the maximum tuned bandwidth (even with parameter errors), the Gopinath observer diverges at a lower bandwidth when errors are present. Since both observers contain a current-feedforward element, you will see nearly zero-lag properties out to the bandwidth of the feedback observer controller. Clearly, disturbances (in the form of modeling errors here) do not influence low frequency estimation (likely due to the addition of integrators in the observer controllers). The Gopinath observer was particularly sensitive to errors in Kt indicating its reliance on the feedforward estimation path. Notice in particular in Figure 7 that zero-lag estimation occurs even with inaccurate K_t (albeit with non-zero estimation frequency response at all frequencies).

Time-response simulations were run with *identically tuned observers* with a sample commanded trajectory (rotation angle) of $\theta^*(t) = \sin(10t)$. Iterations were run to establish the effects of 20% inertia underestimation and the effects of sensor noise *on command tracking accuracy*. Sensor noise was modeled as random numbers with zero-mean and unity variance.

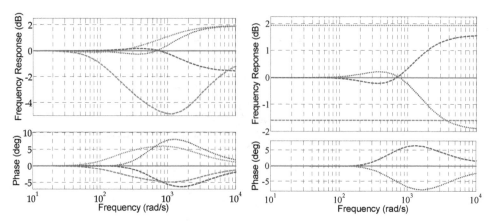

Fig. 7. LEFT: Comparison of estimation accuracy frequency response functions for incorrect $\hat{J}_p$. Luenberger (blue) Gopinath (red); dotted = -20% error, solid = 0% error; dashed = +20% error. RIGHT: Comparison of estimation accuracy frequency response functions for incorrect $K_t=K_e$. Luenberger (blue) Gopinath (red); dotted = -20% error, solid = 0% error; dashed = +20% error.

Figure 9 reveals the methodology for apples-to-apples comparison of effects on command tracking. Manual switches were used to evaluate a given case with the results displayed in Figure 7 and Figure 8. General conclusions may be drawn. Feedback control handles incorrectly estimated just fine, especially since inertia has nothing to do with the feedback control strategy (lacking a feedforward strategy). Using the Luenberger observer performs nearly as well if actual $\theta(t)$ is used for estimation, while it does not perform as well when $\theta^*(t)$ is used for estimation. This is intuitive, since $\theta^*(t)$ does not include the errors and noises associate with the process, while $\theta(t)$ includes these errors and noises. In all cases examined, the Gopinath-styled observer was inferior, which reinforces the earlier revelation of

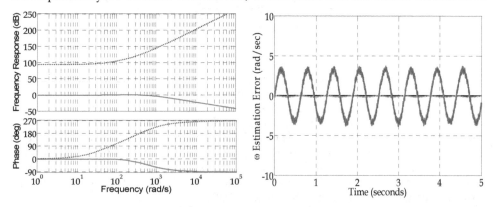

Fig. 8. LEFT: Frequency Response Functions for the motion control system. RIGHT: Estimation errors for $\hat{J}_p = 0.8Jp$ and $\mu=0$, $\lambda^2=1$ sensor noise. Black solid line is Luenberger with $\theta(s)$ input; Green dotted line is Luenberger with $\theta^*(s)$ input; red solid line is Gopinath with $Ia(s)$ input; blue dotted line is Gopinath with $I^*(s)$ input.

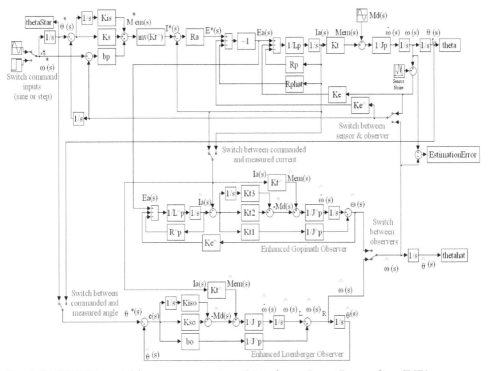

Fig. 9. SIMULINK model for error comparison Disturbance Input Decoupling (DID).

parameter sensitivity (in the discussion of the estimation frequency response functions). In addition to examining the effects on command tracking accuracy, estimation accuracy was plotted from the simulations to confirm the indications garnered from the discussion of figures 1 & 2 (estimation accuracy frequency response functions, FRFs). The single case of 20% inertia underestimation with zero-mean and unity variance sensor noise confirmed that the enhanced Luenberger-styled observer provided superior estimates compared to the Gopinath styled observer for this sinusoidal commanded trajectory.

One suggestion for improved command tracking is to remove *feedback* decoupling as done here replacing it with *feedforward* decoupling permitting the disturbance torque to excite the decoupling. One other thing: Note the maximum phase lag of 90 degrees. Such a maximum would be expected in a system with a command feedforward control scheme. Since the feedforward path would remain nearly zero-lag, the 90-degree phase lag would be creditable to Shannon's sampling-limit theory. Since there is no command feedforward control in this scheme, the lack of a maximum phase shift of 180 degrees (for a double integrator plant) is puzzling.

Observer tuning (not the current loop tuning) determines the maximum frequency for nearly zero-lag accurate estimation. Since the commanded and actual current are nearly identical (also with zero lag) out to the *higher* current loop bandwidth, *it was expected that* the effects of commanded versus actual current are mitigated by feedback decoupling (i.e. we exceed the observer bandwidths before there is an appreciable difference in commanded versus actual current).

Actually, the Luenberger observer was sensitive to output noise associate with actual current. The noisier actual current signal does not pass through a smoothing integrator before going directly into the plant dynamics. On other hand, the Gopinath observer compares the estimated and actual/commanded current (i.e. current estimation error) through a smoothing integrator in the observer controller and also passes a portion through a separate smoothing integrator associate with angular rate estimation. Thus, the Gopinath-styled observer was insensitive to commanded versus measured current due to feedback decoupling. The Luenberger observer may be made less sensitive to the difference between commanded and actual current (and other system noises and errors) by using the actual rotation angle as input to the observer (Figure 4 and Table 2). As a matter of fact, this iteration resulted in the best performance for the evaluated case of sinusoidal sensor noise demonstrating the least mean error.

RECOMMENDATION: Use enhanced Luenberger-styled observers with actual $\theta(s)$.

6. Disturbance Input Decoupling (DID)

This paragraph reformulates the dual observer-based DID system in Yoon, 2007, consistent with physics-based control methods and furthermore evaluates opportunities in the proposed structure, [1]-[7]. Physics-based methods recommend 1) *disturbance input decoupling* followed by 2) *state feedback decoupling* of system cross-coupling, then 3) elimination of *virtual zero references*, and then finally adding *active* state feedback with full state *references*. Note the observer structure in Yoon, 2007 is different where we have added command feedforward (reference [1]) shown in Figure 6 & Figure 10. The [Yoon] paper evaluates the controlled dynamics of a magnetic levitation machine, whose dynamics are similar to a free-floating spacecraft when the cross-product has been decoupled (noting the spacecraft is suspended by gravity while the mag-lev system uses controlled magnetic field instead. Nonetheless, the physics-based decoupling principles remain the same. The main goal of DID is to formally identify the disturbance online, then use feedback to decouple the effects of disturbance input. Although the decoupling signal is actually the disturbance identified at the immediately previous timestep, using this value is far superior than simply treating disturbances as unknown quantities. The disturbance moment $M_d(s)$ is estimated in the observer in the feedforward element $\hat{M}_{em}(s)$.

Emphasize velocity estimation for state feedback of motion controllers. The improvements achieve near-zero lag, accurate velocity estimation are displayed and zoomed in Figure 12 for clarity. The larger scale reveals the advantages over the most recently proposed improved methods. High-frequency roll-off is drastically improved by addition of command feedforward (of the true manipulated input) to the Luenberger observer. Additional inclusion of disturbance input decoupling in the motion control system improves velocity estimates in the observer, essentially eliminating roll-off and estimation lag. This later claim is more clearly displayed in the zoomed response plot in Figure 12.

The cascaded control topology should be eliminated adding full command references. Command feedforward control should be added. The electro-dynamics should not be ignored in the analysis. It causes the illusion that force is the manipulated input as opposed

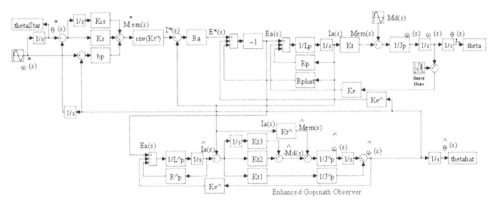

Fig. 10. Decoupled motion control w/DID & Luenberger observer with command feedforward.

to current (the true manipulated input) resulting in lower bandwidth. Neglecting the electrodynamics results in an analysis that is inadequately reinforces the experiments. Yoon refers to "disturbances forces generated by the current controller" to explain the difference between experimentation and analysis. Decoupling the electro-dynamics will improve performance even without full command references. Without manipulated input decoupling (MID), you have an implied zero-reference command for current. Assuming an inductor motor's electronics, decoupling K_e should dramatically increase disturbance rejection isolating the electrical system. The paper utilizes a dual observer to permit individual tuning for disparate purposes (DID and velocity estimation), but then implies using identical observer gains! That makes no sense. Instead of using identical gains, eliminate one of the observers to simplify the algorithmic complexity. Alternatively, utilize different gains optimized respectively for velocity and disturbance estimation. A first step for comparison requires repetition of the Yoon paper results. Equations (3), (4), and (5) in the Yoon paper are plotted in Figure 11, which should duplicate figure (5) in the Yoon paper.

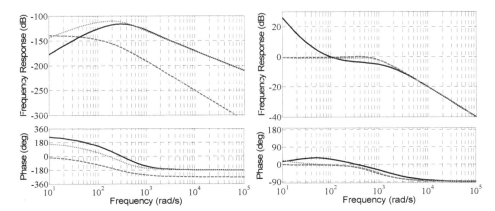

Fig. 11. LEFT: Nominal response comparison: Solid-black line is Luenberger observer; Blue-dashed line is Modified Luenberger observer; Red-dotted line is no compensation. RIGHT: Response comparison: Solid-black line is Luenberger observer; Red-dotted line is Modified Luenberger observer; Blue-dashed line is Dual Observer.

Note the slightly different result was achieved only in the case of modified observer (not the proposed dual-observer method).

Next, equations (6), (7), and (8) in Yoon, 2007 [12] were plotted in Figure 11, which duplicates Yoon's figure 6. Again, notice a slight difference this time with the estimation FRF of the basic Luenberger observer. According to the paper's plots in figure 6, the modified observer estimates more poorly than the nominal observer by dramatically overestimating velocity. This clearly indicates a labeling-error in the paper's figure. Also, the Luenberger observer does not estimate well *within the observer bandwidth*, so my results displayed here seems more credible. The difference is negligible considering the performance to be gained using physics-based reformulation.

$$
F_{dist} = [P(s)-\hat{P}(s)]\left[K_d s+K_p+\frac{K_i}{s}\right] = \frac{s^2\left[\hat{P}(s)-P(s)\right]\left[K_d s+K_p+\frac{K_i}{s}\right]}{s^2} =
$$
$$
= \frac{\left[\hat{V}(s)-V(s)\right]\left[K_d s+K_p+\frac{K_i}{s}\right]}{s^2}
$$
(15)

$$
\hat{V}(s)\left[1+\frac{L_d}{\hat{M}}\frac{1}{s}+\frac{L_p}{\hat{M}}\frac{1}{s^2}+\frac{L_i}{\hat{M}}\frac{1}{s^3}\right] = V(s)\left[\frac{\hat{K}_T}{K_T}\frac{M}{\hat{M}}+\frac{L_d}{\hat{M}}\frac{1}{s}+\frac{L_p}{\hat{M}}\frac{1}{s^2}+\frac{L_i}{\hat{M}}\frac{1}{s^3}\right]+
$$
$$
+\left[\hat{V}(s)-V(s)\right]\left[K_d+\frac{K_p}{s}+\frac{K_i}{s^2}\right]
$$
(16)

$$
\hat{V}(s)\left[1+\frac{L_d}{\hat{M}}\frac{1}{s}+\frac{L_p}{\hat{M}}\frac{1}{s^2}+\frac{L_i}{\hat{M}}\frac{1}{s^3}-K_d-\frac{K_p}{s}-\frac{K_i}{s^2}\right] =
$$
$$
=V(s)\left[\frac{\hat{K}_T}{K_T}\frac{M}{\hat{M}}+\frac{L_d}{\hat{M}}\frac{1}{s}+\frac{L_p}{\hat{M}}\frac{1}{s^2}+\frac{L_i}{\hat{M}}\frac{1}{s^3}-K_d-\frac{K_p}{s}-\frac{K_i}{s^2}\right]
$$
(17)

Multiplying by s^3:

$$
\hat{V}(s)\left[s^3+\frac{L_d}{\hat{M}}s^2+\frac{L_p}{\hat{M}}s+\frac{L_i}{\hat{M}}-K_d s^3-K_p s^2-K_i s\right] =
$$
$$
=V(s)\left[\frac{\hat{K}_T}{K_T}\frac{M}{\hat{M}}s^3+\frac{L_d}{\hat{M}}s^2+\frac{L_p}{\hat{M}}s+\frac{L_i}{\hat{M}}-K_d s^3-K_p s^2-K_i s\right]
$$
(18)

$$
\frac{\hat{V}(s)}{V(s)} = \frac{\left[\frac{\hat{K}_T}{K_T}\frac{M}{\hat{M}}-K_d\right]s^3+\left[\frac{L_d}{\hat{M}}-K_p\right]s^2+\left[\frac{L_p}{\hat{M}}-K_i\right]s+\frac{L_i}{\hat{M}}}{[1-K_d]s^3+\left[\frac{L_d}{\hat{M}}-K_p\right]s^2+\left[\frac{L_p}{\hat{M}}-K_i\right]s+\frac{L_i}{\hat{M}}}
$$
(19)

The reformulation (Figure 10) results in the estimation FRF *with DID and command feedforward* is displayed Figure 12. Immediately notice that addition of the command feedforward to the modified Luenberger observer yields *nearly-zero* lag estimates, far superior to Yoon, 2007 (which omitted the command feedforward path in what they call an observer). It is a premise of the physics-based methodology that the title "observer" implies nearly-zero lag estimation, so one might argue that the Yoon paper really utilizes a state filter rather than a state observer.

The results using the physics-based methodology are clearly superior despite relative algorithmic simplicity. Adding the command feedforward permits accurate, near-zero lag estimation of velocity without a velocity sensor. Furthermore, disturbance input decoupling increases system robustness and permits accurate estimation inaccuracy even when unknown disturbances are present. Certainly, accounting for the electrodynamics should always be done rather than neglecting them as "system noise" as done in Yoon, 2007.

Figure 12 displays a Solid-blue line is Modified Luenberger observer with command feedforward; Red-dashed line is Modified Luenberger observer with command feedforward and disturbance input decoupling. RIGHT: Observer Improvements estimation comparison: Dotted-black line from the Yoon paper (using dual observers). Solid-blue line is Modified Luenberger observer with command feedforward; Red-dashed line is Modified Luenberger observer with command feedforward and disturbance input decoupling; Dashed-black line is Dual Observers.

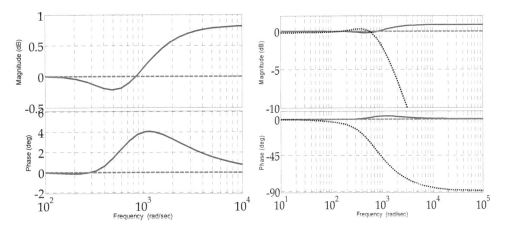

Fig. 12. LEFT: Observer Improvements estimation comparison.

7. Physics-based methods for idealized feedforward control

Feedforward control is a basic starting point for spacecraft rotational maneuver control. Assuming a rigid body spacecraft model in the presence of no disturbances and known inertia [J], an open loop (essentially feedforward) command should exactly accomplish the commanded maneuver. When disturbances are present, feedback is typically utilized to insure command tracking. Additionally, if the spacecraft inertia [J] is unknown, the open command will not yield tracking. Consider a spacecraft that is actually much heavier about it's yaw axis than anticipated in the assumed model. The same open loop command torque would yield less rotational motion for heavier spacecraft. Similarly, if the spacecraft were much lighter than modeled, the open loop command torque would result in excess rotation of the lighter spacecraft. Observe in Figure 13, a rigid spacecraft simulator (TASS2 at Naval Postgraduate School) has been modeled in SIMULINK. An open loop feedforward command has been formulated to produce 10 seconds of regulation followed by a 30° yaw-only rotation in 10 seconds, followed by another 10 seconds of regulation at the new attitude. The assumed inertia matrix is not diagonal, so coupled dynamics are accounted for in the feedforward command.

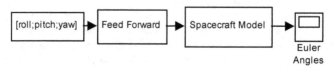

Fig. 13. SIMULINK model of TASS2 Spacecraft Simulator at Naval Postgraduate School.

With no disturbances and a known, correct model, the open loop feedforward command can effectively perform the maneuver.

$$[J]_{\text{modeled}} = [J]_{\text{previously estimated}} = [J]_{\text{feedforward}} = \begin{bmatrix} 119.1259 & -15.7678 & -6.5486 \\ -15.7678 & 150.6615 & 22.3164 \\ -6.5486 & 22.3164 & 106.0288 \end{bmatrix} \quad (20)$$

Recall in the real world systems are not always as we model them, disturbances are presence, and our sensor measurements of the maneuver will also be quite noisy. Nonetheless, the idealized case is a useful place to start, as it gives us confidence that our model has been correctly coded. Proof is easily provided by sending an acceleration command (scaled by the inertia) to the spacecraft model to verify the identical acceleration is produced (Figure 14). We have not yet added noise, disturbances, or modeling errors, so exact following should be anticipated. Next, we will alter the inertia [J] of TASS2. This is real-world, since the spacecraft has recently received its optical payload, so the yaw inertia

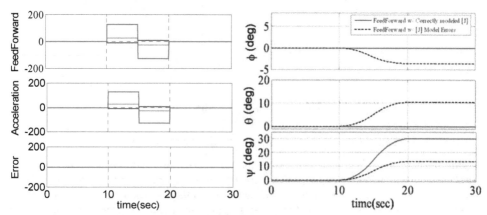

Fig. 14. LEFT: Feedforward input and resultant TASS2 acceleration (note zero error).
RIGHT: Open Loop Feedforward TASS2 Maneuver Simulation.

components have increased significantly. Using the previous experimentally determined inertia [J] in the feedforward command should result in difficulties meeting the open loop pointing command.

Notice in Figure 14 the maneuver is not correctly executed using the identical feedforward command for the assumed, modeled TASS2. The current inertia matrix has not been experimentally determined, so inertia components were varied arbitrarily (making sure to increase yaw inertia dramatically). This new inertia was used in the spacecraft model, but is presumed to be unknown. Thus, the previous modeled open loop feedforward command is used and proven to ineffective. Options to improve system performance include feedback, and adapting the feedforward command to eliminate the tracking error. Since adaptive control is more difficult, we will first examine feedback control with the identical models and maneuver.

8. Feedback control

Feedback control components multiply a gain to the tracking error components in each of the 3-axes. When multiplying gains to the tracking error itself, the control is referred to as proportional control (or P-control). When multiplying gains to the tracking error integral, the control is referred to as integral control (or I-control). Finally, when multiplying gains to the tracking error rate (derivative), the control is referred to as derivative control (or D-control). Summing multiple gained control signals results in combinations such as: PI, PD, PID, etc. PD control is extremely common for Hamiltonian systems, as it is easily veritably a stable control. PD control was augmented to the previous case of feedforward control with inertia modeling errors (Figure 15) dramatically improving performance, while not restoring the ideal case.

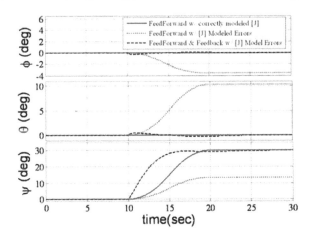

Fig. 15. Demonstration of Feedback Control Effectiveness.

It is clear that feedback control augmentation is a powerful tool to eliminate real world factors like modeling errors. An identical comparison was performed with gravity gradient disturbances associated with an unbalanced TASS2. The comparison is not presented here for brevity's sake, but the results were qualitatively identical.

While feedback appears extremely effective to accomplish the overall tracking maneuver, some missions require faster, more accurate tracking with less error. Such missions often consider augmenting the feedforward-feedback control scheme by adding adaptive control to either signal.

9. Adaptive control

Adaptive control techniques typically adapt control inputs based upon errors tracking commanded trajectories and/or estimation errors. *Direct* adaptive control techniques typically directly adapt the control signal to eliminate tracking errors without estimation of unknown system parameters. *Indirect* adaptive control techniques indirectly adapt the control signal by modifying estimates of unknown system parameters. The adaptation rule is derived using a proof that demonstrates the rapid elimination of tracking errors (the real objective). The proof must also demonstrate stability, since the closed loop system is highly nonlinear with the adaptive control included. Two fields of application of adaptive control is robotic manipulators and spacecraft maneuvers utilizing both approaches [15], [16], [17].

While some adaptive techniques concentrate on adaptation of the feedback control, others have been suggested to modify a feedforward control command retaining a typical feedback controller, such as Proportional-Derivative (PD). Adaptation of the feedforward signal has been suggested in the inertial reference frame [18], [19], but the resulting regression model requires several pages to express for 3-dimensional spacecraft rotational maneuvers. The regression matrix of "knowns" is required in the control calculation, so this approach is computationally inappropriate for spacecraft rotational maneuvers. Subsequently, the identical approach was suggested for implementation in the body reference frame [20]. The method was demonstrated for slip translation of the space shuttle. This method appears promising for practical utilization in 3-dimensional spacecraft rotational maneuvers. A derivation of the Slotine-Fossen approach is derived for 3-dimensional spacecraft rotational maneuvers next, then implementation permits evaluation of the effectiveness of the approach in the context of the previous results for classical feedforward-feedback control of the TASS2 plant with modeling errors.

9.1 Adaptive feedforward command derivation

The equation of motion may be written by various methods (Newton-Euler, Lagrange, Kane's, momentum, etc.) as follows: $[J]\{\ddot{q}\}_B + [C]\{\dot{q}\}_B = \{\tau\}_B$ where $[J]$ is the inertia matrix, $[C]$ is the Coriolis matrix representing the cross-coupling dynamics, τ is the sum of external torques and q is the body coordinates (quaternion, Euler angles, etc.). The body coordinates may be transformed to inertial coordinates via the transformation matrix $[S]$ per the following: $\{\dot{x}\}_i = [S]_{B2i}\{\dot{q}\}_B$. Similarly, we may define a reference trajectory *in the body coordinates*: $\{\dot{x}_r\}_i = [S]_{B2i}\{\dot{q}_r\}_B$. Rewriting the transformation and differentiating:

$\{\dot{q}_r\}_B = [S]^{-1}\{\dot{x}_r\}_i \rightarrow \{\ddot{q}_r\}_B = [S]^{-1}\{\ddot{x}_r\}_i - [S]^{-1}[\dot{S}][S]^{-1}\{\dot{x}_r\}_i$. This may be substitute into the equation of motion allowing us to express the equation of motion in terms of the reference trajectory.

$$[J]\Big[[S]^{-1}\{\ddot{x}_r\}_i - [S]^{-1}[\dot{S}][S]^{-1}\{\dot{x}_r\}_i\Big] + [C]\Big[[S]^{-1}\{\ddot{x}_r\}_i\Big] = \{\tau\}_B \qquad (21)$$

$$[J][S]^{-1}\{\ddot{x}_r\}_i + \Big[-[J][S]^{-1}[\dot{S}][S]^{-1} + [C][S]^{-1}\Big]\{\dot{x}_r\}_i = \{\tau\}_B \qquad (22)$$

Pre-multiplying by $[S]^T[S]^{-T} = 1$ allows us to understand [Slotine]'s original approach in reference [19]:

$$[S]^T[S]^{-T}[J][S]^{-1}\{\ddot{x}_r\}_i + [S]^T[S]^{-T}\left[-[J][S]^{-1}[\dot{S}][S]^{-1} + [C][S]^{-1}\right]\{\dot{x}_r\}_i = [S]^T[S]^{-T}\{\tau\}_B \quad (23)$$

$$[S]^T \underbrace{[S]^{-T}[J][S]^{-1}}\{\ddot{x}_r\}_i + [S]^T\underbrace{\left[-[S]^{-T}[J][S]^{-1}[\dot{S}][S]^{-1} + [S]^{-T}[C][S]^{-1}\right]}\{\dot{x}_r\}_i = [S]^T[S]^{-T}\{\tau\}_B \quad (24)$$

$$\text{Slotine's } J^* \qquad\qquad\qquad \text{Slotine's } C^*$$

$$[S]^T\left[J^*\right]\{\ddot{x}_r\}_i + [S]^T\left[C^*\right]\{\dot{x}_r\}_i = [S]^T\underbrace{\left[J^*\right]\{\ddot{x}_r\}_i + \left[C^*\right]\{\dot{x}_r\}_i}_{\Phi^*(x,x,\ddot{x}_r,\ddot{x}_r)\theta} = \{\tau\}_B \quad (25)$$

Slotine uses the linear regression model to define an equivalent system based on parameter estimates:

$$\Phi^*(x,\dot{x},\dot{x}_r,\ddot{x}_r)\Theta = \underbrace{\Phi^*(x,\dot{x},\dot{x}_r,\ddot{x}_r)}_{\text{knowns}} \underbrace{\hat{\Theta}}_{\text{unknowns}} + \text{error}. \quad (26)$$

The estimates $\hat{\Theta}$ are adapted using an adaption rule that makes the closed loop system stable. The regression model is then used in the control, which is where the complication arises. The $\Phi^*(x,\dot{x},\dot{x}_r,\ddot{x}_r)$ matrix of "knowns" occupies several pages and is used *at each time* step to formulate the adapted control signal making the method computationally impractical. [Fossen] on the other hand formulates the regression model in the body coordinates eliminated the complications seen above with the numerous multiplications with the coordinate transformation matrix [S]. Picking up from [Slotine]'s method above, we can simply express the regression model *including* the transformation matrix:

$$[S]^T\left[J^*\right]\{\ddot{x}_r\}_i + [S]^T\left[C^*\right]\{\dot{x}_r\}_i = [S]^T\underbrace{\left[\left[J^*\right]\{\ddot{x}_r\}_i + \left[C^*\right]\{\dot{x}_r\}_i\right]}_{\Phi(x,\dot{q},\dot{q}_r,\ddot{q}_r)\theta} = \{\tau\}_B \quad (27)$$

noting the $\Phi(x,\dot{x},\dot{x}_r,\ddot{x}_r)$ matrix of "knowns" has no asterisk. Preface [Slotine]'s mathematical trick (pre-multiplication) above:

$$[J][S]^{-1}\{\ddot{x}_r\}_i + \left[-[J][S]^{-1}[\dot{S}][S]^{-1} + [C][S]^{-1}\right]\{\dot{x}_r\}_i = \{\tau\}_B \cdot \quad (28)$$

Continuing here yields [Fossen]'s substantial simplification through the following 3 steps: Solve the earlier defined transformation equations for $\dot{x}_r$ & $\ddot{x}_r$:

$$\{\dot{x}_r\} = [S]\{\dot{q}_r\} \quad (29)$$

$$\{\ddot{x}_r\}_i = [S]\left[\ddot{q}_r + [S]^{-1}[\dot{S}][S]^{-1}\{\dot{x}_r\}\right] = [S]\ddot{q}_r + [\dot{S}][S]^{-1}[S]\{\dot{q}_r\} = [S]\ddot{q}_r + [\dot{S}]\{\dot{q}_r\} \quad (30)$$

Substitute into $[J][S]^{-1}\{\ddot{x}_r\}_i + \left[-[J][S]^{-1}[\dot{S}][S]^{-1} + [C][S]^{-1}\right]\{\dot{x}_r\}_i = \{\tau\}_B$ instead of pre-multiplying.

$$[J][S]^{-1}\underbrace{\left[[S]\ddot{q}_r + [\dot{S}]\{\dot{q}_r\}\right]}_{\{\ddot{x}_r\}_i} + \left[-[J][S]^{-1}[\dot{S}][S]^{-1} + [C][S]^{-1}\right]\underbrace{[S]\{\dot{q}_r\}}_{\{\dot{x}_r\}_i} = \{\tau\}_B \quad (31)$$

Reduce this to linear regression form:

$$[J][S]^{-1}[S]\ddot{q}_r + [J][S]^{-1}[\dot{S}]\{\dot{q}_r\} - [J][S]^{-1}[\dot{S}][S]^{-1}[S]\{\dot{q}_r\} + [C][S]^{-1}[S]\{\dot{q}_r\} = \{\tau\}_B \tag{32}$$

$$[J]\ddot{q}_r + \left[\underbrace{[J][S]^{-1}[\dot{S}] - [J][S]^{-1}[\dot{S}]}^{0} + [C] \right]\{\dot{q}_r\} = \{\tau\}_B \tag{33}$$

$$\boxed{[J]\{\ddot{q}_r\} + [C]\{\dot{q}_r\} = \{\tau\}_B} \tag{34}$$

All that remains now is to multiply this out long-hand and regroup the terms into the linear regression model: $\Phi^*(x, \dot{x}, \dot{x}_r, \ddot{x}_r)\Theta = \underbrace{\Phi^*(x, \dot{x}, \dot{x}_r, \ddot{x}_r)}_{knowns} \underbrace{\hat{\Theta}}_{unknowns} + error$. In order to do this, we must define the reference trajectory. The modifications to the overall feedforward control strategy may be embodied in these two venues: 1) estimate/adapt estimates of inertia in the regression model above, and 2) choose a reference trajectory that addresses system lead/lag when applying the assumed control to a spacecraft with modeling errors, disturbances and noise.

9.2 Reference trajectory

Define the reference trajectory such that the control helps the spacecraft "catch up" to the commanded trajectory. If the spacecraft is actually heavier than modeled, it needs a little extra control to achieve tracking than will be provided by classical feedforward control. If the spacecraft is actually lighter than modeled, the control must be reduced so as not to overshoot the commanded trajectory. Consider defining the reference trajectory as follows:

$$\ddot{q}_r = \ddot{q}_d - \lambda(\dot{q} - \dot{q}_d) \text{ and } \dot{q}_r = \dot{q}_d - \lambda(q - q_d) \tag{35}$$

Note we have scaled the reference acceleration and velocity to add/subtract the velocity and position error respectively scaled by a positive definite constant, λ. This should help the feedforward control component regardless of indirect adaption. Accordingly, subsequent sections will evaluate the effectiveness of the reference trajectory by itself and the also the indirect adaption/estimation by itself as well. First, let's conclude the derivation by multiplying out the linear regression form so that the reader can have the simple equation for spacecraft rotational maneuvers.

9.3 Feedforward & feedback control with reference trajectories

Simplify $[J]\{\ddot{q}_r\} + [C]\{\dot{q}_r\} = \{\tau\}_B$ letting $\ddot{q}_r = \ddot{q}_d - \lambda(\dot{q} - \dot{q}_d)$ and $\dot{q}_r = \dot{q}_d - \lambda(q - q_d)$ and use $\ddot{q}_r = \dot{\omega}_r$ and $\dot{q}_r = \omega_r$:

$$[J]\{\dot{\omega}_r\} + [C]\{\omega_r\} = \{\tau\}_B$$

$$[J]\{\dot{\omega}_r\} = [J]\{\omega_r\} \times \{\omega_r\} + \{\tau\}_B$$

$$[J]\{\dot{\omega}_r\} = [H \times]\{\omega_r\} + \{\tau\}_B \text{ where } [H \times] \text{ is the skew symmetric matrix form of the momentum}$$

vector. Expand $[J]\{\dot{\omega}_r\} - [H \times] = \{\tau\}_B$:

$$\begin{bmatrix} J_{xx} & J_{xy} & J_{xz} \\ J_{yx} & J_{yy} & J_{yz} \\ J_{zx} & J_{zy} & J_{zz} \end{bmatrix}\begin{Bmatrix} \dot{\omega}_x \\ \dot{\omega}_y \\ \dot{\omega}_z \end{Bmatrix} - \begin{bmatrix} 0 & -H_z & H_y \\ H_z & 0 & -H_x \\ -H_y & H_x & 0 \end{bmatrix}\begin{Bmatrix} \omega_x \\ \omega_y \\ \omega_z \end{Bmatrix} = \begin{Bmatrix} \tau_x \\ \tau_y \\ \tau_z \end{Bmatrix}_B = \tag{36}$$

$$= \underbrace{\left[\hat{J}\right]\{\dot{\omega}_r\} - \left[\hat{H}\times\right]\{\omega_r\}}_{\text{Adaptive Feedforward}} - \underbrace{K_d S^{-1}(\dot{x} - \dot{x}_r)}_{\text{Ref Trajectory Feedback}}$$

$$\begin{bmatrix} -H_y\omega_z + H_z\omega_y + J_{xx}\dot{\omega}_x + J_{xy}\dot{\omega}_y + J_{xz}\dot{\omega}_z \\ H_x\omega_z - H_z\omega_x + J_{yx}\dot{\omega}_x + J_{yy}\dot{\omega}_y + J_{yz}\dot{\omega}_z \\ -H_x\omega_y + H_y\omega_x + J_{zx}\dot{\omega}_x + J_{zy}\dot{\omega}_y + J_{zz}\dot{\omega}_z \end{bmatrix} = \begin{Bmatrix} \tau_x \\ \tau_y \\ \tau_z \end{Bmatrix}_B \tag{37}$$

Let $\theta^T = \{J_{xx}\ \ J_{xy}\ \ J_{xz}\ \ J_{yy}\ \ J_{yz}\ \ J_{zz}\ \ H_x\ \ H_y\ \ H_z\}$ and assume $J_{xy} = J_{yx}, J_{xz} = J_{zx}, J_{yz} = J_{zy}$ allowing us to express $[\Phi(\dot{\omega},\ddot{\omega})]_{3x9}\{\theta\}_{9x1}$:

$$\begin{bmatrix} \dot{\omega}_x & \dot{\omega}_y & \dot{\omega}_z & 0 & 0 & 0 & 0 & -\omega_z & \omega_y \\ 0 & \dot{\omega}_x & 0 & \dot{\omega}_y & \dot{\omega}_z & 0 & \omega_z & 0 & \omega_x \\ 0 & 0 & \dot{\omega}_x & 0 & \dot{\omega}_y & \dot{\omega}_z & -\omega_y & \omega_x & 0 \end{bmatrix}\begin{Bmatrix} J_{xx} \\ J_{xy} \\ J_{xz} \\ J_{yy} \\ J_{yz} \\ J_{zz} \\ H_x \\ H_y \\ H_z \end{Bmatrix} = [\Phi(\dot{\omega},\ddot{\omega})]_{3x9}\{\theta\}_{9x1} = \left[\hat{\Phi}\right]\{\theta\} + error$$

$$\{\tau\} = \left[\hat{\Phi}\right]\{\theta\} - K_d S^{-1}(\dot{x} - \dot{x}_r) \leftarrow \boxed{\text{Use this control}} \tag{38}$$

Where $\{\dot{\hat{\theta}}\} = -\Gamma[\Phi]_{[3x9]}[S]^{-1}(\dot{x} - \dot{x}_r) = -\Gamma[\Phi]_{[3x9]}(\dot{q} - \dot{q}_r) \leftarrow \boxed{\text{use this adaption rule}} \tag{39}$

9.4 Adaptive feedforward effectiveness

Especially since typical feedback control deals with modeling errors effectively, we wish to evaluate the effectiveness of indirect adaptive feedforward control with a rigorously disciplined approach. Accordingly, the examination will evaluate the individual effectiveness of each control component in the following paragraphs:

- Reference trajectory without indirect adaption (feedforward, feedback, and both)
- Indirect adaption without a scaled reference trajectory (feedforward, feedback, and both)
- Indirect adaption with reference trajectory (previously derived application of [Fossen] suggested improvement to [Slotine]'s method)

The examination is performed by manually activating switches in the SIMULINK simulation model to insure all aspects of the maneuver are identical with exception of the aspect being switched for investigation. Note the feedback control is configured as a proportional-derivative-integral (PID) controller with the following gains: K_p=100, K_d=300, K_I=0, thus a PD controller.

It seems likely that utilization of the reference trajectory alone should improve system performance without the computational complications of estimation/adaption. Consider the reference trajectory as derived previously: $\ddot{q}_r = \ddot{q}_d - \lambda(\dot{q} - \dot{q}_d)$ and $\dot{q}_r = \dot{q}_d - \lambda(q - q_d)$. This trajectory adds/subtracts a little extra amount (the previous integral scaled by a positive constant). If the system is lagging behind the desired angle for example, that lag is scaled and added to the reference velocity trajectory resulting in more control inputs. Since we use measurements to generate the reference command, it seems intuitively appropriate for feedback control. Nonetheless, it is implemented in feedforward, feedback, and both for completeness sake.

Referencing Figure 16, note that the reference trajectory with feedforward control *only* with a correctly modeled system is not effective. This makes sense, since the feedforward control on a correctly modeled plant with no disturbances was previously demonstrated to perform well (Figure 14) while unrealistic for real world systems.

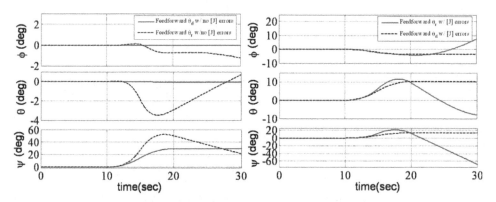

Fig. 16. LEFT: Feedforward (only) control with correctly modeled inertia. RIGHT: Feedforward (only) control with inertia errors.

Next, consider the reference trajectory for a system that is not well modeled. As we saw previously (Figure 15), open loop control when the inertia is increased results in the system falling short of the desired maneuver. The control is designed for a lighter spacecraft. We see in Figure 16 that feedforward control alone with a reference trajectory fairs no better. As a matter of fact, the performance is worse. Addition of feedback control seems appropriate. Before examining feedback control added to feedforward control, first examine feedback control by itself so that we may see the effects of the reference trajectory. Notice in Figure 17 that when the model is well known (correct), feedback control works quite well, and system performance is dramatically improved using the reference trajectory. Again, this is intuitive since the control is given a little something extra to account for tracking errors. This is also important for us to remember when analyzing indirect adaptive control with a reference trajectory. Tracking performance can be improved considerably without the complications of inertia estimation/adaption if the system is the assumed model.

When the model is not known, or has changed considerably from its assumed form, the performance improvement using the reference trajectory is not as pronounced as just seen with a well known model. Figure 17 illustrates that system damping has been reduced by the addition of the reference trajectory. The initial response is much faster, but there is overshoot and oscillatory settling. Notice in this example the two plots settle in similar times, so use of the reference trajectory has not drastically improved or degraded system performance.

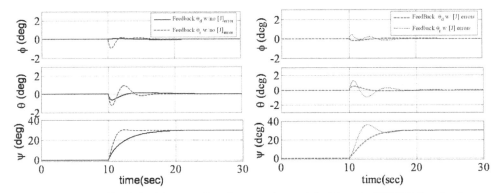

Fig. 17. LEFT: Feedback (only) control with correctly modeled inertia. RIGHT: Feedback (only) control with inertia errors.

Thus far, we see that the reference trajectory does not improve system performance when using feedforward control alone, but can improve performance with feedback control alone especially when the system inertia is known. Next, consider combined feedback & feedforward control. Figure 18 reveals expected results. Feedforward and feedback control with a reference trajectory is superior to using the desired trajectory when the plant model is known (no inertia errors). Similarly to the previous results, the reference trajectory with high inertia errors reduces system damping and exhibits faster response with overshoot and oscillatory settling. To conclude the evaluation of control with the reference trajectory without adaption/estimation, consider using the reference trajectory for feedback only and maintain the desired trajectory to formulate the feedforward control.

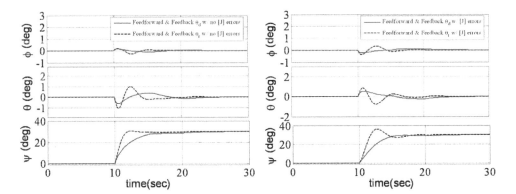

Fig. 18. LEFT: Feedforward & Feedback control with correctly modeled inertia. RIGHT: Feedforward & feedback control with inertia errors.

Notice in Figure 18 the system performance using the reference signal for both feedback and feedforward. This leaves us with a good understanding of how the reference trajectory affects the controlled system. To generalize:

Feedback control may be improved by utilization of a reference trajectory that adds a component scaled on the previous integral tracking error. When the system model is known, performance is improved drastically. In the example, J_{zz} was altered >100% and the reference trajectory still effectively controlled the spacecraft yaw maneuver.

Such reference trajectories are not advisable for feedforward control. Use of the reference trajectory in feedforward control does not improve system performance even in combination with feedback control.

Now that we have a good understanding that reference trajectories can improve system performance without estimation/adaption, let's continue by examining indirect adaptive control without the reference trajectory.

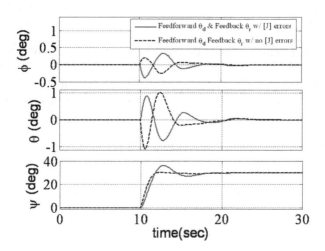

Fig. 19. Feedforward θ_d & Feedback θ_r with and without inertia errors.

9.5 Adaption without reference trajectory

Figure 20 displays a comparison of indirect adaptive control with and without a reference trajectory. In both cases, estimates are used to update a feedforward signal. The former case feeds the reference signal is generated by adding the scaled previous integral (scaled by a positive constant λ) as previously discussed. The latter case sets $\lambda=0$ making the reference trajectory equal to the desired (commanded) trajectory. The figure reveals that adaption//estimation alone does not produce good control. The reference trajectory is a key piece of the control scheme's effectiveness. This is intuitive having established the significance of the reference trajectory in previous sections of this study.

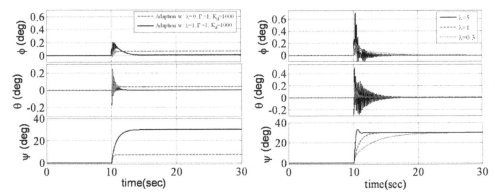

Fig. 20. LEFT: Indirect adaptive control with and without reference trajectory. RIGHT: Effects of scale constant λ on indirect adaptive control with reference trajectory.

9.6 Adaption with reference trajectory

Having established adaptive feedforward control is most effective with a reference trajectory; the following section iterates the design scale constant, λ. As seen in Figure 20, lower values of scale constant, λ result in slower controlled response. As λ is increased, system response is faster, but oscillations are increased. Scale constant value between one and five result in good performance preferring a value closer to one to avoid the oscillatory response.

10. Conclusions

Physics based control is a method that seeks to significantly incorporate the dominant physics of the problem to be controlled into the control design. Some components of the methods include elimination of zero-virtual reference, observers for sensor replacements, manipulated input decoupling, and disturbance-input estimation and decoupling. As pointing requirements have become more stringent to accomplish military missions in space, decoupling dynamic disturbance torques is an attractive solution provided by the physics-based control design methodology. Approaches demonstrated in this paper include elimination of virtual-zero references, manipulated input decoupling, sensor replacement and disturbance input decoupling. This paper compares the performance of the physics-based control to control methods found in the literature typically including cascaded control topology and neglecting factors such as back-emf. Another benefit of using the dynamics derived from the predominant physics of the controlled system lies in that an idealized feedforward results that can easily be augmented with adaptive technique to learn a better command while on-orbit and also assist with system identification. .

11. References

[1] C. J. Kempf and S. Kobayashi, "Disturbance observer and feedforward design of a high-speed direct-drive positioning table", IEEE Trans. On Control Systems Tech., vol. 7, no. 5, Sep., 1999.

[2] Tesfaye, H. S. Lee and M. Tomizuka, "A sensitivity optimization approach to design of a disturbance observer in digital motion control systems", IEEE/ASME Trans. on Mechatronics, vol. 5, no. 1, March, 2000.

[3] K.K. Tan, T. H. Lee, H. F. Dou, S. J. Chin, and Shao Zhao, "Precision motion control with disturbance observer for pulsewidth-modulated-driven permanent-magnet linear motors," IEEE Trans. on Magnetics, vol. 39, no. 3, May, 2003.

[4] C. J. Kempf and S. Kobayashi, "Disturbance observer and feedforward design of a high-speed direct-drive positioning table", IEEE Trans. on Control Systems Tech., vol. 7, no. 5, Sep., 1999.

[5] Tesfaye, H. S. Lee and M. Tomizuka, "A sensitivity optimization approach to design of a disturbance observer in digital motion control systems", IEEE/ASME Trans. on Mechatronics, vol. 5, no. 1, March, 2000.

[6] K.K. Tan, T. H. Lee, H. F. Dou, S. J. Chin, and Shao Zhao, "Precision motion control with disturbance observer for pulsewidth-modulated-driven permanent-magnet linear motors," IEEE Trans. on Magnetics, vol. 39, no. 3, May, 2003.

[7] S. M. Yang and Y. J. Deng, "Observer-based inertia identification for auto-tuning servo motor-drives", in Industry Applications Conference, vol. 2, 2005, pp. 968-972.

[8] M. Z. Liu, T. Tsuji, and T. Hanamato, "Position control of magnetic levitation transfer system by pitch angle," Journal of Power Electronics, vol. 6, no. 3, July 2006.

[9] T. Ohmae, T. Matsuda, K. Kaniyama, and M. Tachikawa, "A microprocessor-controlled high-accuracy wide-range speed regulator for motor drives," IEEE Trans. on Ind. Electron., vol. 29, no. 3, August, 1982.

[10] J. K. Kim, J. W. Choi, and S. K. Sul, "High performance position control of linear permanent magnet synchronous motor for surface mount device in placement system," in Conf. Rec. PCC-Osaka, vol. 1, 2002, pp. 37-42.

[11] Yoo, Y. D. Yoon, S. K. Sul, M. Hisatune, and S. Morimoto, "Design of a current regulator with extended bandwidth for servo motor drive," in Conf. Rec. PCC-Nagoya, 2007.

[12] Y. D. Yoon, E. Jung, A. Yoo, and S. K. Sul, "Dual observers for the disturbance rejection of a motion control system," in Conf. Rec. 42nd IAS Annual Meeting, 2007, pp. 256-261.

[13] Topographies taken from ME746 course notes, University of Wisconsin at Madison

[14] S. M. Yang, Y. J. Deng, "Observer-based inertial identification for auto-tuning servo motor drives," in Industry Application Conference, vol. 2, 2005, pp. 968-972.

[15] Ahmed, J. "Asymptotic Tracking of Spacecraft Attitude Motion with Inertia Identification", AIAA Journal of Guidance, Dynamics and Control, Sep-Oct 1998.

[16] Cristi, R., "Adaptive Quaternion Feedback Regulation for Eigenaxis Rotation", AIAA Journal of Guidance, Dynamics and Control, Nov-Dec 1994.

[17] Sanya, A. "Globally Convergent Adaptive Tracking of Spacecraft Angular Velocity with Inertia Identification", Proceedings of IEEE Conference of Decision and Control, 2003.

[18] Niemeyer, G. and Slotine, J.J.E, "Performance in adaptive manipulator control", Proceedings of 27th IEEE Conference on Decision and Control, Decemebr, 1988.

[19] Slotine, J.J.E. and Benedetto, M.D.Di, "Hamiltonian Adaptive Control of Spacecraft", IEEE Transactions on Automatic Control, Vol. 35, pp. 848-852, July 1990.

[20] Fossen, T. "Comments on 'Hamiltonian Adaptive Control of Spacecraft' ", IEEE Transactions on Automatic Control, Vol. 38., No. 4, April 1993.

2

Spacecraft Relative Orbital Motion

Daniel Condurache
"Gheorghe Asachi" Technical University of IASI
Romania

1. Introduction

The relative orbital motion problem may now be considered classic, because of so many scientific papers written on this subject in the last few decades. This problem is also quite important, due to its numerous applications: spacecraft formation flying, rendezvous operations, distributed spacecraft missions.

The model of the relative motion consists in two spacecraft flying in Keplerian orbits due to the influence of the same gravitational attraction center (see Fig. 1). The main problem is to determine the position and velocity vectors of the Deputy satellite with respect to a reference frame originated in the Leader satellite center of mass. This non-inertial reference frame, traditionally named LVLH (Local-Vertical-Local-Horizontal) is chosen as follows: the C_x axis has the same orientation as the position vector of the Leader with respect to an inertial reference frame originated in the attraction center; the C_z axis has the same orientation as the Leader orbit angular momentum; the C_y axis completes a right-handed frame.

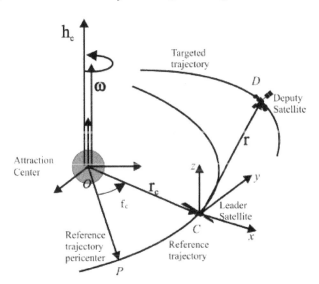

Fig. 1. The model of the relative orbital motion.

Consider $\omega = \omega(t)$ the angular velocity of the LVLH reference frame with respect to an inertial frame originated in the attraction center. By denoting $\mathbf{r}_c$ the Leader position vector with respect to an inertial frame originated in O (the attraction center), $f_c = f_c(t)$ the true anomaly, e_c the eccentricity and p_c the semilatus rectum of the Leader orbit, it follows that vector ω has the expression:

$$\omega = \dot{f}_c \frac{\mathbf{h}_c}{h_c} = \frac{1}{r_c^2}\mathbf{h}_c = \left[\frac{1 + e_c \cos f_c(t)}{p_c}\right]^2 \mathbf{h}_c, \tag{1}$$

where vector $\mathbf{r}_c$ is expressed with respect to the LVLH frame and has the form

$$\mathbf{r}_c = \frac{p_c}{1 + e_c \cos f_c(t)} \frac{\mathbf{r}_c^0}{r_c^0}, \tag{2}$$

and $\mathbf{h}_c$ is the angular momentum of the leader which will be named in the following satellite chief (or chief).

Vector $\mathbf{r}_c^0$ points to the initial position of the Leader spacecraft with respect to the inertial reference frame originated in the attraction center O. The initial value problem that models the motion of the Deputy satellite with respect to the LVLH reference frame is

$$\begin{cases} \ddot{\mathbf{r}} + 2\omega \times \dot{\mathbf{r}} + \omega \times (\omega \times \mathbf{r}) + \dot{\omega} \times \mathbf{r} + \frac{\mu}{|\mathbf{r}_c + \mathbf{r}|^3}(\mathbf{r}_c + \mathbf{r}) - \frac{\mu}{r_c^3}\mathbf{r}_c = 0 \\ \mathbf{r}(t_0) = \Delta\mathbf{r},\ \dot{\mathbf{r}}(t_0) = \Delta\mathbf{v} \end{cases} \tag{3}$$

where $\mu > 0$ is the gravitational parameter of the attraction center and $\Delta\mathbf{r}; \Delta\mathbf{v}$ represent the relative position and relative velocity vectors of the Deputy spacecraft with respect to LVLH at the initial moment of time $t_0 \geq 0$.

The analysis of relative motion began in the early 1960s with the paper of Clohessy and Wiltshire (Clohessy & Wiltshire (1960)), who obtained the equations that model the relative motion in the situation in which the chief spacecraft has a circular orbit and the attraction force is not affected by the Earth oblateness. They linearized the nonlinear initial value problem that models the relative motion by assuming that the relative distance between the two spacecraft remains small during the mission. The Clohessy - Wiltshire equations are still used today in rendezvous maneuvers, but they cannot offer a long-term accuracy because of the secular terms present in the expression of the relative position vector. Independently, Lawden (Lawden (1963)), Tschauner and Hempel (Tschauner & Hempel (1964)), and Tschauner (Tschauner (1966)) obtained the solution to the linearized equations of motion in the situation in which the chief orbit is elliptic, but their solutions still involved secular terms and also had singularities. The singularities in the Tschauner - Hempel equations were removed firstly by Carter (Carter (1990)) and also by Yamanaka and Andersen (Yamanaka & Andersen (2002)). Later on, the formation flying concept began to be considered, and the problem of deriving equations for the relative motion with a long-term accuracy degree raised, together with the need to obtain a more accurate solution to the relative orbital motion problem (Alfriend et al. (2009)). Gim and Alfriend (Gim & Alfriend (2003)) used the state transition matrix in the study of the relative motion.

The main goal was to express the linearized equations of motion with respect to the initial conditions, with applications in formation initialization and reconfiguration. Attempts to offer more accurate equations of motion starting from the nonlinear initial value problem that models the motion were made. Gurfil and Kasdin (Gurfil & N.J.Kasdin (2004)) derived closed-form expression of the relative position vector, but only when the reference trajectory is circular. Similar expressions for the law of relative motion starting from the nonlinear model are presented in (Alfriend et al. (2009); Balaji & Tatnall (2003); Ketema (2006); Lee et al. (2007)). The relative orbital motion problem was also studied from the point of view of the associated differential manifold. Gurfil and Kholshevnikov (Gurfil & Kholshevnikov (2006)) introduced a metric which helps to study the relative distance between Keplerian orbits. Gronchi (Gronchi (2006),Gronchi (2005)) also introduced a metric between two confocal Keplerian orbits and used this instrument in problems of asteroid and comet collisions.

In 2007, Condurache and Martinusi (Condurache & Martinusi (2007b;c)) offered the closed-form solution to the nonlinear unperturbed model of the relative orbital motion. The method led to closed-form vectorial coordinate-free expressions for the relative law of motion and relative velocity and it was based on an approach first introduced in 1995 (Condurache (1995)). It involves the Lie group of proper orthogonal tensor functions and its associated Lie algebra of skew-symmetric tensor functions. Then, the solution was generalized to the problem of the relative motion in a central force field (Condurache & Martinusi (2007e; 2008a;b)). An inedite solution to the Kepler problem by using the algebra of hypercomplex numbers was offered in (Condurache & Martinusi (2007d)). Based on this solution and by using the hypercomplex eccentric anomaly, a unified closed-form solution to the relative orbital motion was determined (Condurache & Martinusi (2010a)).

The present approach offers a tensor procedure to obtain exact expressions for the relative law of motion and the relative velocity between two Keplerian confocal orbits. The solution is obtained by pure analytical methods and it holds for any chief and deputy trajectories, without involving any secular terms or singularities. The relative orbital motion is reduced, by an adequate change of variables, into the classic Kepler problem. It is proved that the relative orbital motion problem is superintegrable. The tensor play only a catalyst role, the final solution being expressed in a vectorial form.

To obtain this solution, one has to know only the inertial motion of the chief spacecraft and the initial conditions (position and velocity) of the deputy satellite in the local-vertical-local-horizontal (LVLH) frame. Both the relative law of motion and the relative velocity of the deputy are obtained, by using the tensor instrument that is developed in the first part of the paper. Another contribution is the expression of the solution to the relative orbital motion by using universal functions, in a compact and unified form. Once the closed form solution is given a comprehensive analysis of the relative orbital motion of satellites is presented. Next the periodicity conditions in the relative orbital motion are revealed and in the end a tensor invariant in the relative motion is highlighted. The tensor invariant is a very useful propagator for the state of the deputy spacecraft in the LVLH frame.

2. Mathematical preliminaries

The key notions that are studied in this Section are proper orthogonal tensorial maps and a Sundman-like vectorial regularization, the latter introduced via a vectorial change of variable. The proper orthogonal tensorial maps are related with the skew-symmetric tensorial maps via the Darboux equation. The results presented in this section appeared for the first time in (Condurache (1995)). The section related to orthogonal tensorial maps after a powerful instrument in the study of the motion with respect to a non-inertial reference frames.

2.1 Proper orthogonal tensorial maps

We denote $SO_3^{\mathbb{R}}$ the set of maps defined on the set of real numbers $\mathbb{R}$ with values in the set of proper orthogonal tensors SO_3:

$$SO_3^{\mathbb{R}} = \left\{ \mathbf{R} : \mathbb{R} \to SO_3 | \mathbf{R}\mathbf{R}^T = \mathbf{I}_3, \det \mathbf{R} = 1 \right\} \tag{4}$$

We denote so_3^{R} the set of maps defined on the set of real numbers $\mathbb{R}$ with values in the set of skew-symmetric tensors so_3^{R}:

$$so_3^{R} = \left\{ \widetilde{\omega} : \mathbb{R} \to so_3 | \widetilde{\omega}^T = -\widetilde{\omega} \right\} \tag{5}$$

We denote $\mathbb{V}_3^{\mathbb{R}}$ to be the set of applications that can be on $\mathbb{R}$ with values in the free vectors set with dimension 3 ($\mathbb{V}_3$).

Theorem 1. *The initial value problem:*

$$\dot{\mathbf{Q}} + \widetilde{\omega}\mathbf{Q} = \mathbf{0}_2, \mathbf{Q}(t_0) = \mathbf{I}_3 \tag{6}$$

has a unique solution $\mathbf{Q} \in SO_3^{\mathbb{R}}$ *for any continuous map* $\widetilde{\omega} \in so_3^{\mathbb{R}}$.

Proof. Denote $\mathbf{Q}^T$ the transpose of tensor $\mathbf{Q}$. Computing:

$$\frac{d}{dt}(\mathbf{Q}\mathbf{Q}^T) = \dot{\mathbf{Q}}\mathbf{Q}^T + \mathbf{Q}\dot{\mathbf{Q}}^T = \mathbf{Q}\widetilde{\omega}\mathbf{Q}^T - \mathbf{Q}\widetilde{\omega}\mathbf{Q}^T = \mathbf{0}_3 \tag{7}$$

it follows that

$$\mathbf{Q}\mathbf{Q}^T = \mathbf{Q}\mathbf{Q}^T(t_0) = \mathbf{I}_3 \tag{8}$$

Since $\mathbf{Q} = \mathbf{Q}(t)$ is a continuous map, $t \geq t_0$, it follows that $\det(\mathbf{Q})$ is a continuous map too. From Eq. (8) it results $\det(\mathbf{Q}) \in [-1, 1]$. Since $\det(\mathbf{Q}(t_0)) = \det \mathbf{I}_3 = 1$, it follows that:

$$\begin{cases} \mathbf{Q}\mathbf{Q}^T = \mathbf{I}_3 \\ \det(\mathbf{Q}) = 1 \end{cases} \tag{9}$$

therefore $\mathbf{Q} \in SO_3^{\mathbb{R}}$ is a proper orthogonal tensor map.

Equation (6) represents the tensor form of the Darboux equation (Condurache & Martinusi (2010b); Darboux (1887)). Its solution will be denoted $\mathbf{R}_{-\omega}$. It models the rotation with instantaneous angular velocity $-\omega$ (ω is the vectorial map associated to the skew-symmetric tensor $\widetilde{\omega}$). The link between them is given by: $\widetilde{\omega}\mathbf{x} = \omega \times \mathbf{x}, \forall \mathbf{x} \in \mathbb{V}_3^{\mathbb{R}}$; where $\mathbb{V}_3$ is the three-dimensional linear space of free vectors and $" \times "$ denotes the cross product.

The inverse (in this case the transpose) of tensor $\mathbf{R}_{-\omega}$ is denoted:

$$\mathbf{R}_{-\omega}^T = \mathbf{F}_\omega \tag{10}$$

Theorem 2. *The tensor map* F_ω *satisfies:*

1. F_ω *is invertible and* $F_\omega^{-1} = F_\omega^T$

2. $F_\omega u \cdot F_\omega u = u \cdot v, \forall u, v \in \mathbb{V}_3^{\mathbb{R}}$

3. $|F_\omega u| = |u|, \forall u \in \mathbb{V}_3^{\mathbb{R}}$

4. $F_\omega(u \times u) = F_\omega u \times F_\omega v, \forall u, v \in \mathbb{V}_3^{\mathbb{R}}$

5. $\frac{d}{dt} F_\omega u = F_\omega(\dot{u} + \omega \times u), \forall u \in \mathbb{V}_3^{\mathbb{R}}$, *differentiable*

6. $\frac{d^2}{dt^2} F_\omega u = F_\omega(\ddot{u} + 2\omega \times \dot{u} + \omega \times (\omega \times u) + \dot{\omega} \times u), \forall u \in \mathbb{V}_3^{\mathbb{R}}$.

If vector ω has fixed direction, given by the unit vector $u; \omega = \omega(t)u$ with ω a continuous real valued map, the Darboux equation (6) has the explicit solution:

$$R_{-\omega} = I_3 - (\sin \varphi)\tilde{u} + (1 - \cos \varphi)\tilde{u}^2 \tag{11}$$

where $\varphi(t) = \int_{t_0}^{t} \omega(s) ds$

Following from Eq (11), if vector ω is constant and nonzero, the solution to the Darboux equation (6) is written as:

$$R_{-\omega} = I_3 - [\sin \omega(t - t_0)]\frac{\tilde{\omega}}{\omega} + [1 - \cos \omega(t - t_0)]\frac{\tilde{\omega}^2}{\omega}. \tag{12}$$

3. Closed-form solution to the relative orbital motion problem

3.1 Vectorial solutions

In this section we present the closed-form exact solution to Eq. (3). In the initial value problem (3), we make the change of variable:

$$r_* = F_\omega(r + r_c) \tag{13}$$

where r_c is the solution of the problem:

$$\begin{cases} \ddot{r}_c + 2\omega \times \dot{r}_c + \omega \times (\omega \times r_c) + \dot{\omega} \times r_c - \frac{\mu}{r_c^3} r_c = 0 \\ r_c(t_0) = r_c^0, \ \dot{r}_c(t_0) = \dot{r}_c^0 \end{cases} \tag{14}$$

After some algebra, it follows that

$$\ddot{r}_* = F_\omega \left\{ (\ddot{r} + \ddot{r}_c) + 2\omega \times (\dot{r} + \dot{r}_c) + \omega \times (\omega \times (r + r_c)) + \dot{\omega} \times (r + r_c) \right\} \tag{15}$$

and furthermore

$$\ddot{r}_* = F_\omega \left\{ \ddot{r} + 2\omega \times \dot{r} + \omega \times (\omega \times r) + \dot{\omega} \times r \right\} + F_\omega \left\{ \ddot{r}_c + 2\omega \times \dot{r}_c + \omega \times (\omega \times r_c) + \dot{\omega} \times r_c \right\} \tag{16}$$

Using Eqs. (3) and (14) we obtain:

$$\ddot{r}_* = F_\omega \left[\frac{\mu}{r_c^3} r_c - \frac{\mu}{|r + r_c|^3}(r + r_c) - \frac{\mu}{r_c^3} r_c \right] = -\frac{\mu}{|r + r_c|^3} F_\omega(r + r_c) \tag{17}$$

which leads to:

$$\ddot{\mathbf{r}}_* + \frac{\mu}{r_*^3}\mathbf{r}_* = 0 \tag{18}$$

The initial conditions for equation (18) are deduced by taking into account that $\mathbf{F}_\omega(t_0) = \mathbf{I}_3$ and Eq. (13):

$$\mathbf{r}_*(t_0) = \mathbf{r}_c^0 + \Delta\mathbf{r} \tag{19}$$

$$\dot{\mathbf{r}}_*(t_0) = \mathbf{v}_c^0 + \Delta\mathbf{v} + \boldsymbol{\omega}(t_0) \times \Delta\mathbf{r} \tag{20}$$

where $\mathbf{r}_c^0 = \mathbf{r}_c(t_0)$, $\mathbf{v}_c^0 = \dot{\mathbf{r}}_c(t_0) + \boldsymbol{\omega}(t_0) \times \mathbf{r}_c^0$.

From (10) and (13) we deduce:

$$\mathbf{r} = \mathbf{R}_{-\omega}\mathbf{r}_* - \mathbf{r}_c \tag{21}$$

The above considerations lead to the main result of this paper. This is stated thus: the solution to the relative orbital motion problem, described by the initial value problem (3) is:

$$\boxed{\mathbf{r} = \mathbf{R}_{-\omega}\mathbf{r}_* - \frac{p_c}{1 + e_c \cos f_c(t)} \frac{\mathbf{r}_c^0}{r_c^0}} \tag{22}$$

where $\mathbf{R}_{-\omega}$ is the solution of Eq. (6) and $\mathbf{r}_*$ is the solution to the initial value problem:

$$\ddot{\mathbf{r}}_* + \frac{\mu}{r_*^3}\mathbf{r}_* = 0; \quad \mathbf{r}_*(t_0) = \mathbf{r}_c^0 + \Delta\mathbf{r}; \quad \dot{\mathbf{r}}_*(t_0) = \mathbf{v}_c^0 + \Delta\mathbf{v} + \boldsymbol{\omega}(t_0) \times \Delta\mathbf{r} \tag{23}$$

and the relative velocity may be computed as:

$$\boxed{\mathbf{v} = \mathbf{R}_{-\omega}\dot{\mathbf{r}}_* - \tilde{\omega}\mathbf{R}_{-\omega}\mathbf{r}_* - \frac{e_c|\mathbf{h}_c|\sin f_c(t)}{p_c} \frac{\mathbf{r}_c^0}{r_c^0}} \tag{24}$$

This result shows a very interesting property of the relative orbital motion problem (3). We have proven that this problem is super-integrable, by reducing it to the classic Kepler problem (23). The solution of the relative orbital motion problem is expressed thus:

$$\mathbf{r} = \mathbf{r}(t, t_0, \Delta\mathbf{r}, \Delta\mathbf{v}); \quad \mathbf{v} = \mathbf{v}(t, t_0, \Delta\mathbf{r}, \Delta\mathbf{v}) \tag{25}$$

The Kepler problem (23) satisfies the prime integral of energy:

$$\frac{\dot{\mathbf{r}}_*^2}{2} - \frac{\mu}{r_*} = \zeta. \tag{26}$$

Taking into account (22), (24) and (26) results that the problem which models the motion of the Deputy satellite with respect to the LVLH frame Eq. (3) has the following prime integral

$$\frac{\mathbf{v}^2}{2} - V(\mathbf{r}, \dot{\mathbf{r}}, t) = \zeta \tag{27}$$

where $V = V(\mathbf{r}, \dot{\mathbf{r}}, t)$ is the generalized potential defined by:

$$V(\mathbf{r}, \dot{\mathbf{r}}, t) = (\boldsymbol{\omega}, \mathbf{r}, \dot{\mathbf{r}}) + \frac{1}{2}(\boldsymbol{\omega} \times \mathbf{r})^2 + \frac{\mu}{|\mathbf{r} + \mathbf{r}_c|} - \frac{\mu}{r_c^3}\mathbf{r} \cdot \mathbf{r}_c \tag{28}$$

and ζ

$$\zeta = \frac{1}{2}|\mathbf{v}_c^0 + \Delta\mathbf{v} + \boldsymbol{\omega}(t_0) \times \Delta\mathbf{r}|^2 - \frac{\mu}{|\mathbf{r}_c^0 + \Delta\mathbf{r}|}. \tag{29}$$

The prime integral (27) generates in the phase space a differential manifold associated to the relative orbital motion. The solutions (22) and (24) are a parametrization of this manifold.

3.2 An unified solution for relative orbital motion

Here, we present another formulation of the solution to the relative orbital motion. Let $U_k, k = \{0,1,2,3\}$, $U_k = U_k(\chi, \alpha)$ be the universal functions defined in (Battin (1999)), pp. 175-179, with

$$\alpha = \frac{2}{|\mathbf{r}_c^0 + \Delta \mathbf{r}|} - \frac{|\mathbf{v}_c^0 + \Delta \mathbf{v} + \boldsymbol{\omega}(t_0) \times \Delta \mathbf{r}|^2}{\mu} = -\mu \zeta \tag{30}$$

and χ a Sudman-like independent universal variable that satisfies

$$\frac{dt}{d\chi} = \frac{1}{\sqrt{\mu}} r_* \tag{31}$$

Then, the solution to the initial value problem (23) may be expressed as Eq. (25):

$$\mathbf{r}_* = \left\{ U_0 + \left[\frac{1}{|\mathbf{r}_c^0 + \Delta \mathbf{r}|} - \frac{|\mathbf{v}_c^0 + \Delta \mathbf{v} + \boldsymbol{\omega}(t_0) \times \Delta \mathbf{r}|^2}{\mu} \right] U_2 \right\} (\mathbf{r}_c^0 + \Delta \mathbf{r}) +$$
$$+ \left[U_1 \frac{|\mathbf{r}_c^0 + \Delta \mathbf{r}|}{\sqrt{\mu}} + U_2 \frac{(\mathbf{r}_c^0 + \Delta \mathbf{r}) \cdot (\mathbf{v}_c^0 + \Delta \mathbf{v} + \boldsymbol{\omega}(t_0) \times \Delta \mathbf{r})}{\mu} \right] \times (\mathbf{v}_c^0 + \Delta \mathbf{v} + \boldsymbol{\omega}(t_0) \times \Delta \mathbf{r}) \tag{32}$$

and the magnitude of the solution is:

$$r_* = |\mathbf{r}_c^0 + \Delta \mathbf{r}| U_0 + \frac{(\mathbf{r}_c^0 + \Delta \mathbf{r}) \cdot (\mathbf{v}_c^0 + \Delta \mathbf{v} + \boldsymbol{\omega}(t_0) \times \Delta \mathbf{r})}{\mu} U_1 + U_2 \tag{33}$$

The velocity of the motion governed by Eq.(23) is

$$\dot{\mathbf{r}}_* = -\frac{\sqrt{\mu}}{r_*} U_1 \frac{\mathbf{r}_c^0 + \Delta \mathbf{r}}{|\mathbf{r}_c^0 + \Delta \mathbf{r}|} + \frac{\sqrt{\mu}}{r_*} \left[U_0 \frac{|\mathbf{r}_c^0 + \Delta \mathbf{r}|}{\sqrt{\mu}} \right.$$
$$\left. + U_1 \frac{(\mathbf{r}_c^0 + \Delta \mathbf{r}) \cdot (\mathbf{v}_c^0 + \Delta \mathbf{v} + \boldsymbol{\omega}(t_0) \times \Delta \mathbf{r})}{\mu} \right] \times (\mathbf{v}_c^0 + \Delta \mathbf{v} + \boldsymbol{\omega}(t_0) \times \Delta \mathbf{r}) \tag{34}$$

Then, using (22) and (24) together with (32) and (34), the solution to the initial value problem (3) may be written as:

$$\mathbf{r} = \mathbf{R}_{-\omega} \left\{ \left\{ U_0 + \left[\frac{1}{|\mathbf{r}_c^0 + \Delta \mathbf{r}|} - \frac{|\mathbf{v}_c^0 + \Delta \mathbf{v} + \boldsymbol{\omega}(t_0) \times \Delta \mathbf{r}|^2}{\mu} \right] U_2 \right\} (\mathbf{r}_c^0 + \Delta \mathbf{r}) + \right.$$
$$\left. + \left[U_1 \frac{|\mathbf{r}_c^0 + \Delta \mathbf{r}|}{\sqrt{\mu}} + U_2 \frac{(\mathbf{r}_c^0 + \Delta \mathbf{r}) \cdot (\mathbf{v}_c^0 + \Delta \mathbf{v} + \boldsymbol{\omega}(t_0) \times \Delta \mathbf{r})}{\mu} \right] \right.$$
$$\left. \times (\mathbf{v}_c^0 + \Delta \mathbf{v} + \boldsymbol{\omega}(t_0) \times \Delta \mathbf{r}) \right\} - \frac{p_c}{1 + e_c \cos f_c(t)} \frac{\mathbf{r}_0}{r_0}. \tag{35}$$

$$\mathbf{v} = \mathbf{R}_{-\omega} \left\{ \frac{\sqrt{\mu}}{r_*} U_1 \frac{\mathbf{r}_c^0 + \Delta \mathbf{r}}{|\mathbf{r}_c^0 + \Delta \mathbf{r}|} + \frac{\sqrt{\mu}}{r_*} \left[U_0 \frac{|\mathbf{r}_c^0 + \Delta \mathbf{r}|}{\sqrt{\mu}} \right. \right.$$
$$\left. \left. + U_1 \frac{(\mathbf{r}_c^0 + \Delta \mathbf{r}) \cdot (\mathbf{v}_c^0 + \Delta \mathbf{v} + \boldsymbol{\omega}(t_0) \times \Delta \mathbf{r})}{\mu} \right] \times (\mathbf{v}_c^0 + \Delta \mathbf{v} + \boldsymbol{\omega}(t_0) \times \Delta \mathbf{r}) \right\}$$
$$- \tilde{\omega} \mathbf{R}_{-\omega} \left\{ \left\{ U_0 + \left[\frac{1}{|\mathbf{r}_c^0 + \Delta \mathbf{r}|} - \frac{|\mathbf{v}_c^0 + \Delta \mathbf{v} + \boldsymbol{\omega}(t_0) \times \Delta \mathbf{r}|^2}{\mu} \right] U_2 \right\} (\mathbf{r}_c^0 + \Delta \mathbf{r}) + \right.$$
$$\left. + \left[U_1 \frac{|\mathbf{r}_c^0 + \Delta \mathbf{r}|}{\sqrt{\mu}} + U_2 \frac{(\mathbf{r}_c^0 + \Delta \mathbf{r}) \cdot (\mathbf{v}_c^0 + \Delta \mathbf{v} + \boldsymbol{\omega}(t_0) \times \Delta \mathbf{r})}{\mu} \right] \right.$$
$$\left. \times (\mathbf{v}_c^0 + \Delta \mathbf{v} + \boldsymbol{\omega}(t_0) \times \Delta \mathbf{r}) \right\} - \frac{e_c |\mathbf{h}_c| \sin f_c(t)}{p_c} \frac{\mathbf{r}_c^0}{r_c^0}. \tag{36}$$

where $\mathbf{R}_{-\omega} = \mathbf{I}_3 - \sin f_c^0 \dfrac{\widetilde{\mathbf{h}}_c}{h_c} + (1 - \cos f_c^0) \dfrac{\widetilde{\mathbf{h}}_c^2}{h_c^2}$ and $f_c^0 = f_c(t) - f_c(t_0)$.

The universal functions U_k are linked by a Kepler-like equation (Battin (1999)):

$$\sqrt{\mu}(t - t_0) = U_1(\chi; \alpha)|\mathbf{r}_c^0 + \Delta \mathbf{r}| + U_2(\chi; \alpha)\frac{(\mathbf{r}_c^0 + \Delta \mathbf{r}) \cdot (\mathbf{v}_c^0 + \Delta \mathbf{v} + \boldsymbol{\omega}(t_0) \times \Delta \mathbf{r})}{\sqrt{\mu}} + U_3(\chi; \alpha) \tag{37}$$

Equations (35) and (36) offer the closed-form compact solution to the relative orbital motion problem. They hold for all types of reference trajectories of the chief and deputy (elliptic, parabolic, hyperbolic).

4. Comprehensive analysis of the relative orbital motion of satellites

By using the results presented in the previous sections, we are about to offer the closed-form solution to the relative orbital motion in all possible particular cases. In this approach, the chief inertial trajectory is less important than the deputy inertial trajectory, and the study will focus on the nature of the latter. We must make here the remark that in fact the initial value problem (23) models the motion of the deputy spacecraft in the inertial frame. This equation is deduced by knowing only the chief motion and the initial conditions of the deputy in the LVLH frame. From this point, when referring to the deputy inertial motion, we refer in fact to the motion governed by the initial value problem (23).

It is possible to obtain a closed-form solution to the nonlinear model of the relative orbital motion (3) in the situation where the inertial deputy trajectory is an ellipse, a parabola, or a hyperbola. These situations are delimited by the sign of the generalized specific energy of the deputy spacecraft (Battin (1999); Condurache & Martinusi (2007a)). It was proven that in the conditions that are given above, the sign of the quantity

$$\zeta = \frac{1}{2}|\mathbf{v}_c^0 + \Delta \mathbf{v} + \boldsymbol{\omega}(t_0) \times \Delta \mathbf{r}|^2 - \frac{\mu}{|\mathbf{r}_c^0 + \Delta \mathbf{r}|} \tag{38}$$

gives the type of the Keplerian inertial trajectory of the deputy spacecraft, i.e., if $\zeta < 0$ the inertial trajectory of the deputy is an ellipse, if $\zeta = 0$ it is a parabola, and if $\zeta > 0$ it is a hyperbola. An accurate observer would remark that the previous phrase is mathematically correct only if the angular momentum $\mathbf{h}$ of the deputy inertial orbit is nonzero, $\mathbf{h} \neq \mathbf{0}$,

$$\mathbf{h} = (\mathbf{r}_c^0 + \Delta \mathbf{r}) \times (\mathbf{v}_c^0 + \Delta \mathbf{v} + \boldsymbol{\omega}(t_0) \times \Delta \mathbf{r}). \tag{39}$$

Only this situation will be taken into consideration in this approach.

In the following the elliptic inertial deputy trajectory ($\zeta < 0, \mathbf{h} \neq \mathbf{0}$) will be analyzed. The inertial trajectory of the deputy spacecraft is an ellipse (or a circle). The motion on this orbit is modeled by the position vector $\mathbf{r}_*$, which is the solution to the initial value problem (23). The expressions for the vectors $\mathbf{r}_*$ and $\dot{\mathbf{r}}_*$ are:

$$\mathbf{r}_* = \mathbf{a}[\cos E(t) - e] + \mathbf{b} \sin E(t) \tag{40}$$

$$\dot{\mathbf{r}}_* = \frac{n}{1 - e \cos E(t)}[-\mathbf{a} \sin E(t) + \mathbf{b} \cos E(t)] \tag{41}$$

where $\mathbf{e}$ represents the vector corresponding to the vectorial eccentricity of the Keplerian motion described by Eq. (23); its expression is

$$\mathbf{e} = \frac{1}{\mu}(\mathbf{v}_c^0 + \Delta\mathbf{v} + \boldsymbol{\omega}(t_0) \times \Delta\mathbf{r}) \times \mathbf{h} - \frac{\mathbf{r}_c^0 + \Delta\mathbf{r}}{|\mathbf{r}_c^0 + \Delta\mathbf{r}|}. \tag{42}$$

If $\mathbf{e} = 0$, the inertial trajectory of the deputy spacecraft is circular; $\mathbf{h}$ is defined in Eq. (31); n is the mean motion of the motion described by Eq. (23); $\mathbf{a}$ and $\mathbf{b}$ represent the vectors that model the semimajor and semiminor axis of the deputy inertial trajectory respectively; their expressions are (Condurache & Martinusi (2007a)):

$$n = \frac{(2|\zeta|)^{\frac{3}{2}}}{\mu}; \mathbf{a} = \begin{cases} \frac{\mu}{2e|\zeta|}\mathbf{e}, & \mathbf{e} \neq 0 \\ \mathbf{r}_c^0 + \Delta\mathbf{r}, & \mathbf{e} = 0 \end{cases}; \mathbf{b} = \begin{cases} \frac{1}{e\sqrt{2|\zeta|}}(\mathbf{h} \times \mathbf{e}), & \mathbf{e} \neq 0 \\ \frac{1}{n}(\mathbf{v}_c^0 + \Delta\mathbf{v} + \boldsymbol{\omega}(t_0) \times \Delta\mathbf{r}), & \mathbf{e} = 0 \end{cases}; \tag{43}$$

$E(t)$ represents the deputy spacecraft eccentric anomaly; it is the solution to the Kepler equation:

$$E(t) - e\sin E(t) = n(t - t_p), \quad t \in [t_0, +\infty] \tag{44}$$

where t_p denotes the time of periapsis passage of the deputy spacecraft and it is computed from (Condurache & Martinusi (2007a))

$$t_p = t_0 - \frac{1}{n}[E(t_0) - e\sin E(t_0)] \tag{45}$$

while

$$\cos E(t_0) = \frac{1}{e}\left(1 - n\frac{|\mathbf{r}_c^0 + \Delta\mathbf{r}|}{\sqrt{2|\zeta|}}\right) \tag{46}$$

$$\sin E(t_0) = n\frac{\Delta\mathbf{v} \cdot (\mathbf{r}_c^0 + \Delta\mathbf{r})}{2e|\zeta|}\left[1 - \frac{\boldsymbol{\omega}(t_0) \cdot \mathbf{h}}{\mu}|\mathbf{r}_c^0 + \Delta\mathbf{r}|\right]. \tag{47}$$

From (22) and (24) combined with (40) and (41) the relative law of motion and the relative velocity are modeled by:

$$\mathbf{r} = [\cos E(t) - e]\left\{\frac{\mathbf{h}_c \cdot \mathbf{a}}{|\mathbf{h}_c|^2}\mathbf{h}_c - \sin f_c^0\frac{\tilde{\mathbf{h}}_c \cdot \mathbf{a}}{|\mathbf{h}_c|} - \cos f_c^0\frac{\tilde{\mathbf{h}}_c^2 \cdot \mathbf{a}}{|\mathbf{h}_c|^2}\right\} \\ + \sin E(t)\left\{\frac{\mathbf{h}_c \cdot \mathbf{b}}{|\mathbf{h}_c|^2}\mathbf{h}_c - \sin f_c^0\frac{\tilde{\mathbf{h}}_c \cdot \mathbf{b}}{|\mathbf{h}_c|} - \cos f_c^0\frac{\tilde{\mathbf{h}}_c^2 \cdot \mathbf{b}}{|\mathbf{h}_c|^2}\right\} - \frac{p_c}{1 + e_c\cos f_c(t)}\frac{\mathbf{r}_0}{r_0} \tag{48}$$

$$\mathbf{v} = \frac{-n\sin E(t)}{1 - e\cos E(t)}\left\{\frac{\mathbf{h}_c \cdot \mathbf{a}}{|\mathbf{h}_c|^2}\mathbf{h}_c - \sin f_c^0\frac{\tilde{\mathbf{h}}_c \cdot \mathbf{a}}{|\mathbf{h}_c|} - \cos f_c^0\frac{\tilde{\mathbf{h}}_c^2 \cdot \mathbf{a}}{|\mathbf{h}_c|^2}\right\} \\ + \frac{n\cos E(t)}{1 - e\cos E(t)}\left\{\frac{\mathbf{h}_c \cdot \mathbf{b}}{|\mathbf{h}_c|^2}\mathbf{h}_c - \sin f_c^0\frac{\tilde{\mathbf{h}}_c \cdot \mathbf{b}}{|\mathbf{h}_c|} - \cos f_c^0\frac{\tilde{\mathbf{h}}_c^2 \cdot \mathbf{b}}{|\mathbf{h}_c|^2}\right\} \\ + \frac{[1 + e_c\cos f_c(t)]^2[\cos E(t) - e]}{p_c^2} \times \left\{\frac{\sin f_c^0(t)}{|\mathbf{h}_c|}\tilde{\mathbf{h}}_c^2\mathbf{a} - \cos f_c^0(t)\tilde{\mathbf{h}}_c\mathbf{a}\right\} \\ + \frac{[1 + e_c\cos f_c(t)]^2[\sin E(t)]}{p_c^2} \times \left\{\frac{\sin f_c^0(t)}{|\mathbf{h}_c|}\tilde{\mathbf{h}}_c^2\mathbf{b} - \cos f_c^0(t)\tilde{\mathbf{h}}_c\mathbf{b}\right\} \\ - \frac{e_c|\mathbf{h}_c|\sin f_c(t)}{p_c}\frac{\mathbf{r}_c^0}{|\mathbf{r}_c^0|} \tag{49}$$

If the deputy trajectory is circular ($e = 0$), Eqs. (43) are taken into account, together with:

$$p = |\mathbf{r}_c^0 + \Delta\mathbf{r}|; \quad E(t) = \frac{|\mathbf{h}|}{|\mathbf{r}_c^0 + \Delta\mathbf{r}|^2}(t - t_0) \tag{50}$$

If the reference trajectory is circular, the closed-form Eqs. (48) and (49) change according to the following expressions:

$$e_c = 0; \quad f_c^0(t) = n_c(t - t_0) \tag{51}$$

It follows that in the situation when the chief spacecraft has an inertial circular trajectory, Eqs. (48) and (49) transform into

$$
\begin{aligned}
\mathbf{r} = [\cos E(t) - e] &\left\{ \frac{\mathbf{h}_c \cdot \mathbf{a}}{|\mathbf{h}_c|^2}\mathbf{h}_c - \sin(n_c(t - t_0))\frac{\widetilde{\mathbf{h}}_c \cdot \mathbf{a}}{|\mathbf{h}_c|} - \cos(n_c(t - t_0))\frac{\widetilde{\mathbf{h}}_c^2 \cdot \mathbf{a}}{|\mathbf{h}_c|^2} \right\} \\
+ \sin E(t) &\left\{ \frac{\mathbf{h}_c \cdot \mathbf{b}}{|\mathbf{h}_c|^2}\mathbf{h}_c - \sin(n_c(t - t_0))\frac{\widetilde{\mathbf{h}}_c \cdot \mathbf{b}}{|\mathbf{h}_c|} - \cos(n_c(t - t_0))\frac{\widetilde{\mathbf{h}}_c^2 \cdot \mathbf{b}}{|\mathbf{h}_c|^2} \right\} - \mathbf{r}_c^0
\end{aligned}
\tag{52}
$$

$$
\begin{aligned}
\mathbf{v} = \frac{-n \sin E(t)}{1 - e \cos E(t)} &\left\{ \frac{\mathbf{h}_c \cdot \mathbf{a}}{|\mathbf{h}_c|^2}\mathbf{h}_c - \sin(n_c(t - t_0))\frac{\widetilde{\mathbf{h}}_c \cdot \mathbf{a}}{|\mathbf{h}_c|} - \cos(n_c(t - t_0))\frac{\widetilde{\mathbf{h}}_c^2 \cdot \mathbf{a}}{|\mathbf{h}_c|^2} \right\} \\
+ \frac{n \cos E(t)}{1 - e \cos E(t)} &\left\{ \frac{\mathbf{h}_c \cdot \mathbf{b}}{|\mathbf{h}_c|^2}\mathbf{h}_c - \sin(n_c(t - t_0))\frac{\widetilde{\mathbf{h}}_c \cdot \mathbf{b}}{|\mathbf{h}_c|} - \cos(n_c(t - t_0))\frac{\widetilde{\mathbf{h}}_c^2 \cdot \mathbf{b}}{|\mathbf{h}_c|^2} \right\} \\
+ \frac{\cos E(t) - e}{|\mathbf{r}_c^0|^2} &\left\{ \frac{1}{|\mathbf{h}_c|}\sin[n_c(t - t_0)]\widetilde{\mathbf{h}}_c^2\mathbf{a} - \cos[n_c(t - t_0)]\widetilde{\mathbf{h}}_c\mathbf{a} \right\} \\
+ \frac{\sin E(t)}{|\mathbf{r}_c^0|^2} &\left\{ \frac{1}{|\mathbf{h}_c|}\sin[n_c(t - t_0)]\widetilde{\mathbf{h}}_c^2\mathbf{b} - \cos[n_c(t - t_0)]\widetilde{\mathbf{h}}_c\mathbf{b} \right\}
\end{aligned}
\tag{53}
$$

We make here the following remark: the equations (48) and (49) represent the generalization to the Tschauner-Hempel (TH) and Lawden solution. While TH and Lawden equations are the solution to the linearized model for the relative motion, the equations deduced here represent the solution to the nonlinear original model of the relative motion. They stand true for any elliptic targeted and reference trajectory. The Eqs. (52) and (53) generalize the Clohessy-Wiltshire model.

In the end of this subsection, we will present the closed-form exact expressions for the relative law of motion and velocity with respect to the eccentric anomalies in the situation when both chief and deputy are satellites (the ellipse-ellipse situation). From the Kepler equations written for both chief and deputy inertial motions

$$E_c - e_c \sin E_c = n_c(t - t_p^c) \tag{54}$$

$$E - e \sin E = n(t - t_p) \tag{55}$$

one may derive the implicit equation that links these anomalies by eliminating the time t from Eqs. (54) and (55):

$$\frac{E_c - e_c \sin E_c}{n_c} + t_p^c = \frac{E - e \sin E}{n} + t_p \tag{56}$$

As the motion of the chief satellite is known, so is function E_c. The eccentric anomaly of the Deputy satellite is then obtained by solving the implicit functional equation:

$$E - e \sin E = \frac{n}{n_c}(E_c - e_c \sin E_c) + n(t_p^c - t_p) \tag{57}$$

By taking into account the relations between the true anomaly and the eccentric anomaly of a Keplerian elliptic orbit

$$\begin{cases} \cos f = \dfrac{\cos E - e}{1 - e \cos E} \\ \sin f = \dfrac{\sqrt{1 - e^2} \sin E}{1 - e \cos E} \end{cases} , \tag{58}$$

equations (49) and (50) are transformed into:

$$\begin{aligned}
\mathbf{r} = [\cos E - e] &\left\{ \frac{\mathbf{h}_c \cdot \mathbf{a}}{|\mathbf{h}_c|^2} \mathbf{h}_c - \sqrt{1 - e_c^2} \frac{\sin(E_c^0 + E_c) - e_c(\sin E_c + \sin E_c^0)}{(1 - e_c \cos E_c)(1 - e_c \cos E_c^0)} \frac{\widetilde{\mathbf{h}}_c \mathbf{a}}{|\mathbf{h}_c|} \right. \\
&\left. - \frac{(\cos E_c - e_c)(\cos E_c^0 - e_c) - (1 - e_c^2) \sin E_c \sin E_c^0}{(1 - e_c \cos E_c)(1 - e_c \cos E_c^0)} \frac{\widetilde{\mathbf{h}}_c^2 \mathbf{a}}{|\mathbf{h}_c|^2} \right\} \\
+ \sin E &\left\{ \frac{\mathbf{h}_c \cdot \mathbf{b}}{|\mathbf{h}_c|^2} \mathbf{h}_c - \sqrt{1 - e_c^2} \frac{\sin(E_c^0 + E_c) - e_c(\sin E_c + \sin E_c^0)}{(1 - e_c \cos E_c)(1 - e_c \cos E_c^0)} \frac{\widetilde{\mathbf{h}}_c \mathbf{b}}{|\mathbf{h}_c|} \right. \\
&\left. - \frac{(\cos E_c - e_c)(\cos E_c^0 - e_c) - (1 - e_c^2) \sin E_c \sin E_c^0}{(1 - e_c \cos E_c)(1 - e_c \cos E_c^0)} \frac{\widetilde{\mathbf{h}}_c^2 \mathbf{b}}{|\mathbf{h}_c|^2} \right\} \\
&- \frac{p_c(1 - e_c \cos E_c)}{1 - e_c^2} \frac{\mathbf{r}_c^0}{|\mathbf{r}_c^0|}
\end{aligned} \tag{59}$$

$$\begin{aligned}
\mathbf{v} = \frac{-n \sin E}{1 - e \cos E} &\left\{ \frac{\mathbf{h}_c \cdot \mathbf{a}}{|\mathbf{h}_c|^2} \mathbf{h}_c - \sqrt{1 - e_c^2} \frac{\sin(E_c^0 + E_c) - e_c(\sin E_c + \sin E_c^0)}{(1 - e_c \cos E_c)(1 - e_c \cos E_c^0)} \frac{\widetilde{\mathbf{h}}_c \mathbf{a}}{|\mathbf{h}_c|} \right. \\
&\left. - \frac{(\cos E_c - e_c)(\cos E_c^0 - e_c) - (1 - e_c^2) \sin E_c \sin E_c^0}{(1 - e_c \cos E_c)(1 - e_c \cos E_c^0)} \frac{\widetilde{\mathbf{h}}_c^2 \mathbf{a}}{|\mathbf{h}_c|^2} \right\} \\
+ \frac{n \cos E}{1 - e \cos E} &\left\{ \frac{\mathbf{h}_c \cdot \mathbf{b}}{|\mathbf{h}_c|^2} \mathbf{h}_c - \sqrt{1 - e_c^2} \frac{\sin(E_c^0 + E_c) - e_c(\sin E_c + \sin E_c^0)}{(1 - e_c \cos E_c)(1 - e_c \cos E_c^0)} \frac{\widetilde{\mathbf{h}}_c \mathbf{b}}{|\mathbf{h}_c|} \right. \\
&\left. - \frac{(\cos E_c - e_c)(\cos E_c^0 - e_c) - (1 - e_c^2) \sin E_c \sin E_c^0}{(1 - e_c \cos E_c)(1 - e_c \cos E_c^0)} \frac{\widetilde{\mathbf{h}}_c^2 \mathbf{b}}{|\mathbf{h}_c|^2} \right\} \\
+ \frac{(1 - e_c^2)(\cos E - e)}{(1 - e_c \cos E_c)p_c^2} &\times \left\{ -\sqrt{1 - e_c^2} \frac{\sin(E_c^0 + E_c) - e_c(\sin E_c + \sin E_c^0)}{(1 - e_c \cos E_c)(1 - e_c \cos E_c^0)} \frac{\widetilde{\mathbf{h}}_c \mathbf{a}}{|\mathbf{h}_c|} \right. \\
&\left. - \frac{(\cos E_c - e_c)(\cos E_c^0 - e_c) - (1 - e_c^2) \sin E_c \sin E_c^0}{(1 - e_c \cos E_c)(1 - e_c \cos E_c^0)} \widetilde{\mathbf{h}}_c \mathbf{a} \right\} \\
+ \frac{(1 - e_c^2)(\sin E)}{(1 - e_c \cos E_c)p_c^2} &\times \left\{ -\sqrt{1 - e_c^2} \frac{\sin(E_c^0 + E_c) - e_c(\sin E_c + \sin E_c^0)}{(1 - e_c \cos E_c)(1 - e_c \cos E_c^0)} \frac{\widetilde{\mathbf{h}}_c \mathbf{b}}{|\mathbf{h}_c|} \right. \\
&\left. - \frac{(\cos E_c - e_c)(\cos E_c^0 - e_c) - (1 - e_c^2) \sin E_c \sin E_c^0}{(1 - e_c \cos E_c)(1 - e_c \cos E_c^0)} \widetilde{\mathbf{h}}_c \mathbf{b} \right\} \\
&- \frac{e_c |\mathbf{h}_c|(1 - e_c^2) \sin E_c}{(1 - e_c \cos E_c)p_c} \frac{\mathbf{r}_c^0}{|\mathbf{r}_c^0|}
\end{aligned} \tag{60}$$

where $E_c^0 = E_c(t_0)$.

4.1 Parametric Cartesian solution of relative orbital motion

In the following we present the scalar Cartesian expressions for the relative position and relative velocity as they are deduced from the expressions presented in this section. By denoting $\mathbf{r} = [x \ y \ z]^T$ the relative position vector, below we present the closed form expressions for $x, y, z, \dot{x}, \dot{y}, \dot{z}$. We denote $\mathbf{u}_x, \mathbf{u}_y, \mathbf{u}_z$ the unit vectors that define the axes of the LVLH frame; their expressions are

$$\mathbf{u}_x = \frac{\mathbf{r}_c^0}{|\mathbf{r}_c^0|}; \quad \mathbf{u}_y = \frac{\tilde{\mathbf{h}}_c \mathbf{r}_c^0}{|\mathbf{h}_c||\mathbf{r}_c^0|}; \quad \mathbf{u}_z = \frac{\mathbf{h}_c^0}{|\mathbf{h}_c^0|} \tag{61}$$

If $\zeta < 0, \mathbf{h} \neq \mathbf{0}$ then using (52) and (53) results:

$$
\begin{aligned}
x(t) = & [\cos E(t) - e] \left\{ (\mathbf{u}_x \cdot \mathbf{a}) \cos f_c^0(t) + (\mathbf{u}_y \cdot \mathbf{a}) \sin f_c^0(t) \right\} \\
& + \sin E(t) \left\{ (\mathbf{u}_x \cdot \mathbf{b}) \cos f_c^0(t) + (\mathbf{u}_y \cdot \mathbf{b}) \sin f_c^0(t) \right\} \\
& - \frac{p_c}{1 + e_c \cos f_c(t)}
\end{aligned}
\tag{62}
$$

$$
\begin{aligned}
y(t) = & [\cos E(t) - e] \left\{ (-\mathbf{u}_x \cdot \mathbf{a}) \sin f_c^0(t) + (\mathbf{u}_y \cdot \mathbf{a}) \cos f_c^0(t) \right\} \\
& + \sin E(t) \left\{ (-\mathbf{u}_x \cdot \mathbf{b}) \sin f_c^0(t) + (\mathbf{u}_y \cdot \mathbf{b}) \cos f_c^0(t) \right\}
\end{aligned}
\tag{63}
$$

$$z(t) = [\cos E(t) - e](\mathbf{u}_z \cdot \mathbf{a}) + \sin E(t)(\mathbf{u}_z \cdot \mathbf{b}) \tag{64}$$

$$
\begin{aligned}
\dot{x}(t) = & \frac{n \sin E(t)}{1 - e \cos E(t)} \left\{ (\mathbf{u}_x \cdot \mathbf{a}) \cos f_c^0(t) + (\mathbf{u}_y \cdot \mathbf{a}) \sin f_c^0(t) \right\} \\
& + \frac{n \cos E(t)}{1 - e \cos E(t)} \left\{ (\mathbf{u}_x \cdot \mathbf{b}) \cos f_c^0(t) + (\mathbf{u}_y \cdot \mathbf{b}) \sin f_c^0(t) \right\} \\
& - \frac{\mu [1 + e_c \cos f_c(t)]^2 [\cos E(t) - e]}{|\mathbf{h}|_c} \left\{ (-\mathbf{u}_x \cdot \mathbf{a}) \sin f_c^0(t) + (\mathbf{u}_y \cdot \mathbf{a}) \cos f_c^0(t) \right\} \\
& - \frac{\mu [1 + e_c \cos f_c(t)]^2 \sin E(t)}{|\mathbf{h}|_c} \left\{ (-\mathbf{u}_x \cdot \mathbf{a}) \sin f_c^0(t) + (\mathbf{u}_y \cdot \mathbf{a}) \cos f_c^0(t) \right\} \\
& - \frac{e_c |\mathbf{h}_c| \sin f_c(t)}{p_c}
\end{aligned}
\tag{65}
$$

$$
\begin{aligned}
\dot{y}(t) = & \frac{n \sin E(t)}{1 - e \cos E(t)} \left\{ (\mathbf{u}_x \cdot \mathbf{a}) \sin f_c^0(t) - (\mathbf{u}_y \cdot \mathbf{a}) \cos f_c^0(t) \right\} \\
& - \frac{n \cos E(t)}{1 - e \cos E(t)} \left\{ -(\mathbf{u}_x \cdot \mathbf{b}) \sin f_c^0(t) + (\mathbf{u}_y \cdot \mathbf{b}) \cos f_c^0(t) \right\} \\
& - \frac{|\mathbf{h}|_c [1 + e_c \cos f_c(t)]^2 [\cos E(t) - e]}{p_c} \left\{ (\mathbf{u}_y \cdot \mathbf{a}) \sin f_c^0(t) + (\mathbf{u}_x \cdot \mathbf{a}) \cos f_c^0(t) \right\} \\
& - \frac{|\mathbf{h}_c| [1 + e_c \cos f_c(t)]^2 \sin E(t)}{p_c} \left\{ (\mathbf{u}_y \cdot \mathbf{b}) \sin f_c^0(t) + (\mathbf{u}_x \cdot \mathbf{b}) \cos f_c^0(t) \right\}
\end{aligned}
\tag{66}
$$

$$\dot{z}(t) = \frac{n}{[1 - e \cos E(t)]} [-\sin E(t)(\mathbf{u}_z \cdot \mathbf{a}) + \cos E(t)(\mathbf{u}_z \cdot \mathbf{b})] \tag{67}$$

When the deputy trajectory is also an ellipse and one expresses the equations of the relative motion with respect to both eccentric anomalies, Eqs. (59-60) are transformed into:

$$
\begin{aligned}
x(t) = [\cos E(t) - e] &\left\{ \frac{(\cos E_c - e_c)(\cos E_c^0 - e_c) - (1 - e_c^2)\sin E_c \sin E_c^0}{(1 - e_c \cos E_c)(1 - e_c \cos E_c^0)}(\mathbf{u}_x \cdot \mathbf{a}) \right. \\
&\left. - \sqrt{1 - e_c^2}\frac{(\sin E_c^0 - e_c) - e_c(\sin E_c + \sin E_c^0)}{(1 - e_c \cos E_c)(1 - e_c \cos E_c^0)}(\mathbf{u}_y \cdot \mathbf{a}) \right\} \\
+ \sin E(t) &\left\{ \frac{(\cos E_c - e_c)(\cos E_c^0 - e_c) - (1 - e_c^2)\sin E_c \sin E_c^0}{(1 - e_c \cos E_c)(1 - e_c \cos E_c^0)}(\mathbf{u}_x \cdot \mathbf{a}) \right. \\
&\left. - \sqrt{1 - e_c^2}\frac{(\sin E_c^0 - e_c) - e_c(\sin E_c + \sin E_c^0)}{(1 - e_c \cos E_c)(1 - e_c \cos E_c^0)}(\mathbf{u}_y \cdot \mathbf{b}) \right\} - \frac{p_c(1 - e_c \cos E_c)}{1 - e_c^2}
\end{aligned}
\tag{68}
$$

$$
\begin{aligned}
y(t) = -[\cos E(t) - e] &\left\{ (\mathbf{u}_x \cdot \mathbf{a})\sin f_c^0(t) \right. \\
&\left. + \frac{(\cos E_c - e_c)(\cos E_c^0 - e_c) - (1 - e_c^2)\sin E_c \sin E_c^0}{(1 - e_c \cos E_c)(1 - e_c \cos E_c^0)}(\mathbf{u}_y \cdot \mathbf{a}) \right\} \\
- \sin E(t) &\left\{ -\sqrt{1 - e_c^2}\frac{(\sin E_c^0 - e_c) - e_c(\sin E_c + \sin E_c^0)}{(1 - e_c \cos E_c)(1 - e_c \cos E_c^0)}(\mathbf{u}_x \cdot \mathbf{b}) \right. \\
&\left. + \frac{(\cos E_c - e_c)(\cos E_c^0 - e_c) - (1 - e_c^2)\sin E_c \sin E_c^0}{(1 - e_c \cos E_c)(1 - e_c \cos E_c^0)}(\mathbf{u}_y \cdot \mathbf{b}) \right\}
\end{aligned}
\tag{69}
$$

$$
z(t) = [\cos E(t) - e](\mathbf{u}_z \cdot \mathbf{a}) + \sin E(t)(\mathbf{u}_z \cdot \mathbf{b})
\tag{70}
$$

$$
\begin{aligned}
\dot{x}(t) = \frac{n \sin E(t)}{1 - e \cos E(t)} &\left\{ \frac{(\cos E_c - e_c)(\cos E_c^0 - e_c) - (1 - e_c^2)\sin E_c \sin E_c^0}{(1 - e_c \cos E_c)(1 - e_c \cos E_c^0)}(\mathbf{u}_x \cdot \mathbf{a}) \right. \\
&\left. + \sqrt{1 - e_c^2}\frac{(\sin E_c^0 - e_c) - e_c(\sin E_c + \sin E_c^0)}{(1 - e_c \cos E_c)(1 - e_c \cos E_c^0)}(\mathbf{u}_y \cdot \mathbf{a}) \right\} \\
+ \frac{n \cos E(t)}{1 - e \cos E(t)} &\left\{ \frac{(\cos E_c - e_c)(\cos E_c^0 - e_c) - (1 - e_c^2)\sin E_c \sin E_c^0}{(1 - e_c \cos E_c)(1 - e_c \cos E_c^0)}(\mathbf{u}_x \cdot \mathbf{b}) \right. \\
&\left. + \sqrt{1 - e_c^2}\frac{(\sin E_c^0 - e_c) - e_c(\sin E_c + \sin E_c^0)}{(1 - e_c \cos E_c)(1 - e_c \cos E_c^0)}(\mathbf{u}_y \cdot \mathbf{b}) \right\} \\
- \frac{\mu[1 + e_c \cos f_c(t)]^2[\cos E(t) - e]}{|\mathbf{h}|_c} &\left\{ -\sqrt{1 - e_c^2}\frac{(\sin E_c^0 - e_c) - e_c(\sin E_c + \sin E_c^0)}{(1 - e_c \cos E_c)(1 - e_c \cos E_c^0)}(\mathbf{u}_x \cdot \mathbf{a}) \right. \\
&\left. + \frac{(\cos E_c - e_c)(\cos E_c^0 - e_c) - (1 - e_c^2)\sin E_c \sin E_c^0}{(1 - e_c \cos E_c)(1 - e_c \cos E_c^0)}(\mathbf{u}_y \cdot \mathbf{a}) \right\} \\
- \frac{\mu[1 + e_c \cos f_c(t)]^2 \sin E(t)}{|\mathbf{h}|_c} &\left\{ -\sqrt{1 - e_c^2}\frac{(\sin E_c^0 - e_c) - e_c(\sin E_c + \sin E_c^0)}{(1 - e_c \cos E_c)(1 - e_c \cos E_c^0)}(\mathbf{u}_x \cdot \mathbf{b}) \right. \\
&\left. + \frac{(\cos E_c - e_c)(\cos E_c^0 - e_c) - (1 - e_c^2)\sin E_c \sin E_c^0}{(1 - e_c \cos E_c)(1 - e_c \cos E_c^0)}(\mathbf{u}_y \cdot \mathbf{b}) \right\} - \frac{e_c|\mathbf{h}_c| \sin f_c(t)}{p_c}
\end{aligned}
\tag{71}
$$

$$\dot{y}(t) = -\frac{n\sin E(t)}{1 - e\cos E(t)}\left\{ -\frac{(\cos E_c - e_c)(\cos E_c^0 - e_c) - (1 - e_c^2)\sin E_c \sin E_c^0}{(1 - e_c\cos E_c)(1 - e_c\cos E_c^0)}(\mathbf{u}_y \cdot \mathbf{a}) \right.$$

$$\left. -\sqrt{1 - e_c^2}\frac{(\sin E_c^0 - e_c) - e_c(\sin E_c + \sin E_c^0)}{(1 - e_c\cos E_c)(1 - e_c\cos E_c^0)}(\mathbf{u}_x \cdot \mathbf{a}) \right\}$$

$$+\frac{n\cos E(t)}{1 - e\cos E(t)}\left\{ -\frac{(\cos E_c - e_c)(\cos E_c^0 - e_c) - (1 - e_c^2)\sin E_c \sin E_c^0}{(1 - e_c\cos E_c)(1 - e_c\cos E_c^0)}(\mathbf{u}_y \cdot \mathbf{b}) \right.$$

$$\left. +\sqrt{1 - e_c^2}\frac{(\sin E_c^0 - e_c) - e_c(\sin E_c + \sin E_c^0)}{(1 - e_c\cos E_c)(1 - e_c\cos E_c^0)}(\mathbf{u}_x \cdot \mathbf{b}) \right\}$$

$$-\frac{|\mathbf{h}_c|[1 + e_c\cos f_c(t)]^2[\cos E(t) - e]}{|\mathbf{h}|_c}\left\{ \sqrt{1 - e_c^2}\frac{(\sin E_c^0 - e_c) - e_c(\sin E_c + \sin E_c^0)}{(1 - e_c\cos E_c)(1 - e_c\cos E_c^0)}(\mathbf{u}_y \cdot \mathbf{a}) \right.$$

$$\left. +\frac{(\cos E_c - e_c)(\cos E_c^0 - e_c) - (1 - e_c^2)\sin E_c \sin E_c^0}{(1 - e_c\cos E_c)(1 - e_c\cos E_c^0)}(\mathbf{u}_x \cdot \mathbf{a}) \right\}$$

$$-\frac{|\mathbf{h}_c|[1 + e_c\cos f_c(t)]^2\sin E(t)}{|\mathbf{h}|_c}\left\{ \sqrt{1 - e_c^2}\frac{(\sin E_c^0 - e_c) - e_c(\sin E_c + \sin E_c^0)}{(1 - e_c\cos E_c)(1 - e_c\cos E_c^0)}(\mathbf{u}_y \cdot \mathbf{b}) \right.$$

$$\left. +\frac{(\cos E_c - e_c)(\cos E_c^0 - e_c) - (1 - e_c^2)\sin E_c \sin E_c^0}{(1 - e_c\cos E_c)(1 - e_c\cos E_c^0)}(\mathbf{u}_x \cdot \mathbf{b}) \right\} - \frac{e_c|\mathbf{h}_c|\sin f_c(t)}{p_c}$$

$$\tag{72}$$

$$\dot{z}(t) = \frac{n}{[1 - e\cos E(t)]}[-\sin E(t)(\mathbf{u}_z \cdot \mathbf{a}) + \cos E(t)(\mathbf{u}_z \cdot \mathbf{b})] \tag{73}$$

An interesting remark is that the motion along the Oz axis of LVLH (the out-of-plane motion) is completely decoupled from the in-plane motion.

5. Periodicity conditions in relative orbital motion

An interesting geometric visualization of the relative motion is illustrated in Fig. 2.

It may be seen as the composition among:

- a classic Keplerian motion in a variable plane $\Pi(t), t \geq t_0$; plane $\Pi(t)$ is formed at moment $t = t_0$ if the inertial motion of the Deputy satellite is not rectilinear; this plane is determined by the initial position and initial velocity vectors of the Deputy;
- a precession of plane $\Pi(t)$ with angular velocity $-\omega$ around the attraction center;
- a rectilinear translation of plane $\Pi(t)$ described by vector $-\mathbf{r}_c$.

This geometric interpretation shows that the relative orbital motion is in fact a Foucault pendulum like motion (Condurache & Martinusi (2008a)). Excluding the situation $\mathbf{h} = 0$, the case $\zeta \leq 0$ is equivalent with the Deputy elliptic inertial motion. If the Leader satellite also has an elliptic motion, then the motion of the Deputy with respect to the LVLH frame might be periodic. In fact, recall that:

$$\mathbf{r} = \mathbf{R}_{-\omega}\mathbf{r}_i - \mathbf{r}_c \tag{74}$$

is the maps $\mathbf{R}_{-\omega}$ and $\mathbf{r}_c$ have the same main period T_c, which is that of the Leader, and $\mathbf{r}_i$ has the main period of the Deputy inertial motion, denoted as T_d. The motion in LVLH is then

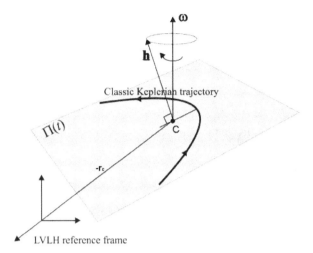

Fig. 2. Geometric interpretation of the relative orbital motion.

periodic if:

$$T_c / T_d \quad \text{is rational number} \tag{75}$$

This leads to a formula that involves the specific energies of the two satellites: the motion is periodic if

$$\left(\frac{\frac{1}{2}(\mathbf{v}_c^0 + \Delta \mathbf{v} + \boldsymbol{\omega}_0 \times \Delta \mathbf{r})^2 - \dfrac{\mu}{|\mathbf{r}_c^0 + \Delta \mathbf{r}|}}{\frac{1}{2}\mathbf{v}_c^{0^2} - \dfrac{\mu}{r_c^0}} \right)^{\frac{3}{2}} = \frac{m}{n} \tag{76}$$

where m and n are relatively prime natural numbers. In spacecraft formations, it can be easily proven that a necessary condition for two or more satellites to remain at a reasonably small distance from one another is that their periods are equal, leading their specific energies to be equal:

$$\frac{1}{2}(\mathbf{v}_c^0 + \Delta \mathbf{v} + \boldsymbol{\omega}_0 \times \Delta \mathbf{r})^2 - \frac{\mu}{|\mathbf{r}_c^0 + \Delta \mathbf{r}|} = \frac{1}{2}\mathbf{v}_c^{0^2} - \frac{\mu}{r_0} \tag{77}$$

Written with respect to the initial conditions Eq. (72) becomes:

$$\frac{1}{2}\Delta \mathbf{v}^2 + \mathbf{v}_c^0 \cdot \Delta \mathbf{v} + \frac{(\mathbf{v}_c^0 + \Delta \mathbf{v}, \mathbf{h}_c, \Delta \mathbf{r})}{r_c^{0^2}} + \frac{1}{2}\frac{(\mathbf{h}_c \times \Delta \mathbf{r})^2}{r_c^{0^4}} - \frac{\mu}{\sqrt{r_c^{0^2} + \Delta \mathbf{r}^2 + 2\mathbf{r}_c^0 \cdot \Delta \mathbf{r}}} + \frac{\mu}{r_c^0} = 0 \tag{78}$$

If the conditions from Eq. (78) are fulfilled the relative orbital motion trajectory is a closed curve.

6. A tensor invariant in the relative motion

In this Section, we will refrain to apply the state flow operator approach to the entire problem which models the relative motion in a gravitational field, but rather to apply it to a part of its solution. We will reveal a very interesting invariance relation, which relates the motion of the deputy and the motion of the attraction center, both referred to LVLH, as well as a very useful propagator for the state of the deputy spacecraft in the same frame.

Consider the relative motion in a gravitational field, where the relative state of the deputy spacecraft in the LVLH frame associated to the chief is expressed like:

$$\begin{cases} \mathbf{r}(t) = \mathbf{R}_{-\boldsymbol{\omega}}\mathbf{r}_*(t) - \mathbf{r}_c(t) \\ \dot{\mathbf{r}}(t) = \mathbf{R}_{-\boldsymbol{\omega}}[\dot{\mathbf{r}}_*(t) - \widetilde{\boldsymbol{\omega}}\mathbf{r}_*(t)] - \dot{\mathbf{r}}_c(t) \end{cases} \tag{79}$$

where $\mathbf{r}_* = \mathbf{r}_*(t)$ is the solution to the initial value problem and it models a Keplerian motion. Equation (79) can be written as:

$$\begin{bmatrix} \mathbf{r}_* \\ \dot{\mathbf{r}}_* \end{bmatrix} = \begin{bmatrix} \boldsymbol{\Phi} & \mathbf{I}_3 \\ \mathbf{I}_3 & \boldsymbol{\Phi} \end{bmatrix} \begin{bmatrix} \mathbf{r}_c(t_0) + \Delta\mathbf{r} \\ \dot{\mathbf{r}}_c(t_0) + \Delta\mathbf{v} + \boldsymbol{\omega}(t_0) \times (\Delta\mathbf{r} + \mathbf{r}_c(t_0)) \end{bmatrix} \tag{80}$$

where $\boldsymbol{\Phi}$ is the unique tensor established by the conditions:

$$\begin{cases} \boldsymbol{\Phi}[\mathbf{r}_*(t_0)] = \mathbf{r}_*(t) \\ \boldsymbol{\Phi}[\dot{\mathbf{r}}_*(t_0)] = \dot{\mathbf{r}}_*(t) \\ \boldsymbol{\Phi}[\mathbf{h}(t_0)] = \mathbf{h}(t) \end{cases} \tag{81}$$

If the Deputy trajectory is elliptic ($\zeta \leq 0$ and $\mathbf{h} \neq 0$), $\boldsymbol{\Phi}$ can be computed as (Condurache & Martinusi (2011); Martinusi (2010)) :

$$\begin{aligned} \boldsymbol{\Phi}(E) =& \left[\frac{\cos E_0(\cos E - e)}{1 - e\cos E_0} + \frac{\sin E_0 \sin E}{1 - e\cos E} \right] \widehat{\mathbf{a}} \otimes \widehat{\mathbf{a}} \\ &+ \left[\frac{\sin E_0(\cos E - e)}{1 - e\cos E_0} - \frac{(\cos E_0 - e)\sin E}{1 - e\cos E} \right] \widehat{\mathbf{a}} \otimes \widehat{\mathbf{b}} \\ &+ \left[\frac{\cos E_0 \sin E}{1 - e\cos E_0} + \frac{\sin E_0 \cos E}{1 - e\cos E} \right] \widehat{\mathbf{b}} \otimes \widehat{\mathbf{a}} \\ &+ \frac{n^3 a^3}{h^2} \left[\frac{\sin E_0 \sin E}{1 - e\cos E_0} - \frac{(\cos E_0 - e)\cos E}{1 - e\cos E} \right] \widehat{\mathbf{b}} \otimes \widehat{\mathbf{b}} + \widehat{\mathbf{h}} \otimes \widehat{\mathbf{h}} \end{aligned} \tag{82}$$

where $E_0 = E(t_0)$ and $\widehat{\mathbf{v}}$ is the unity vector attached to $\mathbf{v}$.

If we denote by $\mathbf{X}(t)$ the state vector attached to the Deputy

$$\mathbf{X}(t) = \begin{bmatrix} \mathbf{r}(t) \\ \dot{\mathbf{r}}(t) \end{bmatrix}, \tag{83}$$

equation (79) may be rewritten like:

$$\mathbf{X}(t) = \boldsymbol{\Psi}(t)\mathbf{Y}_0 + \mathbf{X}_c(t) \tag{84}$$

where:

$$\begin{aligned} \boldsymbol{\Psi}(t) &= \begin{bmatrix} \mathbf{R}_{-\boldsymbol{\omega}}\boldsymbol{\Phi} & \mathbf{0}_3 \\ -\widetilde{\boldsymbol{\omega}}\mathbf{R}_{-\boldsymbol{\omega}}\boldsymbol{\Phi} & \mathbf{R}_{-\boldsymbol{\omega}}\boldsymbol{\Phi} \end{bmatrix} \\ \mathbf{Y}_0 &= \begin{bmatrix} \mathbf{r}_c(t_0) + \Delta\mathbf{r} \\ \dot{\mathbf{r}}_c(t_0) + \Delta\mathbf{v} + \boldsymbol{\omega}(t_0) \times \Delta(\mathbf{r} + \mathbf{r}_c(t_0)) \end{bmatrix} \\ \mathbf{X}_c(t) &= \begin{bmatrix} -\mathbf{r}_c(t) \\ -\dot{\mathbf{r}}_c(t) \end{bmatrix} \end{aligned} \tag{85}$$

Notice that $\mathbf{X}_c(t)$ models the state of the attraction center with respect to the LVLH frame associated to the chief spacecraft. After some manipulations, it follows that the constant vector $\mathbf{Y}_0$ may be rewritten like:

$$\mathbf{Y}_0 = \begin{bmatrix} \mathbf{I}_3 & \mathbf{0}_3 \\ \widetilde{\omega} & \mathbf{I}_3 \end{bmatrix} \left\{ \begin{bmatrix} \Delta\mathbf{r} \\ \Delta\mathbf{v} \end{bmatrix} + \begin{bmatrix} \mathbf{r}_c(t_0) \\ \dot{\mathbf{r}}_c(t_0) \end{bmatrix} \right\} \tag{86}$$

Denote:

$$\Gamma_0 = \begin{bmatrix} \mathbf{I}_3 & \mathbf{0}_3 \\ \widetilde{\omega} & \mathbf{I}_3 \end{bmatrix}; \quad \mathbf{X}_0 = \mathbf{X}(t_0); \quad \mathbf{X}_c^0 = \mathbf{X}_c(t_0); \quad \Lambda(t) = \Psi(t)\Gamma_0 \tag{87}$$

From the above considerations, it follows that:

$$\mathbf{X}(t) - \Lambda(t)\mathbf{X}_0 = \mathbf{X}_c(t) - \Lambda(t)\mathbf{X}_c^0 \tag{88}$$

where the closed form expression of $\Lambda(t)$ is determined by taking into account Equations (85) and (87):

$$\Lambda(t) = \begin{bmatrix} \mathbf{R}_{-\omega}\Phi & \mathbf{0}_3 \\ \mathbf{R}_{-\omega}[\Phi, \widetilde{\omega}] & \mathbf{R}_{-\omega}\Phi \end{bmatrix} = \begin{bmatrix} \mathbf{R}_{-\omega} & \mathbf{0}_3 \\ \mathbf{0}_3 & \mathbf{R}_{-\omega} \end{bmatrix} \begin{bmatrix} \Phi & \mathbf{0}_3 \\ [\Phi, \widetilde{\omega}] & \Phi \end{bmatrix} \tag{89}$$

where $[\,,]$ denotes the comutator brackets:

$$[\mathbf{A}, \mathbf{B}] = \mathbf{AB} - \mathbf{BA}. \tag{90}$$

Note that Eq. (88) is very similar to the velocity invariant expression in rigid body kinematics (Condurache & Matcovschi (2001)). The relative state of the deputy spacecraft in LVLH is propagated by:

$$\mathbf{X}(t) = \mathbf{X}_c(t) + \Lambda(t)(\mathbf{X}_0 - \mathbf{X}_c^0) \tag{91}$$

The above formula is the complete exact solution of relative orbital motion nonlinear problem (3).

7. Conclusions

The tensor approach used in this paper allows us to obtain closed-form exact expressions for the relative law of motion and the relative velocity. This instrument is only a catalyst, and it helps introduce a change of variable which transforms the relative orbital motion problem into the classic Kepler problem. Thus, the problem of the relative orbital motion is super-integrable. The shape of the chief inertial trajectory does not impose special problems, as it does in the linearized approaches. The deputy trajectory does not impose problems either, allowing us to derive exact equations of relative motion in any situation and for any initial conditions. The equations that describe the state of the deputy spacecraft in LVLH depend only on time and the initial conditions. Also all the computational stages needed by this solution are conducted on board in the LVLH frame. The long-term accuracy offered by this solution allows the study of the relative motion for indefinite time intervals, and with no restrictions on the magnitude of the relative distance. The solution may be used in the study of satellite constellations from the point of view of the relative motion. The solution offered in this paper gives a parameterization of the manifold associated to the relative motion. Perturbation techniques may be now used in order to derive more accurate equations of motion when assuming small perturbations on the relative trajectory, due to Earth oblateness, solar wind, moon attraction, and atmospheric drag. Based on this solution, a study of the full-body relative motion might be a subject for future work.

8. Nomenclature

$\mathbf{A}^T$ = transpose of tensor (matrix) $\mathbf{A}$

$\mathbf{r}$ = position vector

$\mathbf{r}_1 \cdot \mathbf{r}_2$ = dot product of vector $\mathbf{r}_1$ and $\mathbf{r}_2$

$\mathbf{r}_1 \times \mathbf{r}_2$ = cross product of vector $\mathbf{r}_1$ and $\mathbf{r}_2$

$\hat{\mathbf{r}}$ = the unity vector attached to $\mathbf{r}$

a = semimajor axis

$\mathbf{a}$=vectorial semimajor axis

b = semimajor axis

$\mathbf{b}$=vectorial semimajor axis

e = eccentricity

$\mathbf{e}$ = vectorial eccentricity

$\mathbf{h}$=specific angular momentum

n=mean motion

p=semilatus rectum (conic parameter)

$\mathbf{R}_\omega$=rotation tensor with angular velocity ω

t= time

u = magnitude of vector $\mathbf{u}$

$\mathbf{v}$ = velocity vector

μ = gravitational parameter

ξ = specific energy

$\boldsymbol{\omega}$ = angular velocity of the rotating reference frame

$\tilde{\omega}$ = skew-symmetric tensor associated with vector $\boldsymbol{\omega}$

9. References

Alfriend, K., Vadali, S., Gurfil, P., How, J. & Breger, L. (2009). *Spacecraft Formation Flying: Dynamics, Control, and Navigation*, Elsevier, Oxford.

Balaji, S. K. & Tatnall, A. (2003). Precise modeling of relative motion for formation flying spacecraft, *Proceedings of 54th International Astronautical Congress of the International Astronautical Federation, the International Academy of Astronautics, and the International Institute of Space Law*, Bremen.

Battin, R. (1999). *An Introduction to the Mathematics and Methods of Astrodynamics*, AIAA Education Series.

Carter, T. (1990). New form for the optimal rendezvous equations near keplerian orbit, *Journal of Guidance, Control, and Dynamics* 13(1): 183–186.

Clohessy, W. & Wiltshire, R. (1960). Terminal guidance system for satellite rendezvous, *Journal of the Aerospace Sciences* 27(9): 653–658.

Condurache, D. (1995). *New Symbolic Methods in the Study of Dynamic Systems - Phd Thesis*, Gheorghe Asachi Technical University of Iasi, Iasi, Romania.

Condurache, D. & Martinusi, V. (2007a). A complete closed form solution to the Kepler problem, *Meccanica* 42(5): 465–476.

Condurache, D. & Martinusi, V. (2007b). Kepler's problem in rotating reference frames. part I: Prime integrals. vectorial regularization, *Journal of Guidance, Control, and Dynamics* 30(1): 192–200.

Condurache, D. & Martinusi, V. (2007c). Kepler's problem in rotating reference frames. part II: Relative orbital motion, *Journal of Guidance, Control, and Dynamics* 30(1): 201–213.

Condurache, D. & Martinusi, V. (2007d). A novel hypercomplex solution to Kepler's problem, *PADEU* 19: 201–213.

Condurache, D. & Martinusi, V. (2007e). Relative spacecraft motion in a central force field, *Journal of Guidance, Control, and Dynamics* 30(3): 873–876.

Condurache, D. & Martinusi, V. (2008a). Foucault pendulum-like problems: A tensorial approach, *International Journal of Non-linear Mechanics* 43(8): 743–760.

Condurache, D. & Martinusi, V. (2008b). Exact solution to the relative orbital motion in a central force field, *Proceedings ofthe 2nd International Symposium on Systems and Control in Aeronautics and Astronautics*, Shenzhen, China.

Condurache, D. & Martinusi, V. (2010a). Hypercomplex eccentric anomaly in the unified solution of the relative orbital motion, *Advances in Astronautical Sciences* 135: 281–300.

Condurache, D. & Martinusi, V. (2010b). Quatenionic exact solution to the relative orbital motion problem, *Journal of Guidance, Control, and Dynamics* 33(4): 1035–1047.

Condurache, D. & Martinusi, V. (2011). State space analysis for the relative spacecraft motion in geopotential fields, *Proceedings of AIAA Guidance, Navigation, and Control Conference*, Portland.

Condurache, D. & Matcovschi, M. (2001). Computation of angular velocity and acceleration tensors by direct measurements, *Acta Mechanica* 153(3-4): 147–167.

Darboux, G. (1887). *Lecons sur la Theorie Generale des Surfaces et les Applications Geometriques du Calcul Infinitesimal*, Gauthier-Villars, PAris.

Gim, D.-W. & Alfriend, K. (2003). State transition matrix of relative motion for the perturbed noncircular reference orbit, *Journal of Guidance, Control, and Dynamics* 26(6): 956–971.

Gronchi, G. (2005). On the uncertainty of the minimal distance between two confocal keplerian orbits, *Discrete and Continuous Dynamical Systems Series B* 7(4): 295–329.

Gronchi, G. (2006). An algebraic method to compute the critical points of the distance function between two keplerian orbits, *Celestial Mechanics and Dynamical Astronomy* 93(1): 295–329.

Gurfil, P. & Kholshevnikov, K. (2006). Manifolds and Metrics in the Relative Spacecraft Motion Problem, *Journal of Guidance, Control, and Dynamics* 29(4): 1004–1010.

Gurfil, P. & N.J.Kasdin (2004). Nonlinear modeling of spacecraft relative motion in the configuration space, *Journal of Guidance, Control, and Dynamics* 27(1): 154–157.

Ketema, Y. (2006). An analytical solution for relative motion with an elliptic reference orbit, *Journal of Astronautical Sciences* 53(4): 373–389.

Lawden, D. (1963). *Optimal Trajectories for Space Navigation*, Butterworth, London.

Lee, D., Cochran, J. & Jo, J. (2007). Solutions to the variational equations for relative motion of satellites, *Journal of Guidance, Control, and Dynamics* 30(3): 671–678.

Martinusi, V. (2010). *Lagrangian and Hamiltonian Formulations in Relative Orbital Dynamics. Applications to Spacecraft Formation Flying and Satellite Constellations - Phd Thesis,* Gheorghe Asachi Technical University of Iasi, Iasi, Romania.

Tschauner, J. (1966). The elliptic orbit rendezvous, *Proceedings of AIAA 4th Aerospace Sciences Meeting,* AIAA, Los Angeles.

Tschauner, J. & Hempel, P. (1964). Optimale beschleunigeungsprogramme fur das rendezvous-manouever, *Acta Astronautica* 10: 296–307.

Yamanaka, K. & Andersen, F. (2002). New state transition matrix for relative motion on an arbitrary elliptical orbit, *Journal of Guidance, Control, and Dynamics* 25(1): 60–66.

3

The Middle Atmosphere: Discharge Phenomena

Cheng Ling Kuo

Department of Physics, National Cheng Kung University, Taiwan

1. Introduction

The layer between 10 and 100 km altitude in the Earth atmosphere is generally categorized as the middle atmosphere (Brasseur & Solomon, 1986). The boosting development of rocket and satellite technologies during the past 50 years has made it possible to directly probe the middle atmosphere (Brasseur & Solomon, 1986). Recently, transient luminous events (TLEs) open up another window; through observing the discharge phenomena in the middle atmosphere from both the ground and the space, the physical processes in this region can be inferred. Besides the present satellite missions (ISUAL, Tatiana-2, SPRITE-SAT, Chibis-M mission), future orbit missions include JEM-GLIMS, ASIM, TARANIS will soon join the efforts. These space missions provide the unique platforms to explore the plasma chemistry and atmospheric electricity in the middle atmosphere, and also investigate the possible TLE impact on spacecrafts.

2. Discharge phenomena in the middle atmosphere

The discharge phenomena in the middle atmosphere collectively carry the name of the transient luminous events (TLEs), owing to their fleeting nature (sub-milliseconds to tens of milliseconds) and high luminosity over the thunderstorms; see Fig. 1. The transient luminous events were accidentally observed in the ground observation (Franz et al., 1990) and Earth orbit observation (Boeck et al., 1992), and were soon recognized as the manifestations of the electric coupling between atmospheric lightning and the middle atmosphere/ionosphere. The thunderstorm plays the role of an electric battery in the atmosphere-ionosphere system. The thunderstorms, ~3000 of them at any time on Earth, generate a total electric current of 1.5 kA flowing into the ionosphere, and sustain the electric potential ~200 MV of the ionosphere (Volland, 1987). With the thunderstorms, the electric energy gradually accumulates in the middle atmosphere and a part of the deposited energy later is released as the luminous TLEs, in a way similar to the capacitor discharge. However, how the light emission and electric current distribute in those discharge phenomena will lead to different varieties of transient luminous events between the cloud top and the ionosphere.

2.1 Transient luminous events

Thunderstorm-induced transient optical emissions near the lower ionosphere and in the middle atmosphere are categorized into several types of transient luminous events (TLEs),

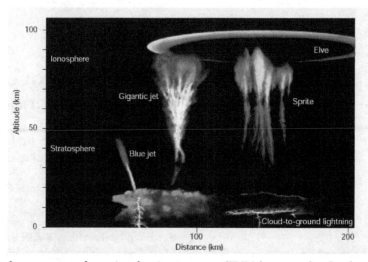

Fig. 1. The known types of transient luminous events (TLEs) between the cloud top and the ionosphere. The causes of TLEs are generally attributed to the activity of cloud discharges. The current known species of TLEs include sprite, elves, blue jet and gigantic jet (Pasko, 2003).

including sprites (Sentman et al., 1995), elves (Fukunishi et al., 1996), blue jets and gigantic jets (Wescott et al., 1995; Pasko et al., 2002; Su et al., 2003).

2.2 Sprites

The University of Minnesota group was the first to obtain the evidence for the existence of the upward electrical discharge on the night of 22, Sep, 1989 (Franz et al., 1990). The first color image of sprites was taken from an aircraft in 1994 that helps to elucidate the luminous structure and its fleeting existence (< 16 ms): a red main body which spans the latitude range between 50 – 90 km, and faint bluish tendrils that extends downward and occasionally reaches the cloud top. The first 0.3 - 1 ms high-speed imaging of sprites, halos and elves were reported by Stanley et al. (1999), Barrington-Leigh et al. (2001) and Moudry et al. (2002, 2003). The high-speed photograph showed that sprites usually initiated at an altitude of ~ 75 km and developed simultaneously upward and downward from the original point (Stanley et al., 1999). In more detail, McHarg et al. (2007) analyzed sub-millisecond (5000 and 7200 frames/s) images of sprites and compiled statistics on velocities of streamer heads. The streamer speeds vary between 10^6 and 10^7 m/s. Additionally, Cummer et al. (2006) showed that the long-persisting sprite beads are formed as the tips of the downward moving sprite streamers are attracted to and, sometimes, collide with other streamer channels. Theoretically, higher-speed dynamic evolutions of the fine structure (streamers) in sprites are also predicted by theoretical streamer models (Pasko et al., 1998; Liu & Pasko, 2004; Liu et al., 2006; Liu et al., 2009), which are well consistent with sprite observations.

Among the main groups of emissions in sprites, the molecular nitrogen first positive band (N_2 1P) was the first to be identified by using an intensifier CCD spectrograph (Mende et al., 1995). Then, the follow-up works further determined the vibrational exciting states of N_2 1P

(Green et al., 1996; Hampton et al., 1996) and obtained evidences that support the existence of N$_2$+ Meinel band emission (Bucsela et al., 2003). Recently, 1 ms time-resolution spectrograph observation has been achieved (Stenbaek-Nielse & McHarg, 2004) and the altitude-resolved spectrum (3 ms temporal and ~3 nm spectral resolution between 640 to 820 nm) showed that the population of the upper vibrational state of the N$_2$ 1P bands, $B^3\Pi_g$, varies with altitude and is similar to that of the laboratory afterglow at high pressure (Kanmae et al., 2007; Kanmae et al., 2010).

2.3 Elves

The enhanced airglow emission (elves) was first discovered in the images recorded by the space shuttle's cargo-bay cameras (Boeck et al., 1992; Boeck et al., 1998), and were subsequently observed (and termed as "ELVES" - Emissions of Light and VLF perturbations due to EMP Sources) in the ground observation using a multi-channel high-speed photometer and image intensified CCD cameras (Fukunishi et al., 1996). The Stanford University group built and used an array of photomultipliers called Fly's Eve to resolve the rapid lateral expansion of optical emissions in elves and the observational results were consistent with those were predicted by the elve model (Inan et al., 1991; Inan et al., 1996; Inan et al., 1997; Barrington-Leigh & Inan, 1999; Barrington-Leigh et al., 2001). The ISUAL experiment on the FORMOSAT-2 satellite, the first spacecraft TLE experiment, then successfully confirmed the existence of ionization and the Lyman-Birge-Hopfield (LBH) band emissions in elves (Mende et al., 2005); the satellite images were used to study their spatial-temporal evolutions and the numerical simulation results of the elve model (Kuo et al., 2007) have beautifully reproduced the observed elve morphology.

2.4 Halos

Halos are pancake-like objects with diameters of ~ 80 km, occurring at altitudes of ~ 80 km (Wescott et al., 2001). Halos were initially thought to be elves by most ground observers using conventional cameras with 30 frames per second until Barrington-Leigh et al. (2001) first proved that halos are distinct from elves (with a much larger diameter of~300 km and a shorter luminous duration of <1ms). The evolution of halo and elves recorded by a high-speed (3000 frames per second) camera were found to be consistent with the modeling results (Barrington-Leigh et al., 2001). Frey et al.(2007) also showed that ~50 % of halos are unexpectedly associated with negative cloud-to-ground (-CG) lightning while nearly 99% sprites are induced by positive cloud-to-ground (-CG) lightning. Wescott et al. (2001) compared the maximum brightness geometry of halos with lightning location using triangulation measurement. They found that the maximum brightness of halo is very close to the location of the parent lightning while the sprite structure can be displaced as far as several tens of kilometers.

2.5 Blue jets and gigantic jets

Blue jets are electric jets that appear to emerge directly from the cloud tops (~ 16-18 km) and shoot upwardly to the final altitudes of ~ 40-50 km (Wescott et al., 1995). Gigantic jets (GJs) are largest discharges in the middle atmosphere, which have been reported by several ground campaigns (Pasko et al., 2002; Su et al., 2003). Based on the monochrome images with a time resolution of 16.7 ms, the temporal optical evolution of the GJs typically contains three stages: the leading jet, the fully-developed jet (FDJ) and the trailing jet (TJ) (Su et al., 2003). The upward propagating leading jet maybe considers being the pre-stage of the FDJ, playing a role similar to that of a stepped leader in the conventional lightning. In the FDJ stage, the GJ optically links the cloud top and the lower ionosphere. The trailing jet shows features similar to those of the blue jets (BJs) and propagates from the cloud top up to ~ 60 km altitude. The optical emission of the trailing jet lasts for more than 0.3 second, and the overall duration of the GJs is ~ 0.5 second (Su et al., 2003).

3. Lightning effect in the middle atmosphere

Since TLEs always occur over active thunderstorms, the electromagnetic radiations from thundercloud discharges being the root cause behind these upper atmospheric luminous phenomena are implied. Thus, to deal with these phenomena, the first two questions should be addressed are "what is the frequency spectrum of a lightning flash and what is the absorption frequency range of the upper atmosphere?" After that one should resolve how the radiation field attenuates in the upper atmosphere and how it reflects at the lower boundary ionosphere.

3.1 Electromagnetic field by lightning current

The lightning frequency spectrum exhibits a peak at 1-10 kHz (Rakov & Uman, 2003, p. 158 and references therein). If we assume that a lightning has a peak current of 60 kA with a channel resistance of 1 Ω (Rakov & Uman, 2003, p. 398 and references therein) and radiates all the electromagnetic energy at 5 kHz. The radiated power can be readily computed to be $P = I^2 R = 3.6 GW$ and power flux at 87 km altitude is 0.0378 W/m². The equivalent energy flux density of the electromagnetic field can be expressed as $c\varepsilon_0 E^2$, where c is the speed of light and ε_0 is the permittivity of free space. Hence the E strength at 87 km altitude is deduced to be ~3.8 V/m, which is ~ 0.25 times of the conventional breakdown field (E_k) where 1 E_k~117.2 Td and 1 Townsend (Td) = 10^{-21} V-m², also refer to the definition of Eq. 1 and Fig. 9). At this altitude, 0.25 Ek corresponds to 14.7 V/m. The reduced E-field is defined as E/N (V-m²) where E is the magnitude of the E-field and N is the neutral density. The reduced E-field for E = 3.8 V/m and N = 1.25×10²⁰ /m³ at an altitude of 87 km is ~ 30.5×10⁻²¹ V-m² or 30.5 Td. The magnitude of E-field (3.8 V/m) is not small comparing to the value of the breakdown E-field (~0.25 E_k), and is sufficient to excite N_2 1P (Veronis et al., 1999). As it will be shown, our calculation indicates that a lightning with peak current of 60 kA or higher will generate a sufficiently strong E-field at 87 km elevation to excite the N_2 1P emissions of molecular nitrogen. Our results are also consistent with the observational fact that any lightning with peak current > 57 kA will have an accompanying elve (Barrington-Leigh & Inan, 1999).

The wavelength of the lightning radiation field in the VLF frequency range is ~ 10-100 km, which is much longer than the electron mean free path of 1 m at the mesospheric elevation (Rakov & Tuni, 2003). Hence, the lightning electromagnetic field can be approximately thought as a DC field. Those DC electric fields can also be caused by the accumulated charges inside the thundercloud. The quasi-electrostatic field will accelerate ambient electrons. The energized electrons excite the neutral particles (N_2 or O_2) to higher excited states. The electronically excited neutral particles tend to return to their low energy states, and rapidly de-excite by emitting photons. In next section, we will discuss the atmospheric discharge in the middle atmosphere with an external quasi-electrostatic field.

3.2 Quasi electrostatic field by charge in the thunderstorm

The critical condition of atmospheric discharge occurrence is whether an electron avalanche process in the upper atmosphere does happen in a short time. The time-varying electron number density can be written as

$$\frac{dN_e}{dt} = (\nu_i - \nu_a)N_e \tag{1}$$

where N_e, ν_i and ν_a are the electron density, the ionization rate and dissociative attachment ($O_2 + e^- \rightarrow O^- + O$) rate, respectively. The values of ν_i and ν_a, which are functions of E-field strength, will be derived from the Boltzmann transport equation and also shown in section 4.5. The criterion of the neutral gas breakdown is decided by Eq. 1 between the electron creation and electron loss term. As $\nu_i > \nu_a$, the gas breakdown is accompanied by electron avalanche. An avalanche process can begin with a small number of seed electrons, due to existing free electrons or second electrons by cosmic ray, even being triggered by a single electron (Raizer, 1991, pp. 128-130).

The E-field strength at the breakdown threshold ($\nu_i = \nu_a$) is characterized by the breakdown E-field or termed as the conventional breakdown E-field, to distinct it from the positive and the negative streamer breakdown E-fields. The positive and negative streamer breakdown E-fields are the minimum E-fields necessary for the propagation of positive and negative streamer in air at ground pressure. The streamer structures in sprite have been confirmed in TLEs campaign and their radius is roughly several hundreds meter (Gerken & Inan, 2003; Gerken & Inan, 2004). The streamer theory has successfully explained the fine structure of sprite column emissions (Pasko et al., 1998; Liu & Pasko, 2004, 2005; Liu et al., 2006; Liu et al., 2009). The typical values of the conventional breakdown (E_k), the positive (E_{cr}^+) and the negative (E_{cr}^-) streamer breakdown E-fields respectively are $E_k \approx 32 \times 10^5$ (Raizer, 1991, p. 135), 12.5×10^5 (Babaeva & Naidis, 1997), 4.4×10^5 V/m (Allen & Ghaffar, 1995) in air at ground pressure.

Pasko et al. (1997) has proposed the quasi-electrostatic model to account for the electrostatic interaction between thunderstorm charge and the middle atmosphere. Fig. 2 shows the altitude profile of applied E-field corresponding to the charge remove in the thunderstorm that is defined by charge moment, $\Delta M = \Delta Q \cdot L$ where ΔQ is charge in units of coulomb and L is distance above the ground in units of km. The effect of removing total charge ΔQ at altitude $z_0 = 10$ km is equivalent to add the charge in cloud with Gaussian spatial distribution of $e^{-[r^2/a^2 + (z-z_0)^2/b^2]}$ where $a = 10$km and $b = 5$ km. Two solid lines are the applied E-fields equivalent to the charge moment changes of 1000 (100C x 10 km) and 3000 C-km (300C x 10 km). The three dashed lines denote the criterion for the conventional breakdown and the propagation of negative/positive streamers. The conventional breakdown field represent large scale (several hundred km wide) gas

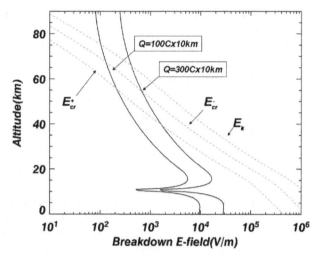

Fig. 2. The altitude profile of the resulting E-field corresponds to the charge removal in the thunderstorm. Two solid lines are the E-fields equivalent to the charge moment changes of 1000 and 3000 C-km. Three dashed lines denote the criterion for the conventional breakdown and the propagation of the negative/the positive streamers.

discharge (Pasko, 2006, and reference therein) and has the strictest condition among the three breakdown mechanisms. The required minimum E-field for the propagation of the positive and the negative streamers are lower because a relatively smaller scale (hundred or tens of meter) space charge is sufficient to enhance the local E-field and cause the streamers to form. For a charge moment change of 1000 C-km, the lowest altitude for the conventional breakdown, the negative and the positive streamers are 75, 70 and 55 km, respectively. For a more extreme case of 300 C x 10 km, the corresponding values will be 65, 55 and 40 km.

4. Microscopic physics in TLEs

Discharge phenomena in the upper atmosphere, e.g., sprites, elves, halo, occur under the physical conditions of low-pressure and low-density atmophsere. We use Boltzmann equation including the collision terms as a method to describe the behaviors of a weakly ionized gas. For the atmospheric discharges, we consider all the important collisional processes in the atmospheric discharge system. We want to calculate the macroscopic quantities (reaction rates, drift velocity and average electron energy) to the physical quantities in the microscopic system of the gas system. From solving the Boltzmann equation, we can derive the physical quantities (chemical reaction rates, drift velocity and average electron energy) and reaction rate of the collisional processes in the atmospheric discharge at the TLE altitudes.

4.1 Electron distribution function to describe the weakly ionized gas

Boltzmann transport equation, which was first devised by Ludwig Boltzmann, is often used to describe the statistical properties of a many-particle system with collision processes. We employ the solver for the Boltzmann transport equation (Morgan & Penetrante, 1990) for this work. The general form of the Boltzmann transport equation is

$$\{\frac{\partial}{\partial t} + \vec{v} \cdot \nabla_{\vec{x}} + \frac{q}{m}[\vec{E}(\vec{x},t) + \vec{v} \times \vec{B}(\vec{x},t)] \cdot \nabla_{\vec{v}}\} f(\vec{x},\vec{v},t) = (\frac{\partial f}{\partial t})_{collision} \qquad (2)$$

where $f(\vec{x},\vec{v},t)$ is the velocity distribution function, defined such that $f(\vec{x},\vec{v},t)dv$ is the number density of finding the particle in a unit volume located at position $\vec{x}$ and at time t with velocity in the range from v to $v+dv$; q is the charge and m is the mass of electron; $\vec{E}$ is the applied electromagnetic field. The right-hand side, which is the collision term, represents changes in the distribution function due to collision processes.

4.2 Collision process in molecular nitrogen

We consider the following collision processes between electrons and molecular nitrogen:

1. Momentum Transfer: $e^- + N_2 \rightarrow e^- + N_2$
2. Rotational Excitation: $e^- + N_2 \rightarrow e^- + N_2^*$
3. Vibrational Excitation: $e^- + N_2 \rightarrow e^- + N_2^*$ (v=1, 2, 3, 4, 5, 6, 7, and 8)
4. Electronic Excitaion: $e^- + N_2 \rightarrow e^- + N_2^*$
                              ($A^3\Sigma_u^+$, $B^3\Pi_g$, $W^3\Delta_u$, $B'\ ^3\Sigma_u^-$, $a'\ ^1\Sigma_u^-$, $a_1\Pi_g$, $w^1\Delta_u$, $C_3\Pi_u$,
                              $E^3\Sigma_g^+$, $a''^1\Sigma_g^+$, Singlet State)
5. Ionization: $e^- + N_2 \rightarrow e^- + N_2^+ + e^-$
                              ($X^2\Sigma_g^+$, $A^2\Pi_u$, $B^2\Sigma_u^+$)

The collision processes between electrons and molecular nitrogen include momentum transfer, rotational excitation, vibrational excitation, electronic excitation and ionization of molecular nitrogen. The total cross section for the electron collision with molecular nitrogen is shown in Fig. 3. The most dominant contribution toward the total cross section is from the momentum transfer process below 100 eV. The differential cross section $\frac{d\sigma}{d\Omega}$ is defined that the number of electrons which is scattered elastically per second into the solid angle $d\Omega$, $\frac{d\sigma}{d\Omega} = \frac{1}{N_e v_e}\frac{dN_e}{d\Omega}$. The total cross section σ can be obtained by integrating $\frac{d\sigma}{d\Omega}$ over the 4π solid angle, $\sigma_m = \int \frac{d\sigma}{d\Omega}d\Omega$ (cm²). The momentum transfer cross section for elastic collisions is defined as $\sigma_m = \int \frac{d\sigma}{d\Omega}(1 - \cos\theta)d\Omega$, which is called the effective cross section. For inelastic collision processes, the cross section could include contributions from the rotational excitation, the vibrational excitation, the electronic excitation and the ionization. The cross sections for inelastic electron collisions in molecular nitrogen are shown in Fig. 3 and the corresponding initiating energies in molecular nitrogen are also shown in Fig. 3, which was compiled by *A. V. Phelps*.

In the energy level diagram of Fig. 4, the electronic excited states are enumerated and labeled by Roman letters, A, B, C, ...or a, b, c,... with X indicating the ground state energy level. In the symbol $X^{2s+1}\Lambda_{u/g}$, X is the ground state; S is the spin angular momentum from the unpaired electrons and 2S+1 is the spin multiplicity; Λ is the total angular momentum quantum number. The parities g and u stand for gerade (even in German) or ungerade (odd). For higher total angular momentum Λ, the Greek symbols Σ, Π, Δ and Φ are used.

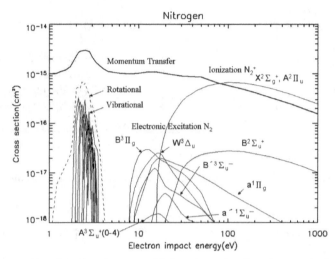

Fig. 3. The cross section data for electron collisions in molecular nitrogen including the momentum transfer, the rotational excitation, the vibrational excitation, the electronic excitation and the ionization processes.

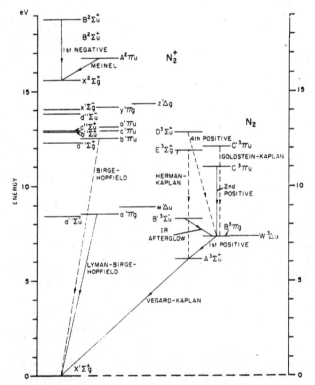

Fig. 4. The energy level diagram of molecular nitrogen (Vallance-Jones, 1974).

The band systems set in the bold-face fonts in Table 1 are N_2 1P, N_2 2P, N_2 LBH, N_2^+ Meniel and N_2^+ 1N band systems; these are the shortest lifetime nitrogen band systems, which also have the highest photon emission rates. Although the upper state (B' $^3\Sigma_u^-$) has a lifetime of 25μs, its excitation cross section (as shown in Fig. 3) is not significant. Hence, the band emission from this state has never been confirmed in sprite ground campaigns.

	State	Transition	Band System	Wave Length (nm)	Energy	Lifetime
N_2	$X^1\Sigma_g^+$	--------	-------	-------	0	-------
	$A^3\Sigma_u^+$	A-X	Vegard-Kapian	125-532.5	6.18	1.9s
	$B^3\Pi_g$	**B-A**	**First Positive(*)**	**478-2531**	**7.37**	**8μs**
	$W^3\Delta_u$	W-B	Wu-Benesch	2200-4300	7.38	2ms-54μs
	$B'\ ^3\Sigma_u^-$	B'-B	IR afterglow	605-890	8.19	25μs
		B'-X	Ogawa-Tanaka-Wilkinson	112-224		
	$a'\ ^1\Sigma_u^-$	A'-X	Ogawa-Tanaka-Wilkinson-Mulliken	108-200	8.42	0.5s
	$a^1\Pi_g$	a-a'	McFarlane IR	3000-8500	8.57	80μs(***)
		a-X	**Lyman-Birge-Hopfield(**)**	**100-240**		
	$w^1\Delta_u$	w-a	McFarlane IR	3000-8500	8.91	100-500μs
		w-X	Tanaka	114-140		
	$C_3\Pi_u$	**C-B**	**Second Positive(*)**	**268-546**	**11.05**	**36.6ns**
	$E^3\Sigma_g^+$	E-A	Herman-Kaplan	213-274	11.90	190μs
	$a''^1\Sigma_g^+$	A'-X	Dressler-Lutz	101	16.74	
N_2^+	$X^2\Sigma_g^+$	--------	--------	--------	15.62	--------
	$A^2\Pi_u$	**A-X**	**Meniel(*)**	**550-1770**	**16.74**	**13.9μs**
	$B^2\Sigma_u^+$	**B-X**	**First Negative(*)**	**286-587**	**18.80**	**62.5ns**

Table 1. Electronic state, transition, band system, wavelength, energy and mean lifetime of N_2 and N_2^+ (*: These band systems have been observed for TLEs. **: Especially for ISUAL spectrophotometer. ***: From William, 1989) (Lofthus & Krupenie, 1977).

4.3 Collision process in molecular oxygen

For electron collisions with molecular oxygen, the following reactions were considered.

1. Momentum Transfer: $e^- + O_2 \rightarrow e^- + O_2$
2. Rotational Excitation: $e^- + O_2 \rightarrow e^- + O_2^*$
3. Vibrational Excitation: $e^- + O_2 \rightarrow e^- + O_2^*$ (v=1, 2, 3, 4)
4. Meta-stable Excitation: $e^- + O_2 \rightarrow e^- + O_2^*$
   [$a^1\Delta_g$(0.98ev), $b^1\Sigma_g^+$(1.63ev), $c^1\Sigma_u^-$, $A^3\Sigma_u^+$(4.5ev)]
5. Dissociative Attachment: $e^- + O_2 \rightarrow O + O^-$ (4.2ev)
6. Dissociation: $e^- + O_2 \rightarrow e^- + O(^3P) + O(^3P)$ [6.0ev]
   $e^- + O_2 \rightarrow e^- + O(^3P) + O(^1D)$ [8.4ev]
   $e^- + O_2 \rightarrow e^- + O(^3D) + O(^1D)$ [9.97ev]
7. Dissociative Excitation: $e + O_2 \rightarrow e + O + O^*$ ($3p^3P$) [14.7ev]

The electron collisional cross sections of molecular oxygen are shown in Fig. 5 and the energy levels of molecular oxygen are shown in Fig. 6.The dissociation reaction with threshold energies of 6.0, 8.4 and 9.97 eV are listed above. The cross sections of molecular oxygen in Fig. 5 are for a smaller set of discrete levels comparing with those of molecular nitrogen in Fig. 3, because the direct transition probability from ground state to the upper states in molecular oxygen is high for dissociation than for excitation. One of the dissociation reactions, e + O_2 → e + $O(^3P)$ + $O(^1D)$ [8.4eV], is used to study the production rate of the metastable oxygen atoms $O(^1D)$ in the mesosphere and is often associated with sprite halo. The dissociation reaction in sprite halo can be one of major sources for $O(^1D)$ at nighttime (Hiraki et al., 2004). The $O(^1D)$ can also be produced by direct impacting of $O(^3P)$ by a several eV electron through the reaction e- (T≥2eV) + $O(^3P)$ → $O(^1D)$ + e-. The excited $O(^1D)$ decays into $O(^3P)$ with a lifetime 110 s with a companying emission of 630 nm, which is a prominent red line in the middle atmosphere. The dissociative recombination, e- + O_2^+ → O + $O(^1D, ^1S)$, could be one of the key processes that produces $O(^1S)$. The excited state $O(^1S)$ decays into $O(^1D)$ with a lifetime 0.7s and emits a 557.7 nm photon, which is a green line emission of the middle atmosphere. The quenching heights for the emission lines 630 and 557.7 nm from upper states $O(^1D)$ and $O(^1S)$ are 250-350 km and ~95 km, respectively. The quenching height is defined as the elevation that an emission rate is reduced to one half of its unquenching value (Vallance-Jones, 1974, p. 118).

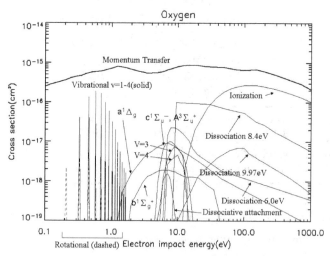

Fig. 5. The cross sections of oxygen include the momentum transfer, the rotational excitation, the vibrational excitation, the electronic excitation and the ionization.

The electronic states, the transition pathways, the energy and the mean lifetime for the related states of O_2 and O_2^+ are listed in Table 2 (Krupenie, 1972). The upper states of IR Atmospheric and Atmospheric bands in molecular oxygen are $b^1\Sigma_g^+$ and $a^1\Delta_g$, with lifetimes of 60 minutes and 12 seconds. The lifetimes of Herzberg I, II bands with upper states $A^3\Sigma_u^+$ and $c^1\Sigma_u^-$ are longer than 1 ms. Hence, the major emission bands of O_2^+ are the first negative and the second negative bands which have short lifetimes and originate from the upper states $b^4\Sigma_g^-$ and $A^2\Pi_u$.

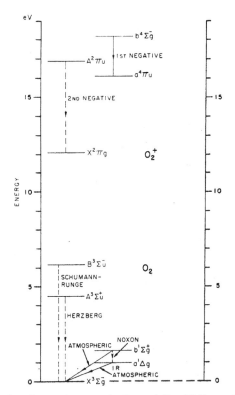

Fig. 6. The energy levels for electronic states in O_2 and O_2^+ (Vallance-Jones, 1974).

	State	Transition	Band System	Wavelength(nm)	Energy	Lifetime
O_2	$X^3\Sigma_g^-$	--------	-------	-------	0	-------
	$a^1\Delta_g$	a-X	IR Atmospheric	924-1580	0.98	60 min
	$b^1\Sigma_g^+$	b-X	Atmospheric	538-997	1.63	12 s
	$c^1\Sigma_u^-$	c-X	Herzberg II	449-479 254-271	4.06	>1 ms
	$A^3\Sigma_u^+$	A-X	Herzberg I	243-488	4.35	1-1000 s
O_2^+	$b^4\Sigma_g^-$	b-a	First negative	499-853	---	1.1-1.2 µs
	$A^2\Pi_u$	A-X	Second negative	-------	---	0.67-0.68µs

Table 2. The electronic state, the transition pathways, the band systems, the wavelength, the energy and the mean lifetime of O_2 and O_2^+ (Krupenie, 1972).

4.4 Electron energy distribution function

The electron energy distribution function (EEDF), which can be numerically solved by ELENDIF (Morgan & Penetrante, 1990), is shown in Fig. 7. The EEDFs calculated by ELENDIF code for several values of the reduced E-field E/N are shown. Recently, the results of ELENDIF

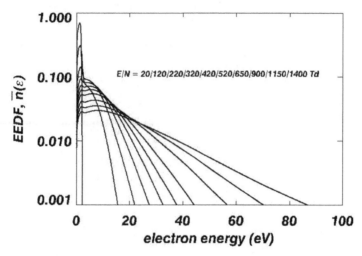

Fig. 7. The electron energy distribution functions calculated using ELNEDF for reduced E-field E/N ranging from 20 to 1400 Td.

are compared to those from the studies using a Monte Carlo model of thermal runaway electrons; the results agree well for electric fields up to ~20 E_k (~2400 Td) (Moss et al., 2006). Hence, even if ELENDIF is strictly valid for weak E-field cases (Morgan & Penetrante, 1990), however its applicable range could be extended to higher E-field cases.

We compare the EEDF by ELENDIF code with EEDF computed via the Maxwell-Boltzmann and Druyvesteyn models in Fig. 8. The mathematical formula of Maxwell-Boltzmann EEDF can be expresses as $\bar{n} = \dfrac{2\varepsilon^{1/2}}{\pi^{1/2}\varepsilon_0^{3/2}} e^{-\frac{\varepsilon}{\varepsilon_0}}$. The Druyvesteyn EEDF is $\bar{n} = C\varepsilon^{1/2} e^{-\frac{3m}{M}\frac{\varepsilon^2}{\varepsilon_0^2}}$ where C is a constant that satisfies the normalization condition $\int_0^\infty \bar{n}(\varepsilon)d\varepsilon = 1$; m and M are the electron mass and the mean air mass (28.6 amu). In Fig. 8, for the high-energy domain (> 5 eV), the Druyvesteyn EEDF ($e^{-\frac{\varepsilon^2}{\varepsilon_0^2}}$) has a steeper decreasing high-energy tail than that of the Maxwell EEDF ($e^{-\frac{\varepsilon}{\varepsilon_0}}$). Whereas in the low-energy domain (< 5 eV), the Druyvesteyn EEDF calls for a large percentage of low energy electron than that in the Maxwell EEDF. The ELENDIF EEDF straddles the middle ground between Druyvesteyn and Maxwell EEDFs. The Druyvesteyn and ELENDIF EEDF are classified as the "non-Maxwellian" EEDFs. Even though both have the same average electron energy, the major difference between the Druyvesteyn EEDF and the Maxwell distribution function is in the high-energy tail; because of in the Druyvesteyn model, the collision frequency is not a constant but is proportional to velocity. Whereas, the ELENDIF EEDF accounts almost for all the important elastic and inelastic processes in air and should provide the best EEDF comparing with the other two models.

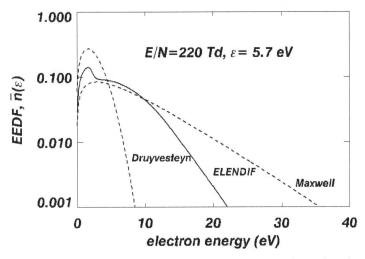

Fig. 8. The EEDF (solid line) calculated by ELENDIF for the case with a reduced E-field of 220 Td, derived from sprite observation (Kuo et al., 2005), and corresponding average electron energy ~5.7 eV. For comparison, the EEDFs computed using the Druyvesteyn and the Maxwell-Boltzmann models are also shown (dashed lines). The EEDFs are all normalized to satisfy $\int_0^\infty \overline{n}(\varepsilon)d\varepsilon = 1$.

4.5 Ionization, attachment and excitation rates in N_2/O_2

The ionization and attachment rates ($K_{s,j}$ is reaction rate for process j and gas density of s-th species) can be calculated using

$$\frac{K_{s,j}}{N_s} = \frac{1}{N_e}(\frac{2}{m_e})^{\frac{1}{2}}\int_0^\infty \varepsilon^{\frac{1}{2}}\sigma_{s,j}(\varepsilon)n(\varepsilon)d\varepsilon \qquad (3)$$

where N_s and N_e are the gas density of s-th species and the electron density, m_e is the electron mass, $n(\varepsilon)$ is the electron energy distribution as a function of electron energy ε, and $\sigma_{s,j}(\varepsilon)$ is the cross section of the process j (ionization, attachment and excitation). The ionization rates of molecular oxygen and nitrogen are represented by the dotted lines in Fig. 9, and the air ionization rate is denoted by the solid line. The breakdown E-field is the field that the curve of the air ionization rate crossing the curve of the dissociative attachment. The breakdown E-field in our calculation is ~ 117.2 Td, which is very close to the published values of 118.5 Td (Papadopoulos et al., 1993).

4.6 Electron-impact processes in N_2/O_2

Numerical results from the Boltzmann transport equation link the underlying microscopic collision processes and the reaction rates of the collisional processes discussed in Section 4.5. The ionization and attachment rates are calculated using Eq. 3, from microscopic collision processes. The derived parameters (v_i and v_a) can be used to calculate the electron number

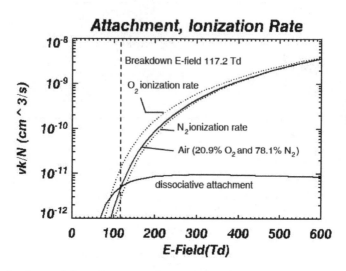

Fig. 9. The ionization and dissociative attachment rates of air.

density in a macroscopic environment where the electron multiplication process is represented by Eq. 1 in Section 3.2.

The reaction rates, which are functions of the reduced E-field, can be directly derived by the ELENDIF after incorporating the experimental cross sections of molecular nitrogen and oxygen. With the collisional processes, e.g. excitation rates in N_2/O_2, properly accounting for, the excitation rates can be used to predict the optical emissions of TLEs. The optical emission model, the population and depopulation equation (Sipler & Biondi, 1972), can be expressed as

$$\frac{\partial n_k}{\partial t} = v_k N_e - n_k (A_k + k_{q,N_2} N_{N_2} + k_{q,O_2} N_{O_2}) + \sum_m n_m A_m \tag{4}$$

where v_k, A_k, $k_{q,,N2}$, $k_{q,O2}$ are the excitation rate from ELENDIF, the Einstein coefficient, and the collisional quenching rates for the k-th electronic state of molecular nitrogen and molecular oxygen as listed in Tables 1 and 2; N_e, N_{N2}, and N_{O2} are the number densities of electrons, molecular nitrogen and molecular oxygen. The last term in the right hand side is a sum over all the cascade terms into the specified k-th excited state. As an example, for $B^3\Pi_g$, we consider major cascading terms from $C^3\Pi_u$ and $B'^3\Sigma_u^-$ (Milikh et al., 1998). The photon emission intensity per volume in units of the number of photons per cubic centimeter per second (ph/cm^3/s) at coordinate r, z and time t for the k-th excited state is represented by $I_k(r,z,t) = n_k(r,z,t)A_k$, in the N_2 1P, N_2 2P, N_2 LBH, N_2^+ Meinel and N_2^+ 1N band systems in our model. The emission lines of the v-th vibrational state of the k-th excited state into the v'-th vibrational state of the k'-th excited state is calculated by

$$I_{k,v;k',v'}(\lambda) = n_k q_{x,0;k,v} A_{k,v;k'.v'} \tag{5}$$

where n_k is the number density of the ambient molecular nitrogen or molecular oxygen; $q_{x,0;k,v}$ is the Franck-Condon factor, which represents the relative population of the v-th vibrational level of the k-th excited state from the 0-th vibrational state of the ground state; $A_{k,v;k',v'}$ is the Einstein coefficient from the v-th vibrational state of k-th electronic state to the v'-th vibrational state of k'-th electronic state, and the adopted values for molecular nitrogen is from (Gilmore et al., 1992) and those values for molecular oxygen from (Krupenie, 1972).

4.7 Full kinetic scheme in discharge gas

Besides electron-impact processes between electron and N_2/O_2, other plasma chemistry reactions also needed to be considered for the discharge processes in TLEs. Recently, Sentman et al. (2008) proposed the full-kinetic plasma chemistry model to compute the involved plasma processes in TLEs; in all, +80 species and +500 chemical reactions are considered in their zero dimensional plasma chemistry model. Sentman et al. (2008) pointed out that the optical emissions in the tail of the sprite streamer may be due to chemiluminescent processes, which follow the electron-impact processes in the head of the sprite streamer. Kuo et al. (Kuo et al., 2011) adopted the Sentman kinetic scheme but include a few corrected chemical processes for a similar but independent plasma chemistry study of TLEs. Kuo et al. (Kuo et al., 2011) found that the modelling intensity ratios N_2 1P/N_2 2P, N_2^+ 1N/N_2 2P were in good agreement with ISUAL optical measurements. Moreover, they also reported, for the first time, the evidence for the existence of O_2 atmosphere (0-0) band in sprites that was predicted by the plasma chemistry model (Sentman et al., 2008; Sentman & Stenbaek-Nielsen, 2009).

5. Space shuttle observations of TLEs

Recently, a few review articles on the TLE orbit missions and their results have been available (Yair, 2006; Lefeuvre et al., 2009; Neubert, 2009; Panasyuk et al., 2009; Pasko, 2010; Pasko et al., 2011). Here, only the relevant orbit missions are revisited and summarized. Before the first satellite mission (ISUAL) for the TLE survey, pioneer quests for the TLE observations had been performed on the space shuttles and on the International Space Station (ISS). The first TLE observed from space was termed as an enhanced airglow emission (Boeck et al., 1992); an elve in the current term. They performed post-reviews of the video tapes recorded by the cargo-bay television cameras from the STS-41 mission of the shuttle Discovery, and identified the enhanced transient luminosity events in the airglow altitude of $\sim$ 95 km. Boeck et al. (Boeck et al., 1992) concluded that the enhanced airglow emission suddenly appeared after a lightning flash, and provided the evidence on the direct coupling between atmospheric lighting and enhanced airglow emission in the bottom of ionosphere (Boeck et al., 1992; Boeck et al., 1995; Boeck et al., 1998).

The Mediterranean Israeli Dust Experiment (MEIDEX) sprite campaign was conducted on board the space shuttle Columbia (Yair et al., 2003; Yair et al., 2004; Yair, 2006) during the STS-107 mission in January 2003. Using an image-intensified Xybin IMC-201 camera, 17 TLEs were identified (7 sprites and 10 elves) along with additional 20 probable events. Their brightness in the 665 nm filter is determined to be in the range of 0.3-1.7 MR and 1.44-1.7 MR in the 860-nm filter (Israelevich et al., 2004; Yair et al., 2004).

The LSO (Lightning and Sprite Observations) on board ISS (International Space Station) is first experiment dedicated to a nadir observation of sprites from the Earth orbit (Blanc et al., 2004; Blanc et al., 2006; Blanc et al., 2007). The first LSO measurements were conducted on the ISS in October 2001. Blanc et al. (Blanc et al., 2004) utilized the emission differentiation method designed for the nadir observation of TLEs to distinguish the sprite emissions from the lighting emissions. The LSO is a pilot experiment for the upcoming TARANIS satellite mission. The configuration of the nadir observation is necessary for a simultaneous study of the optical, the relativistic runaway electron, and the X-gamma emissions from the TLEs.

5.1 The first satellite mission for the survey of TLEs: ISUAL

ISUAL (Imager of Sprites and Upper Atmospheric Lightnings) onboard the FORMOSAT-2 satellite is first satellite payload dedicating for the survey of TLEs (Chern et al., 2003; Mende et al., 2005; Su et al., 2005; Hsu et al., 2009). The FORMOSAT-2 is a sun-synchronized satellite with fourteen daily-revisiting 891 km altitude orbits. The FORMOSAT-2 was successfully launched on 21 May 2004. The ISUAL experiment is an international collaboration between the National Cheng Kung University, Taiwan, Tohoku University, Japan and the instrument development team from the University of California, Berkeley. The ISUAL consists of three sensor packages including an intensified CCD imager, a six-channel spectrophotometer and a dual-band array photometer. The imager is equipped with 6 selectable filters (N_2 1P, 762, 630, 557.7, 427.8 nm filters, and a broadband filter) mounted on a rotatable filter wheel. The spectrophotometer contains six filter photometer channels, their bandpasses ranging from the far ultraviolet to the near infrared regions. The dual-channel AP is fitted with broadband blue and red filters. The mission objectives are to perform a global survey of lightning-induced TLEs, to determine the occurrence rate of TLEs above thunderstorm, to investigate their spatial, temporal and spectral properties, and to investigate of the global distribution of airglow intensity as a function of altitude. ISUAL have completed the first phase (2004-2009) of the orbital mission. Due to a successful five-year mission and the significant scientific achievements, an additional funding has been granted to the ISUAL team for an extended mission of +3 years (2010-2014) from the National Space Organization in Taiwan.

The first sprite from ISUAL was recorded on July 4, 2004/21:31:15.451. From analyzing the ISUAL spectrophotometer data, Kuo et al. (2005) and Liu et al. (2006) estimated the strength of electric field at the streamer tips to be 2-4 E_k; through analyzing the ISUAL array photometer data, Adachi et al. (2006) concluded that the electric field is 1-2 E_k in the diffuse region of the sprite streamer. Recently, based on sprite streamer simulations, (Celestin & Pasko, 2010) pointed out that the electric fields derived basing on the ISUAL spectrophotometer/photometer data were lower-limiting values, since the time of the highest electric field precedes that of detected emission peak for the N_2 excited emission bands. The highest band emission source spatially is behind the electric field peak in streamer simulations, and they estimated that reported electric field strengths should be corrected by multiplying a factor of ~1.5 (Celestin & Pasko, 2010).

Mende et al. (2005) analyzed ISUAL elves whose parent lightning were behind the Earth limb and hence the lightning emissions were blocked by the solid Earth; they reported that the elves contained significant 391.4 nm emission of $1NN_2^+$. Mende et al. (2005) also estimated that reduced electric field in elves was > 200 Td by comparing the ratio of elve-emissions registered by different channels of the ISUAL spectrophotometer and that of the theoretically-derived emission intensity ratio. Using the inferred reduced electric field and the total ionization derived from the registered $1NN_2^+$ emission, they also found that, on average, the free electron density is 210 electrons cm^{-3} in elves; in the region occupied by an elve, the free electron density increases by nearly 100% over the ambient E-layer ionospheric value. Moreover, the FUV emission (Lyman-Birge-Hopfield band) in TLEs was also detected for the first time (Mende et al., 2005). Kuo et al. (2007) developed an elve model using finite difference time domain method to simulate the expected geometry of ISUAL recorded elves and the expected photometric intensities of elves. Their simulation results were in excellent agreement with ISUAL observed events. Kuo et al. (2007) also found there is an exponential relationship between the causative lighting current and the elve emissions. Based on their results, the peak current of the elve-parent lightning can be inferred from the ISUAL photometric intensity data.

Using the ISUAL TLE data, Chen et al. (2008) constructed the first global TLE distribution map and obtained the global TLE occurrence rates. The map indicates that there are six elve hot zones over: the Caribbean Sea, the South China Sea, the east Indian Ocean, the central Pacific Ocean, the west Atlantic Ocean, and the southwest Pacific Ocean. Unlike sprites mostly occur over the lands; elves appear predominately over oceans. Chen et al. (2008) compiled the global occurrence rate of elves and concluded that elve occurrence rate jumps as the sea surface temperature exceeds 26 degrees Celsius. Their finding clearly confirms the existence of an ocean-atmosphere-ionosphere coupling. Kuo et al. (2008) analyzed the photometric and the imagery brightness of TLEs (sprites, halos and elves), and found that total energy deposition rate of TLEs is ~1 GJ/min in the middle atmosphere. Hsu et al. (2009) re-examined a more complete set of ISUAL recorded TLEs, and discovered that the global TLE occurrence rates should be 72, 3.7, and ~1 events/minute, respectively, for elvess, halos, and sprites. Comparing with the results from the first three years of the ISUAL experiment reported in Chen et al. (2008), the global occurrence rates for elves and halos are higher due to the adoption of different correction factors. Using these updated TLE rates, the free electron content over an elve hot zone is estimated to be elevated by more than 10%. Deposited energy in the upper atmosphere by sprites, halos, and elves was found to be 22, 14, and 19 MJ per event, respectively. After factoring in the occurrence rates, in each minute, sprites, halos and elves deliver 22, 52 and 1370 MJs of the troposphere energy to the upper atmosphere.

Using ISUAL recorded gigantic jets, Kuo et al. (2009) performed the first high time resolution analysis of these spectacular events. They reported that the velocity of the upward propagating fully-developed jet of the gigantic jets was ~10^7 m/s, which is in line with that for the downward sprite streamers. Analysis of the spectral ratios of the fully-developed jet emissions gives a reduced E field of > 5 E_k and average electron energy of 8.5–12.3 eV in the gigantic jets. These values are higher than those in the sprites but are similar to those predicted by streamer models (Kuo et al., 2005), which implies the existence of streamer tips in fully-developed jets.

Chou et al. (2010) found that the gigantic jets (GJs) can actually be categorized into three types from their generating sequence and spectral properties. Type I GJs resembles that reported previously in (Su et al., 2003): after the fully-developed jet (FDJ) established the discharge channel, the ISUAL photometers registered a peak that was from a return stroke-like-process between the ionosphere and the cloud-top. The associated ULF (ultra low frequency) sferics indicates that they are negative cloud-to-ionosphere discharges (-CIs). Type II GJs begin as blue jets and then developed into GJs in ~100 ms. Blue jets also frequently occurred at the same region before and after the type II GJs. No identifiable ULF sferics of the type II Gjs were found, though an extra event with +CI ULF is probably a type II GJ. Thus for the type II GJs, the energy and the charge may not accumulate high enough to initiate a bright gigantic jet. Type III GJs were preceded by lightning and a GJ occurred near this preceding lightning. The spectral data of the type III GJs are dominated by lightning signals and the ULF data have high background noise. The average brightness of the type III GJs falls between those of the other two types of GJs. Therefore, they proposed that the discharge polarity of the type III GJs can be either negative or positive, depending on the type of the charge imbalance left by the trigger lightning (Chou et al., 2010).

After analyzing the N_2 1P brightness of the ISUAL elves and their FUV intensity and performing modeling work of elves, Chang et al. (2010) shown that ISUAL-FUV intensity in an elve could be used to infer the peak current of the causative CG lightning. The ISUAL detection rate of elves is also can be improved since the sensitivity of ISUAL FUV photometer is 16 times higher than that of ISUAL N_2 1P-filtered Imager. Hence, FUV photometer can be used to perform a global elve survey and to obtain the peak current of the elve-producing lighting and other salient parameters. Besides, the existences of multi-elves, which are FUV events from the M-components or the multiple strokes in lighting flashes, were also reported.

Lee et al. (2010) analyzed the distribution of the TLEs registered by ISUAL, and deduced the synoptic-scale factors that control the occurrence of TLEs. Two different distribution patterns are found. For the low-latitude tropical regions (25°S ~ 25°N), 84% of the TLEs were found to occur over the Intertropical Convergence Zone (ITCZ) and the South Pacific Convergence Zone. The distribution of TLEs exhibited a seasonal variation that migrates north and south with respect to the equator. For the mid-latitude regions (latitudes beyond ±30°), 88% of the northern winter TLEs and 72% of the southern winter TLEs occurred near the mid-latitude cyclones. The winter TLE occurrence density and the storm-track frequency share similar trends with the distribution of the winter TLEs offset by 10°–15°.

5.2 Other present orbital missions of TLEs

Besides ISUAL mission (2004-) (Chern et al., 2003; Mende et al., 2005; Su et al., 2005; Hsu et al., 2009) for the global survey of TLEs, Tatiana-1 (2005-7) mission performed a similar function; Tatiana is a Moscow State University research educational microsatellite Tatiana. Tatiana mission was carried out in the period between January 2005 and March 2007 (Garipov et al., 2005; Garipov et al., 2006; Shneider & Milikh, 2010). With the Tatiana-1 data, Shneider and Milikh (2010) studied the atmospheric electricity phenomena that can serve as sources for short millisecond range flashes; they reported that the UV flashes in the millisecond scale detected by Tatiana-1 may have been generated by gigantic blue jets (GBJ).

Tatiana-2 (2009-) satellite was launched on 17 September 2009 to a solar-synchronized orbit of 820 km altitude with a inclination angle 98.8°(Garipov et al., 2010). The Tatiana-2 satellite have upgraded their instrument to achieve an higher performance than Tatiana-1 mission in several ways: UV (300-400 nm)- and red (600-700 nm)-filtered photomultiplier tube (PMT) micro-electro-mechanical telescope for extreme lighting (MTEL), photo spectrometer and electron flux detector (Panasyuk et al., 2009; Garipov et al., 2010).

SPRITE-SAT (2010-) is a Japanese micro satellite with a size of 50 cm cube and with a weight of 45 kg, that were designed and developed by Tohoku University, Japan (Takahashi et al., 2010). SPRITE-SAT has a sun-synchronous polar orbit of 670 km altitude. The main scientific goal of SPRITE-SAT satellite is to simultaneously observe TLEs and terrestrial gamma-ray flashes in nadir direction and to study the relationship and generation mechanisms of TLEs and TGFs. SPRITE-SAT has equipped the lightning Imager-1 and Imager-2 with narrow- and wide-band 762 nm filters; the payload include a wide field-of-view camera with a FOV of 140°, a terrestrial gamma-ray counter with a FOV of 134x180°, a high-sensitivity star sensor, and a VLF receiver and antenna. The SPRITE-SAT has been successful launched on 23 January 2009 and is currently operating by the Tohoku University group.

Chibis-M mission (Klimov et al., 2009) (see http://chibis.cosmos.ru/) is another ISS module with the goal to study TLEs and TGFs. Scientific instruments of Chibis-M include X-ray and γ-ray detectors with an energy range of 50-500 eV & a time resolution of 30 ns, an UV detector sensitive in the wavelength band of 180-800 nm, a digital photo camera with a fixed exposure time if 0.2 second, a radiofrequency sensor with a frequency passing band of 20 – 50 Hz, and an ULF-VLF antenna. On January 25, 2012 the micro-satellite Chibis (lapwing) was successfully detached from the transportation vehicle <Progress M-13> and started its mission. Video of the "Chibis-M" detachment from "Progress" can be seen on http://www.roscosmos.ru/main.php?id=216. This mission is dedicated to studies of Terrestrial Gamma-ray Flashes (TGFs) and accompanying emissions above thunderstorms in the upper atmosphere. The multi-instrument technique, covering nearly the whole spectrum of electromagnetic emissions (radio, optical, UV, X-ray and gamma bands), will monitor the lightning discharges with higher time resolution.

5.3 Future orbital missions of TLEs

Global Lightning and sprIte MeasurementS on JEM-EF (JEM-GLIMS, 2011-) is a space mission to observe lightning and TLEs from the Exposure Facility (EF) of the Japanese Experiment Module (JEM) on the International Space Station (ISS). The JEM-GLIMS mission uses two CMOS cameras, two photometers, one spectro-imager, and two VHF receivers to achieve the mission goals of studying the generation mechanism of transient luminous events (TLEs) and identifying the relationship between lightning, TLEs, and terrestrial γ-ray flashes (TGFs) (Sato et al., 2009).

ASIM (Atmpshere-Space Interactions Monitor, 2014-) (Neubert, 2009) is an instrument suite, mounted on the external platform of the European Columbus module for the International Space Station (ISS). The scientific objectives are to understand the global occurrences of TLEs and TGFs, to study the physical mechanism of TLEs and TGFs, and their relationships. The ASIM will further coordinates with the ground EuroSprite campaigns (Neubert et al., 2001; Neubert et al., 2008; Neubert, 2009).

TARANIS (Tool for the Analysis of Radiations from lightNIngs and Sprites, 2016-) (Blanc et al., 2006; Lefeuvre et al., 2009) is a CNES satellite project with a goal to study of the impulsive transfer of energy between the Earth atmosphere and the space environment. TARANIS have a very broad range of scientific objectives for simultaneously probing the

TLEs and Terrestrial Gamma-ray Flashes (TGFs). Therefore, TARANIS instruments including micro-cameras, photometers, X-ray, γ-ray detectors, energetic electron detectors, and radio band antenna. The TRANIS mission aims are: (1) to advance the physical understanding of the links between TLEs, TGFs, (2) to clarify the potential signatures of impulsive transfers of energy, verified by physical mechanism, and (3) to elucidate the physical parameters in TLEs and TGFs (Blanc et al., 2006; Lefeuvre et al., 2009).

6. The impact of TLEs on space shuttle

Space shuttle uses 76 miles (122 km) as their re-entry altitude, which roughly marks the boundary where atmospheric drag becomes important. Below re-entry altitude, space shuttle switches from steering with thrusters to maneuvering with air surfaces. At lower altitude, space shuttle enters the TLE region (10 – 100 km). The magnitude of electric field can as high as 2-3 E_k (10 – 40 V/m) in elves altitudes of 80 - 100 km (Kuo et al., 2007). The average energy of accelerated electrons in elves can as high as several to tens of eV (Kuo et al., 2005; Kuo et al., 2007). In the high tail of electron energy distribution, runaway electron may be up to several kilo electron volt of electron energy. Besides, these energetic electron avalanches in gas breakdown may cause the plasma erosion on the heat shield of space shuttle. Therefore, it is necessary to have space missions to investigate the possible damages on re-entry of space shuttle.

7. Conclusion

Discharge phenomena in the middle atmosphere are one of the hottest research fields for satellite missions; currently with the ISUAL, the Tatiana-2, the SPRITE-SAT, Chibis-M missions perform daily observations of TLEs from space. Other upcoming orbit missions including JEM-GLIMS, ASIM, TARANIS will soon join in to carry out further investigations of these interesting phenomena. These space missions will continue hunting TLEs over the thunderstorm and exploring the associated plasma physics, plasma chemistry, and atmospheric electricity in middle atmosphere. Besides, high electric field pulses and energized electron-impact process may cause the damage as space shuttle flies back to the TLE altitudes (10-100 km).

8. Acknowledgments

We thank Profs. Rue-Ron Hsu and Han-Tzong Su at the National Cheng Kung University, Prof. Lou-Chuug Lee at the National Central University, Prof. Alfred Chen at the National Cheng Kung University, Prof. J. L. Chern at the National Chiao Tung University, Drs. Harald Frey and Stephen Mende at the Space Science Laboratory-University of California, Profs. Horoshi Fukunishi and Yukihiro Takahashi at Tohoku University for helpful discussions and comments. We are also grateful to the National Center for High-performance Computing in Taiwan and Center for Computational Geophysics at National Central University for computer time and facilities. This work was supported in part by grants (NSC 98-2111-M-008-001, NSC 99-2111-M-006-001-MY3, NSC 99-2112-M-006-006-MY3, NSC 099-2811-M-006-004, NSC 100-2119-M-006-015, NSC 100-2811-M-006-004) from National Science Council in Taiwan.

9. References

Adachi, T., et al. (2006). Electric field transition between the diffuse and streamer regions of sprites estimated from ISUAL/array photometer measurements. *Geophys. Res. Lett.*, Vol.*33*, (September 1, 2006), pp. 17803, DOI: 10.1029/2006GL026495.

Allen, N. L., & Ghaffar, A. (1995). The conditions required for the propagation of a cathode-directed positive streamer in air. *J. Phys. D.*, Vol.*28*, (February 1, 1995), pp. 331-337, DOI: 10.1088/0022-3727/28/2/016.

Babaeva, N. Y., & Naidis, G. V. (1997). Dynamics of positive and negative streamers in air in weak uniform electric fields. *IEEE Trans. Plasma Sci.*, Vol.*25*, No.2, pp. 375-379, DOI: 10.1109/27.602514.

Barrington-Leigh, C. P., & Inan, U. S. (1999). Elves triggered by positive and negative lightning discharges. *Geophys. Res. Lett.*, Vol.*26*, (March 1, 1999), pp. 683-686, DOI: 10.1029/1999GL900059.

Barrington-Leigh, C. P.; Inan, U. S., & Stanley, M. (2001). Identification of sprites and elves with intensified video and broadband array photometry. *J. Geophys. Res.*, Vol.*106*, (February 1, 2001), pp. 1741-1750, DOI: 10.1029/2000JA000073.

Blanc, E., et al. (2004). Nadir observations of sprites from the International Space Station. *J. Geophys. Res.*, Vol.*109*, No.A2, pp. A02306, 0148-0227, DOI: 10.1029/2003ja009972.

Blanc, E., et al. (2006).Observations of sprites from space at the nadir: the lso (lightning and sprite observations) experiment on board of the international space station In: *Sprites, Elves and Intense Lightning Discharges*, pp.151-166, Springer Netherlands, ISBN 978-1-4020-4629-2.

Blanc, E., et al. (2007). Main results of LSO (Lightning and sprite observations) on board of the international space station. *Microgravity Science and Technology*, Vol.*19*, No.5, pp. 80-84, 0938-0108, DOI: 10.1007/bf02919458.

Boeck, W. L., et al. (1992). Lightning induced brightening in the airglow layer. *Geophys. Res. Lett.*, Vol.*19*, (January 1, 1992), pp. 99-102, DOI:10.1029/91GL03168.

Boeck, W. L., et al. (1995). Observations of lightning in the stratosphere. *J. Geophys. Res.*, Vol.*100*, (January 1, 1995), pp. 1465-1476, DOI: 10.1029/94JD02432.

Boeck, W. L., et al. (1998). The role of the space shuttle videotapes in the discovery of sprites, jets and elves. *J. Atmos. Sol. Terr. Phys.*, Vol.*60*, (May 1, 1998), pp. 669-677, DOI: 10.1016/S1364-6826(98)00025-X.

Brasseur, G., & Solomon, S. (1986). *Aeronomy of the middle atmosphere : chemistry and physics of the stratosphere and mesosphere* (2nd), D. Reidel Pub. Co., ISBN 902-7723-44-3, Dordrecht.

Bucsela, E., et al. (2003). $N_2(B^3\Pi_g)$ and $N_2^+(A^2\Pi_u)$ vibrational distributions observed in sprites. *J. Atmos. Sol. Terr. Phys.*, Vol.*65*, (March 2003), pp. 583-590, ISSN 1364-8826, DOI: 10.1029/96GL02071.

Celestin, S., & Pasko, V. P. (2010). Effects of spatial non-uniformity of streamer discharges on spectroscopic diagnostics of peak electric fields in transient luminous events. *Geophys. Res. Lett.*, Vol.*37*, No.7, pp. L07804, 0094-8276, DOI: 10.1029/2010gl042675.

Chang, S. C., et al. (2010). ISUAL far-ultraviolet events, elves, and lightning current. *J. Geophys. Res.*, Vol.*115*, pp. A00E46, 0148-0227, DOI: 10.1029/2009ja014861.

Chen, B., et al. (2008). Global distributions and occurrence rates of transient luminous events. *J. Geophys. Res.*, Vol.*113*, pp. A08306, DOI:10.1029/2008JA013101.

Chern, J. L., et al. (2003). Global survey of upper atmospheric transient luminous events on the ROCSAT-2 satellite. *J. Atmos. Sol. Terr. Phys.*, Vol.*65*, (March 1, 2003), pp. 647-659, DOI: 10.1016/S1364-6826(02)00317-6.

Chou, J. K., et al. (2010). Gigantic jets with negative and positive polarity streamers. *J. Geophys. Res.*, Vol.*115*, pp. A00E45, 0148-0227, DOI: 10.1029/2009ja014831.

Cummer, S. A., et al. (2006). Submillisecond imaging of sprite development and structure. *Geophys. Res. Lett.*, Vol.*33*, (February 1, 2006), pp. 04104, DOI: 10.1029/2005GL024969.

Franz, R. C.; Nemzek, R. J., & Winckler, J. R. (1990). Television Image of a Large Upward Electrical Discharge Above a Thunderstorm System. *Science*, Vol.*249*, (July 1, 1990), pp. 48-51, DOI: 10.1126/science.249.4964.48.

Frey, H. U., et al. (2007). Halos generated by negative cloud-to-ground lightning. *Geophys. Res. Lett.*, Vol.*34*, No.18, pp. L18801, 0094-8276, DOI: 10.1029/2007gl030908.

Fukunishi, H., et al. (1996). Elves: Lightning-induced transient luminous events in the lower ionosphere. *Geophys. Res. Lett.*, Vol.*23*, (January 1, 1996), pp. 2157-2160, DOI: 10.1029/96GL01979.

Garipov, G., et al. (2005). Ultraviolet flashes in the equatorial region of the Earth. *JETP Letters*, Vol.*82*, No.4, pp. 185-187, 0021-3640, DOI: 10.1134/1.2121811.

Garipov, G., et al. (2006). Ultraviolet radiation detector of the MSU research educational microsatellite *Universitetskii-Tat'yana*. *Instruments and Experimental Techniques*, Vol.*49*, No.1, pp. 126-131, 0020-4412, DOI: 10.1134/s0020441206010180.

Garipov, G. K., et al. (2010). Program of transient UV event research at Tatiana-2 satellite. *J. Geophys. Res.*, Vol.*115*, pp. A00E24, 0148-0227, DOI: 10.1029/2009ja014765.

Gerken, E. A., & Inan, U. S. (2003). Observations of decameter-scale morphologies in sprites. *J. Atmos. Sol. Terr. Phys.*, Vol.*65*, (March 1, 2003), pp. 567-572, DOI: 10.1016/S1364-6826(02)00333-4.

Gerken, E. A., & Inan, U. S. (2004). Comparison of photometric measurements and charge moment estimations in two sprite-producing storms. *Geophys. Res. Lett.*, Vol.*31*, (February 1, 2004), pp. 03107, DOI: 10.1029/2003GL018751.

Gilmore, F. R.; Laher, R. R., & Espy, P. J. (1992). Franck-Condon Factors, r-Centroids, Electronic Transition Moments, and Einstein Coefficients for Many Nitrogen and Oxygen Band Systems. *J. Phys. Chem. Ref. Data.*, Vol.*21*, (September 1, 1992), pp. 1005-1107, DOI: 10.1063/1.555910.

Green, B. D., et al. (1996). Molecular excitation in sprites. *Geophys. Res. Lett.*, Vol.*23*, (Nov 1, 1996), pp. 2161-2164, DOI: 10.1029/96GL02071.

Hampton, D. L.; Heavner, M. J.; Wescott, E. M., & Sentman, D. D. (1996). Optical spectral characteristics of sprites. *Geophys. Res. Lett.*, Vol.*23*, (Nov 1, 1996), pp. 89-92, DOI: 10.1029/95GL03587.

Hiraki, Y., et al. (2004). Generation of metastable oxygen atom $O(1D)$ in sprite halos. *Geophys. Res. Lett.*, Vol.*31*, (July 1, 2004), pp. 14105, DOI: 10.1029/2004GL020048.

Hsu, R.-R., et al. (2009). On the Global Occurrence and Impacts of Transient Luminous Events (TLEs). *AIP Conference Proceedings*, Vol.*1118*, No.1, pp. 99-107, DOI: 10.1063/1.3137720.

Inan, U. S.; Bell, T. F., & Rodriguez, J. V. (1991). Heating and ionization of the lower ionosphere by lightning. *Geophys. Res. Lett.*, Vol.*18*, (April 1, 1991), pp. 705-708, DOI: 10.1029/91GL00364.

Inan, U. S.; Sampson, W. A., & Taranenko, Y. N. (1996). Space-time structure of optical flashes and ionization changes produced by lighting-EMP. *Geophys. Res. Lett.*, Vol.*23*, (January 1, 1996), pp. 133-136, DOI: 10.1029/95GL03816.

Inan, U. S., et al. (1997). Rapid lateral expansion of optical luminosity in lightning-induced ionospheric flashes referred to as `elves'. *Geophys. Res. Lett.*, Vol.*24*, No.5, (March 1, 1997), pp. 583-586, DOI: 10.1029/97GL00404.

Israelevich, P. L., et al. (2004). Transient airglow enhancements observed from the space shuttle Columbia during the MEIDEX sprite campaign. *Geophys. Res. Lett.*, Vol.*31*, No.6, pp. L06124, 0094-8276, DOI: 10.1029/2003gl019110.

Kanmae, T.; Stenbaek-Nielsen, H. C., & McHarg, M. G. (2007). Altitude resolved sprite spectra with 3 ms temporal resolution. *Geophys. Res. Lett.*, Vol.*34*, (April 1, 2007), pp. 07810, DOI: 10.1029/2006GL028608.

Kanmae, T.; Stenbaek-Nielsen, H. C.; McHarg, M. G., & Haaland, R. K. (2010). Observation of sprite streamer head's spectra at 10,000 fps. *J. Geophys. Res.*, Vol.*115*, pp. A00E48, 0148-0227, DOI: 10.1029/2009ja014546.

Klimov, S. I.; Sharkov, E. A., & Zelenyi, L. M. (2009). The Tropical Cyclones as the Possible Sources of Gamma Emission in the Earth's Atmosphere. *AGU Fall Meeting Abstracts*, Vol.*33*, (December 1, 2009), pp. 0295

Krupenie, P. H. (1972). The spectrum of molecular oxygen. *J. Phys. Chem. Ref. Data.*, Vol.*1*, pp. 423-534, DOI: 10.1063/1.3253101.

Kuo, C.-L., et al. (2005). Electric fields and electron energies inferred from the ISUAL recorded sprites. *Geophys. Res. Lett.*, Vol.*32*, (October 1, 2005), pp. 19103, DOI: 10.1029/2005GL023389.

Kuo, C. L., et al. (2007). Modeling elves observed by FORMOSAT-2 satellite. *J. Geophys. Res.*, DOI: 10.1029/2007JA012407.

Kuo, C. L., et al. (2008). Radiative emission and energy deposition in transient luminous events. *J. Phys. D.*, Vol.*41*, (December 1, 2008), pp. 4014, DOI: 10.1088/0022-3727/41/23/234014.

Kuo, C. L., et al. (2009). Discharge processes, electric field, and electron energy in ISUAL recorded gigantic jets. *J. Geophys. Res.*, Vol.*114*, No.A4, pp. A04314, 0148-0227, DOI: 10.1029/2008ja013791.

Kuo, C. L., et al. (2011). The 762 nm emissions of sprites. *J. Geophys. Res.*, Vol.*116*, No.A1, pp. A01310, 0148-0227, DOI: 10.1029/2010ja015949.

Lee, L.-J., et al. (2010). Controlling synoptic-scale factors for the distribution of transient luminous events. *J. Geophys. Res.*, Vol.*115*, pp. A00E54, 0148-0227, DOI: 10.1029/2009ja014823.

Lefeuvre, F.; Blanc, E., & Team, J. L. P. T. (2009). TARANIS---a Satellite Project Dedicated to the Physics of TLEs and TGFs. *AIP Conference Proceedings*, Vol.*1118*, No.1, pp. 3-7, DOI: 10.1063/1.3137711.

Liu, N., & Pasko, V. P. (2004). Effects of photoionization on propagation and branching of positive and negative streamers in sprites. *J. Geophys. Res.*, Vol.*109*, (April 1, 2004), pp. 04301, DOI: 10.1029/2003JA010064.

Liu, N., & Pasko, V. P. (2005). Molecular nitrogen LBH band system far-UV emissions of sprite streamers. *Geophys. Res. Lett.*, Vol.*32*, (March 1, 2005), pp. 05104, DOI: 10.1029/2004GL022001.

Liu, N., et al. (2006). Comparison of results from sprite streamer modeling with spectrophotometric measurements by ISUAL instrument on FORMOSAT-2 satellite. *Geophys. Res. Lett.*, Vol.33, (January 1, 2006), pp. 01101, DOI: 10.1029/2005GL024243.

Liu, N. Y., et al. (2009). Comparison of acceleration, expansion, and brightness of sprite streamers obtained from modeling and high-speed video observations. *J. Geophys. Res.*, Vol.114, (March 1, 2009), DOI: 10.1029/2008JA013720.

Lofthus, A., & Krupenie, P. H. (1977). The spectrum of molecular nitrogen. *J. Phys. Chem. Ref. Data.*, Vol.6, (January 1, 1977), pp. 113-307, DOI: 10.1063/1.555546.

McHarg, M. G.; Stenbaek-Nielsen, H. C., & Kammae, T. (2007). Observations of streamer formation in sprites. *Geophys. Res. Lett.*, Vol.34, (March 1, 2007), pp. 06804, DOI: 10.1029/2006GL027854.

Mende, S. B.; Rairden, R. L.; Swenson, G. R., & Lyons, W. A. (1995). Sprite spectra; N_2 1 PG band identification. *Geophys. Res. Lett.*, Vol.22, (n/a 1, 1995), pp. 2633-2636, DOI: 10.1029/95GL02827.

Mende, S. B., et al. (2005). D region ionization by lightning-induced electromagnetic pulses. *J. Geophys. Res.*, Vol.110, (November 1, 2005), pp. 11312, DOI: 10.1029/2005JA011064.

Milikh, G.; Valdivia, J. A., & Papadopoulos, K. (1998). Spectrum of red sprites. *J. Atmos. Sol. Terr. Phys.*, Vol.60, (May 1, 1998), pp. 907-915, DOI: 10.1016/S1364-6826(98)00032-7.

Morgan, W. L., & Penetrante, B. M. (1990). ELENDIF: A time-dependent Boltzmann solver for partially ionized plasmas. *Comput. Phys. Commun.*, Vol.58, (February 1, 1990), pp. 127-152, DOI: 10.1016/0010-4655(90)90141-M.

Moss, G. D.; Pasko, V. P.; Liu, N., & Veronis, G. (2006). Monte Carlo model for analysis of thermal runaway electrons in streamer tips in transient luminous events and streamer zones of lightning leaders. *J. Geophys. Res.*, Vol.111, (February 1, 2006), pp. 02307, DOI: 10.1029/2005JA011350.

Moudry, D. R.; Stenbaek-Nielsen, H. C.; Sentman, D. D., & Wescott, E. M. (2002). Velocities of sprite tendrils. *Geophys. Res. Lett.*, Vol.29, (October 1, 2002), pp. 53-51, DOI: 10.1029/2002GL015682.

Moudry, D. R.; Stenbaek-Nielsen, H. C.; Sentman, D. D., & Wescott, E. M. (2003). Imaging of elves, halos and sprite initiation at 1ms time resolution. *J. Atmos. Sol. Terr. Phys.*, Vol.65, (March 1, 2003), pp. 509-518, DOI: 10.1016/S1364-6826(02)00323-1.

Neubert, T.; Allin, T. H.; Stenbaek-Nielsen, H. C., & Blanc, E. (2001). Sprites over Europe. *Geophys. Res. Lett.*, Vol.28, (September 1, 2001), pp. 3585-3588, DOI: 10.1029/2001GL013427.

Neubert, T., et al. (2008). Recent Results from Studies of Electric Discharges in the Mesosphere. *Surveys in Geophysics*, Vol.29, No.2, pp. 71-137, 0169-3298, DOI: 10.1007/s10712-008-9043-1.

Neubert, T. (2009). ASIM---an Instrument Suite for the International Space Station. *AIP Conference Proceedings*, Vol.1118, No.1, pp. 8-12, DOI: 10.1063/1.3137718.

Panasyuk, M. I., et al. (2009). Energetic Particles Impacting the Upper Atmosphere in Connection with Transient Luminous Event Phenomena: Russian Space Experiment Programs. *AIP Conference Proceedings*, Vol.1118, No.1, pp. 108-115, DOI: 10.1063/1.3137702.

Papadopoulos, K., et al. (1993). Ionization rates for atmospheric and ionospheric breakdown. *J. Geophys. Res.*, Vol.*98*, (October 1, 1993), pp. 17593-17596, DOI: 10.1029/93JA00795.

Pasko, V. P.; Inan, U. S.; Bell, T. F., & Taranenko, Y. N. (1997). Sprites produced by quasi-electrostatic heating and ionization in the lower ionosphere. *J. Geophys. Res.*, Vol.*102*, No.A3, pp. 4529-4562, DOI: 10.1029/96JA03528.

Pasko, V. P.; Inan, U. S., & Bell, T. F. (1998). Spatial structure of sprites. *Geophys. Res. Lett.*, Vol.*25*, (June 1, 1998), pp. 2123-2126, DOI: 10.1029/98GL01242.

Pasko, V. P., et al. (2002). Electrical discharge from a thundercloud top to the lower ionosphere. *Nature*, Vol.*416*, pp. 152-154, DOI :10.1038/416152a.

Pasko, V. P. (2003). Atmospheric physics: Electric jets. *Nature*, Vol.*423*, No.6943, pp. 927-929, 0028-0836, DOI: 10.1038/423927a.

Pasko, V. P. (2006).Theoretical modeling of sprites and jets, in Sprites, Elves and Intense Lightning Discharges In: *NATO Science Series II: Mathematics, Physics and Chemistry*, pp.253-311, Springer, Heidleberg, Germany.

Pasko, V. P. (2010). Recent advances in theory of transient luminous events. *J. Geophys. Res.*, Vol.*115*, pp. A00E35, 0148-0227, 10.1029/2009ja014860.

Pasko, V. P.; Yair, Y., & Kuo, C.-L. (2011). Lightning Related Transient Luminous Events at High Altitude in the Earth's Atmosphere: Phenomenology, Mechanisms and Effects. *Space Science Reviews*, pp. 1-42, 0038-6308, DOI: 10.1007/s11214-011-9813-9.

Raizer, Y. P. (1991). *Gas discharge physics*, Springer-Verlag, ISBN 0387194622, New York.

Rakov, V. A., & Tuni, W. G. (2003). Lightning electric field intensity at high altitudes: Inferences for production of elves. *J. Geophys. Res.*, Vol.*108*, No.D20, (October 1, 2003), pp. 4639, DOI: 10.1029/2003JD003618.

Rakov, V. A., & Uman, M. A. (2003). *Lightning: physics and effects*, Cambridge University Press, ISBN 0521583276, Cambridge, UK.

Sato, M., et al. (2009). Science Goal and Mission Status of JEM-GLIMS. *AGU Fall Meeting Abstracts*, Vol.*23*, (December 1, 2009), pp. 03

Sentman, D. D., et al. (1995). Preliminary results from the Sprites94 aircraft campaign: 1. Red sprites. *Geophys. Res. Lett.*, Vol.*22*, (May 1, 1995), pp. 1205-1208, DOI: 10.1029/95GL00583.

Sentman, D. D.; Stenbaek-Nielsen, H. C.; McHarg, M. G., & Morrill, J. S. (2008). Plasma chemistry of sprite streamers. *J. Geophys. Res.*, Vol.*113*, DOI: 10.1029/2007jd008941.

Sentman, D. D., & Stenbaek-Nielsen, H. C. (2009). Chemical effects of weak electric fields in the trailing columns of sprite streamers. *Plasma Sources Science and Technology*, Vol.*18*, No.3, pp. 034012, 0963-0252, DOI: 10.1088/0963-0252/18/3/034012.

Shneider, M. N., & Milikh, G. M. (2010). Analysis of UV flashes of millisecond scale detected by a low-orbit satellite. *J. Geophys. Res.*, Vol.*115*, pp. A00E23, 0148-0227, DOI: 10.1029/2009ja014685.

Sipler, D. P., & Biondi, M. A. (1972). Measurements of O^1 D quenching rates in the F region. *J. Geophys. Res.*, Vol.*77*, (1 Nov. 1972), pp. 6202-6212, DOI: 10.1029/JA077i031p06202.

Stanley, M., et al. (1999). High speed video of initial sprite development. *Geophys. Res. Lett.*, Vol.*26*, (October 1, 1999), pp. 3201-3204, DOI: 10.1029/1999GL010673.

Stenbaek-Nielse, H., & McHarg, M. G. (2004). Sprite Spectra at 1000 fps. *AGU Fall Meeting Abstracts*, Vol.*51*, (December 1, 2004), pp. 08

Su, H., et al. (2005). Key results from the first fourteen months of ISUAL experiment. *AGU Fall Meeting Abstracts*, Vol.*11*, (December 1, 2005), pp. 01

Su, H. T., et al. (2003). Gigantic jets between a thundercloud and the ionosphere. *Nature*, Vol.*423*, pp. 974-976, DOI: 10.1038/nature01759.

Takahashi, Y.; Yoshida, K.; Sakamoto, Y., & Sakamoi, T. (2010).SPRITE-SAT: A University Small Satellite for Observation of High-Altitude Luminous Events, pp.197-206.

Vallance-Jones, A. (1974). *Aurora*, D. Reidel Publishing Co., Dordrecht.

Veronis, G.; Pasko, V. P., & Inan, U. S. (1999). Characteristics of mesospheric optical emissions produced by lighting discharges. *J. Geophys. Res.*, Vol.*104*, (June 1, 1999), pp. 12645-12656, DOI: 10.1029/1999JA900129.

Volland, H. (1987). Electromagnetic Coupling Between Lower and Upper Atmosphere. *Physica Scripta*, Vol.*1987*, No.T18, pp. 289, 1402-4896, DOI: 10.1088/0031-8949/1987/T18/029.

Wescott, E. M., et al. (1995). Preliminary results from the Sprites94 aircraft campaign: 2. Blue jets. *Geophys. Res. Lett.*, Vol.*22*, (May 1, 1995), pp. 1209-1212, DOI: 10.1016/0032-0633(92)90056-T.

Wescott, E. M., et al. (2001). Triangulation of sprites, associated halos and their possible relation to causative lightning and micrometeors. *J. Geophys. Res.*, Vol.*106*, No.A6, pp. 10467-10477, 0148-0227, DOI: 10.1029/2000ja000182.

Yair, Y., et al. (2003). Sprite observations from the space shuttle during the Mediterranean Israeli dust experiment (MEIDEX). *Journal of Atmospheric and Solar-Terrestrial Physics*, Vol.*65*, No.5, pp. 635-642, 1364-6826, DOI: 10.1016/s1364-6826(02)00332-2.

Yair, Y., et al. (2004). New observations of sprites from the space shuttle. *J. Geophys. Res.*, Vol.*109*, No.D15, pp. D15201, 0148-0227, DOI: 10.1029/2003jd004497.

Yair, Y. (2006). Observations of Transient Luminous Events from Earth Orbit. *IEEJ Transactions on Fundamentals and Materials*, Vol.*126*, (Nov 1, 2006), pp. 244-249, DOI: 10.1541/ieejfms.126.244.

4

Research on the Method of Spacecraft Orbit Determination Based on the Technology of Dynamic Model Compensation

Pan Xiaogang, Wang Jiongqi and Zhou Haiyin
National University of Defence Technology,
China

1. Introduction

For the limit of dynamic model knowledge, model error will exist objectively in satellite dynamic model, especially for non-cooperation satellite, whose quality and shapes are unknown. While the precision of orbit determination results depends on the satellite dynamic model error, because dynamic model error will transmit to measurement data, and then mixed with measurement data error itself and result in the produce of the new system error data named Mixed Error(ME) in measurement model, which cannot be accurately described with parameters. The ME can be popularly dealt with by three methods: 1. to regard ME as stochastic error. By analyzing the characteristic of error, it is chosen the corresponding estimator[1]. 2. Geometrical method for orbit determination. The dynamic model will not attend the process of orbit determination, so the ME doesn't exist [2]. 3. Reduced-dynamic orbit determination method. Process noise is added to dynamic model, which can absorb the dynamic model error[3.4]. Many scholars had researched the above orbit determination methods, and gained great success under given circumstance. For the first method, it is very difficult to determine the distribution of error data, so the precision of orbit determination will be influenced badly. Satellite dynamic model was avoided with the second method, the precision will not be affected by dynamic model, but it cannot forecast the satellite orbit and cannot analyze the orbit characteristic for geometrical method. For the third method, it was the most effective method to restrain the dynamic model error, but it needs precise measurement data (GPS) to restrict the model. It can be summarized from the above analysis that a new dynamic model error compensation method must be established to deal with the common orbit determination question. Semi-parametric model of orbit determination was established in this paper. Parametric function was used to describe the exact dynamic model, and non parametric function was founded to describe the ME. Because of the rationality of semi-parametric model, it gained great development since it was produced by Engle in 1980s. Many solve method of semi-parametric was researched such as regularization methd, B-spline compensation method and two stage method [5]. In traditional two stage method, nonlinear model regression was usually used to estimate the non parametric parts in semi-parametric model, but it can be distorted by the gross error, so the data depth theory was applied to semi-parametric model called Stahel –Donoho Kernel Estimator [6].

The precision of orbit determination depends on the precision of measurement data and the precision of dynamic model. In modern times, the degree of measurement can be limited in millimeter degree, so the key method to improve the precision of orbit determination is to increase the exactness of dynamic model, or to compensate the model error. In this chapter, some mathematic method are proposed to compensate the uncertainty model error, all those method is focus on the mathematics models, at last, a new orbit determination method based on model error compensation is put forward to deal with directly the dynamic force.

2. Three different method of orbit determination based on dynamic model

The dynamic model can describe the orbit character of satellite, and the position and velocity at time t can be got by integral of dynamic model with condition of initial parameters. Based on the different method of state transfer and dynamic model ,the method of orbit determination is different too.

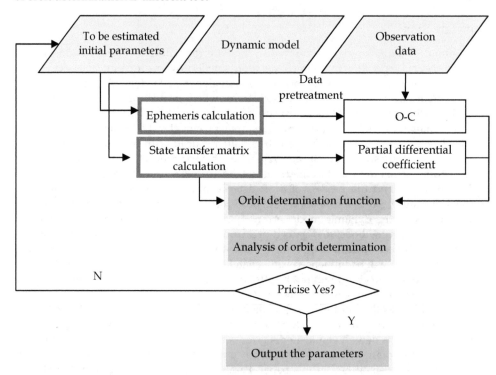

Fig. 1. Flow chart of orbit determination based on dynamic model.

2.1 Theory of orbit determination based on transcendental information

Based on the function of satellite, the dynamic model of satellite can be noted as

$$\begin{cases} \dot{X}(t) = f(X(t),t) \\ X(t_0) = X_0 \end{cases} \tag{1}$$

Where, $X = (r, \dot{r}, p_c, p_l)^T$ is to be improved state vector and model parameters. $r, \dot{r}$ is the position and velocity vector of satellite or the ephemeris of satellite, p_c is the vector of dynamic model parameters including atmosphere drag coefficient and light pressure coefficient and coefficients of experiential acceleration and others, p_l is other to be estimated parameters of other model.

Orbit state at any time can be get by integral the dynamic model(1), so the parameters to be estimated at time t can be transformed to initial time t_0, noted as $X_0 = (r_0, \dot{r}_0, p_c, p_l)^T$, then the observation function can be described as

$$H = G(X_0, b) + \varepsilon \tag{2}$$

Where, b is the parameter with no relation to dynamic model, such as index of system error.

In the JPL/ODP and GTDS, there are some parts of parameters meaning considering vector, which is known with low precision. Dividing the all parameters to two parts, named estimating vector χ and considering vector z, and dividing the estimating vector χ to dynamic model parameters X_0 and observation model parameters b. The observation function can be rewritten as

$$H = G(\chi, z) + \varepsilon \tag{3}$$

Let the transcendental value of estimating vector χ and considering vector z be χ_0, z_0.

Where, $E(\chi_0) = \chi, E(z_0) = z, \text{cov}(\chi_0) = P_{\chi 0}, \text{cov}(z_0) = P_{z 0}$.

Define 1: suppose the loss function be,

$$\tilde{Q}(\chi) = (H - G(\chi, z_0))^T W (H - G(\chi, z_0))^T + (\chi - \chi_0)^T P_\chi^{-1} (\chi - \chi_0) \tag{4}$$

Where W is weight matrix of observation.

The second item of the right parts in the above formula(4) is to restrict the estimating value to the transcendental value with the covariance matrix P_χ. Based on the iterative method of Gauss-Newton, unwrap the observation function at χ_i,

$$G(\chi, z) = G(\chi_i, z_0) + B_1 \Delta \chi_i + B_2 \Delta z_i \tag{5}$$

Where, $\Delta \chi_i = \chi - \chi_i, \Delta z_i = z - z_0, B_1 = \left(\dfrac{\partial G}{\partial \chi} \right)\bigg|_{\chi = \chi_i, z = z_0}, B_2 = \left(\dfrac{\partial G}{\partial z} \right)\bigg|_{\chi = \chi_i, z = z_0}.$

The linear loss function can be obtain when put the observation function (5)to loss function(4),

$$Q(\Delta \chi_i) = (\Delta H_i - B_{1i} \Delta \chi_i)^T W (\Delta H_i - B_{1i} \Delta \chi_i) + (\Delta \chi_i - \Delta \tilde{\chi}_i)^T P_\chi^{-1} (\Delta \chi_i - \Delta \tilde{\chi}_i)$$

Where, $\Delta H_i = H - G(\chi_i, z)$ is the residual error of calculation measurement data and actual measurement data, named OC residual error. $\Delta \tilde{\chi}_i = \chi_0 - \chi_i$ is the error of transcendental value and iterative value at i.

Based on the least square estimation theory, the value of $\Delta \chi_i$ to minimize Q,

$$\Delta \hat{\chi}_{i+1} = \left(B_{1i}^T W B_{1i} + P_\chi^{-1}\right)^{-1} \left(B_{1i}^T W \Delta H_i + P_\chi^{-1} \Delta \tilde{\chi}_i\right) \tag{6}$$

Then the estimation is,

$$\hat{\chi}_{i+1} = \chi_0 + \sum_{k=1}^{i+1} \Delta \hat{\chi}_k = \hat{\chi}_i + \Delta \hat{\chi}_{i+1} \tag{7}$$

The orbit determination method is different with different method to solve the dynamic model and state transfer matrix.

2.2 Orbit determination based on analytical method

In analytical method, the ephemeris value can be calculated by mean elements, simple and with high efficiency. And it ignore up two rank chronically items and all the periodic items, for the degree of all the items is $O(J_2)$,with conditions of $n(t-t_0) = 1 / J_2$.

The partial differential coefficient matrix of observation vector to dynamic model parameters can be got by chain principle,

$$B = \frac{\partial G}{\partial (r, \dot{r})} \cdot \frac{\partial (r, \dot{r})}{\partial X} \cdot \frac{\partial X}{\partial X_0} \tag{8}$$

1.   the first kind partial differential coefficient matrix $\dfrac{\partial G}{\partial (r, \dot{r})}$

The sensitive degree of observation value at time t to position and velocity vector is reflected by the first kind partial differential coefficient matrix

2.   the second kind partial differential coefficient matrix $\dfrac{\partial (r, \dot{r})}{\partial X}$

The ephemeris of satellite and the position and velocity can be transformed each other, so the second kind partial differential coefficient is transfer matrix.

3.   the third kind partial differential coefficient matrix $\dfrac{\partial X}{\partial X_0}$

The third kind partial differential coefficient matrix is the state transfer matrix, it shows the transform from vector at time t to initial time. In the process of orbit determination based on analytical method, the matrix can be predigested.

2.3 Orbit determination based on numerical method

Although the dynamic model can be solved by analytical method, it generally cannot obtain the exact value, for the complicated dynamic model. With numerical method, the position and velocity of satellite at time t obtained by integral. The formula (1) can transform as

$$\frac{\partial \ddot{r}}{\partial X_0} = \frac{\partial \ddot{r}}{\partial r}\frac{\partial r}{\partial X_0} + \frac{\partial \ddot{r}}{\partial \dot{r}}\frac{\partial \dot{r}}{\partial X_0} + \left(\frac{\partial \ddot{r}}{\partial X_0}\right)_e$$

$$\frac{d^2}{dt^2}\left(\frac{\partial r}{\partial X_0}\right) = \frac{\partial \ddot{r}}{\partial r}\frac{\partial r}{\partial X_0} + \frac{\partial \ddot{r}}{\partial \dot{r}}\frac{d}{dt}\left(\frac{\partial r}{\partial X_0}\right) + \left(\frac{\partial \ddot{r}}{\partial X_0}\right)_e$$

$$A(t) = \frac{\partial \ddot{r}(t)}{\partial r}, C(t) = \frac{\partial \ddot{r}(t)}{\partial X_0}, B(t) = \frac{\partial \ddot{r}(t)}{\partial \dot{r}}, Y(t) = \frac{\partial r(t)}{\partial X_0} \tag{9}$$

$$\ddot{Y} = A(t)Y + B(t)\dot{Y} + C(t) \tag{10}$$

Where Y is state transfer matrix.

The value of formula (10) can be get by Adams-Cowell integral method.

2.4 Orbit determination based on difference method

Let the orbit determination be

$$\begin{cases} \dot{X}(t) = f(X(t),t) \\ X(t_0) = X_0 \end{cases} \tag{11}$$

Where, $X(t) = (r, \dot{r}, p_c)^T$ is orbit state at time t, $(r, \dot{r}) = F(r_0, \dot{r}_0, p_c)$.

Let the dynamic model parameters are,

$$p_c = \left(C_{D1}, \cdots C_{Dn_1}; C_{R1}, \cdots, C_{Rn_2}; C_{E1}, \cdots, C_{En3}\right)$$

where $C_{Di} (i = 1,2,\cdots,n_1)$ is the atmosphere coefficients of ith zone, $C_{Ri} (i = 1,2,\cdots,n_2)$ is the sunlight pressure coefficient of ith zone, $C_{Ei} (i = 1,2,\cdots,n_3)$ is experiential acceleration coefficient of ith zone. The total number of parameters to be estimated is $m = 6 + n_1 + n_2 + n_3$, then the state transfer matrix is,

$$\frac{\partial(r,\dot{r})}{\partial(r_0,\dot{r}_0,p_c)} = \frac{\partial F(r_0,\dot{r}_0,t_0,t,p_c)}{\partial(r_0,\dot{r}_0,p_c)} = \begin{pmatrix} \dfrac{\partial x}{\partial x_0} & \dfrac{\partial x}{\partial y_0} & \cdots & \dfrac{\partial x}{\partial v_{z0}} & \cdots & \dfrac{\partial x}{\partial C_{Di}} & \cdots & \dfrac{\partial x}{\partial C_{Ri}} & \cdots & \dfrac{\partial x}{\partial C_{En_3}} \\ \dfrac{\partial y}{\partial x_0} & \dfrac{\partial y}{\partial y_0} & \cdots & \dfrac{\partial y}{\partial v_{z0}} & \cdots & \dfrac{\partial y}{\partial C_{Di}} & \cdots & \dfrac{\partial y}{\partial C_{Ri}} & \cdots & \dfrac{\partial x}{\partial C_{En_3}} \\ \vdots & \vdots & & \vdots & & \vdots & & \vdots & \ddots & \vdots \\ \dfrac{\partial v_z}{\partial x_0} & \dfrac{\partial v_z}{\partial y_0} & \cdots & \dfrac{\partial v_z}{\partial v_{z0}} & \cdots & \dfrac{\partial v_z}{\partial C_{Di}} & \cdots & \dfrac{\partial v_z}{\partial C_{Ri}} & \cdots & \dfrac{\partial v_z}{\partial C_{En_3}} \end{pmatrix}_{6\times m} \tag{12}$$

Where, $\frac{\partial Y}{\partial \Theta}$ ($Y = x,y,\cdots v_z$, $\Theta = x_0,y_0,\cdots,C_{Di},\cdots,C_{Ri},\cdots,C_{Ei},\cdots$) is effect of the initial state Θ to state Y at time t, so the value can be simply got,

$$\frac{\partial \Upsilon}{\partial \Theta} = \frac{\Upsilon(\Theta + \Delta) - \Upsilon(\Theta)}{\Delta}$$

Where, Δ is little value, $\Upsilon(\Theta + \Delta)$ is the state calculated by the initial state Θ add Δ.

The other more concise method based on difference is directly to deal with the Jacobi matrix

$$B = \frac{\partial G}{\partial (r_0, \dot{r}_0, p_c)} = \begin{pmatrix} \dfrac{\partial h_1}{\partial x_0} & \dfrac{\partial h_1}{\partial y_0} & \cdots & \dfrac{\partial h_1}{\partial v_{z0}} & \cdots & \dfrac{\partial h_1}{\partial C_{Di}} & \cdots & \dfrac{\partial h_1}{\partial C_{Ri}} & \cdots & \dfrac{\partial h_1}{\partial C_{En_3}} \\[2ex] \dfrac{\partial h_2}{\partial x_0} & \dfrac{\partial h_2}{\partial y_0} & \cdots & \dfrac{\partial h_2}{\partial v_{z0}} & \cdots & \dfrac{\partial h_2}{\partial C_{Di}} & \cdots & \dfrac{\partial h_2}{\partial C_{Ri}} & \cdots & \dfrac{\partial h_2}{\partial C_{En_3}} \\[2ex] \vdots & \vdots & & \vdots & & \vdots & & \vdots & \ddots & \vdots \\[2ex] \dfrac{\partial h_{n_k}}{\partial x_0} & \dfrac{\partial h_{n_k}}{\partial y_0} & \cdots & \dfrac{\partial h_{n_k}}{\partial v_{z0}} & \cdots & \dfrac{\partial h_{n_k}}{\partial C_{Di}} & \cdots & \dfrac{\partial h_{n_k}}{\partial C_{Ri}} & \cdots & \dfrac{\partial h_{n_k}}{\partial C_{En_3}} \end{pmatrix}_{n_k \times m} \tag{13}$$

Where, n_k is the number of observation elements.

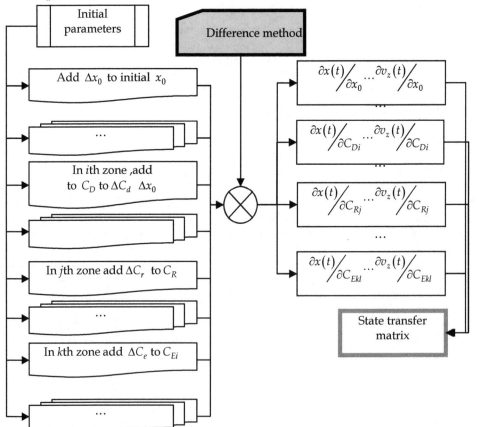

Fig. 2. Flow chart of orbit determination based on difference method.

3. The uniform method of orbit determination with uncertainty model error

The satellite dynamic model can be descriped more and more subtle, but the actual force still can not be entirely expressed by the model. There are some unknown or inexplicable perturbation force and modled pertubation force with unkonwn parameter, for example, the atmosphere consistency parameter in the atmosphere pertubation force modle. So the model error of satellite dynamic model is the inevitable fact. If the orbit determination with no model error compensation is processed, the degree of orbit error will be equal with the degree of dynamic model error, see figure 3.

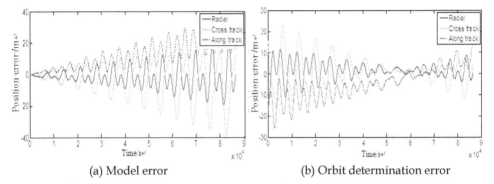

(a) Model error   (b) Orbit determination error

Fig. 3. Modle error and orbit determination error with no model compensation.

The dynamic model in the J2000 reference frame can be described as the following:

$$\begin{cases} \dot{r} = F(r,t) + F_\varepsilon(r,t) \\ r(t_0) = r_0 \end{cases} \tag{14}$$

where $r = (x, y, z, v_x, v_y, v_z)^T$ is the satellite state vector, $F_\varepsilon(r,t)$ is dynamic model error.
The formal function of state transfer can be described as

$$\begin{cases} \dot{\Phi}(t,t_0) = A(t)\Phi(t,t_0) \\ \Phi(t_0,t_0) = I \end{cases} \tag{15}$$

Note, $X_0 = (r_0, \dot{r}_0, p_c)^T$. Then the formula of dynamic model can be showed as the following,

$$\frac{d^2}{dt^2}\left(\frac{\partial r}{\partial X_0}\right) = \frac{\partial \ddot{r}}{\partial r}\frac{\partial r}{\partial X_0} + \frac{\partial \ddot{r}}{\partial \dot{r}}\frac{d}{dt}\left(\frac{\partial r}{\partial X_0}\right) + \left(\frac{\partial \ddot{r}}{\partial X_0}\right)_e$$

$$A(t) = \frac{\partial \ddot{r}(t)}{\partial r}, C(t) = \frac{\partial \ddot{r}(t)}{\partial X_0}, B(t) = \frac{\partial \ddot{r}(t)}{\partial \dot{r}}, Y(t) = \frac{\partial r(t)}{\partial X_0}$$

Where, $A(t) = \dfrac{\partial \ddot{r}}{\partial r} = \dfrac{\partial(F + F_\varepsilon)}{\partial r} \triangleq A_0(t) + A_\varepsilon(t)$.

Based on the differential equation theory, the form root of state transfer function (15) is

$$\Phi = Ce^{(A_0 + A_\varepsilon)} = Ce^{(A_0)} \cdot Ce^{(A_\varepsilon)} = \Phi_0 \Phi_\varepsilon \tag{16}$$

Where, Φ_0 is the state transfer matrix of perturbation F with known model, while Φ_ε is the state transfer matrix of F_ε

The orbit function can be deployed as:

$$\dot{r}(t) = F(r,t) = F\left(r^*,t\right) + \left(\frac{\partial F + F_\varepsilon}{\partial r}\right)\Bigg|_{r=r^*} \Delta r + F_\varepsilon\left(r^*,t\right) \cdots \tag{17}$$

let $\Delta r(t) = r(t) - r^*(t)$,

$$A(t) + A_\varepsilon(t) = \left(\frac{\partial(F + F_\varepsilon)}{\partial r}\right)\Bigg|_{r=r^*}$$

The formula(17)can be noted:

$$\Delta \dot{r}(t) = \left(A(t) + A_\varepsilon(t)\right)\Delta r(t) \tag{18}$$

The formula (18) can be solved by the differential equations theory, which can be expressed as following:

$$\Delta r(t) = \left(\Phi(t,t_0) + \Phi_\varepsilon(t,t_0)\right)\Delta r(t_0) \tag{19}$$

Where $\Phi(t,t_0)$ is state transferring matrix toward precise parameter perturbation force, and $\Phi_\varepsilon(t,t_0)$ is state transferring matrix toward perturbation force error.

3.1 Measasure function with on system error

The measure function can be noted as the following,

$$H = \tilde{G}(r,\dot{r},p_c,t) + \varepsilon = G\left(X_0,t_0,t\right) + \varepsilon \text{ , } X_0 = \left(r_0,\dot{r}_0,p_c\right)^{\mathrm{T}}.$$

Where, the relation of satellite position and velocity at time t and other to be estimated parameters is described by the first equation, and the second equation shows the relation at initial time t_0.

Then, the orbit function with uncertainty dynamic model error can be described as,

$$\begin{cases} \dot{X} = f(X,t) + f_\varepsilon \\ H = G(X_0,t_0,t) + \varepsilon \end{cases}$$

Where, $f_\varepsilon = (0,F_\varepsilon,0)^{\mathrm{T}}$.

Based on the differential coefficient improve process, we can get

$$H - G\left(X_0^*, t_0, t\right) = \frac{\partial G\left(X_0^*, t_0, t\right)}{\partial X_0} \Delta X_0 + \varepsilon = B\Delta X_0 + \varepsilon$$

Note,

$$Y = B\Delta X_0 + \varepsilon \tag{20}$$

Where, Y is OC residual error, $B = \frac{\partial G}{\partial(r_c, \dot{r}_c)} \cdot \frac{\partial(r_c, \dot{r}_c)}{\partial(r, \dot{r})} \cdot \frac{\partial(r, \dot{r})}{\partial X_0}$, ε is error vector.

Based on the formula(20), the matrix satellite state transfer is: $\Phi = \Phi_0 \Phi_\varepsilon$, then

$$\begin{aligned}
\tilde{B} &= \frac{\partial G}{\partial(r_c, \dot{r}_c)} \cdot \frac{\partial(r_c, \dot{r}_c)}{\partial(r, \dot{r})} \cdot \left[\Phi, \frac{\partial(r, \dot{r})}{\partial p_c} \right] \\
&= \frac{\partial G}{\partial(r_c, \dot{r}_c)} \cdot \frac{\partial(r_c, \dot{r}_c)}{\partial(r, \dot{r})} \cdot \left[\Phi_0 \Phi_\varepsilon, \frac{\partial(r, \dot{r})}{\partial p_c} \right]
\end{aligned}$$

$$Y = H - G\left(X_0^*, t_0, t\right) = H - \tilde{G}\left(\Phi X_0^*, t\right) = H - \tilde{G}\left(\Phi_0 \Phi_\varepsilon X_0^*, t\right) \tag{21}$$

In fact the matrix Φ_ε is unkown, So formula (21) should be designed as

$$Y = H - \tilde{G}\left(\Phi_0 X_0^*, t\right) - S \tag{22}$$

Where $S = \tilde{G}\left(\Phi_0 \Phi_\varepsilon X_0^*, t\right) - \tilde{G}\left(\Phi_0 X_0^*, t\right)$, expressed the error caused by dynamic model error. The error S is uncertainty for the unknown matrix Φ_ε.

So the orbit determination function (22) should be changed as

$$Y = \tilde{B}\Delta X_0 + S + \varepsilon \tag{23}$$

For the low precision need of B matrix and the minuteness value of Φ_ε, the formula (23) can be changed as the following,

$$Y = B\Delta X_0 + S + \varepsilon \tag{24}$$

The observation system error S_d caused by dynamic model error can be decomposed from uncertainty model error S .Then,

$$S_d = \tilde{G}\left(\Phi_0 \Phi_\varepsilon X_0^*, t\right) - \tilde{G}\left(\Phi_0 X_0^*, t\right)$$

$$\tilde{G}\left(\Phi_0 \Phi_\varepsilon X_0^*, t\right) = \tilde{G}\left(\Phi_0 \left(\Phi_\varepsilon - I\right) X_0^* + \Phi_0 X_0^*, t\right) \tag{25}$$

Unwrap the above formula (25) at $\Phi_0 X_0^*$

$$\tilde{G}\left(\Phi_0\Phi_\varepsilon X_0^*,t\right) = \tilde{G}\left(\Phi_0\left(\Phi_\varepsilon - I\right)X_0^* + \Phi_0 X_0^*,t\right)$$

$$= \tilde{G}\left(\Phi_0 X_0^*,t\right) + \frac{\partial\tilde{G}}{\partial\Phi_0 X_0^*}\Phi_0\left(\Phi_\varepsilon - I\right)X_0^*$$

$$S_d = \frac{\partial\tilde{G}}{\partial\Phi_0 X_0^*}\Phi_0\left(\Phi_\varepsilon - I\right)X_0^*$$

$$S_d + \frac{\partial\tilde{G}}{\partial\Phi_0 X_0^*}\Phi_0 X_0^* = \frac{\partial\tilde{G}}{\partial\Phi_0 X_0^*}\Phi_0\Phi_\varepsilon X_0^*$$

(26)

In fact, if there are no any dynamic model error, then $\Phi_\varepsilon = I\ S_d = 0$.

$$
\begin{pmatrix} S_{d1} \\ S_{d2} \\ \vdots \\ S_{dn} \end{pmatrix}
+
\begin{pmatrix}
\frac{\partial\tilde{G}_1}{\partial r(t_1)} & & & \\
& \frac{\partial\tilde{G}_2}{\partial r(t_2)} & & \\
& & \ddots & \\
& & & \frac{\partial\tilde{G}_n}{\partial r(t_n)}
\end{pmatrix}
\begin{pmatrix} \Phi_0(t_0,t_1) \\ \Phi_0(t_0,t_2) \\ \vdots \\ \Phi_0(t_0,t_n) \end{pmatrix} X_0^*
$$

$$
=
\begin{pmatrix}
\frac{\partial\tilde{G}_1}{\partial r(t_1)}\Phi_0(t_0,t_1) & & & \\
& \frac{\partial\tilde{G}_2}{\partial r(t_2)}\Phi_0(t_0,t_2) & & \\
& & \ddots & \\
& & & \frac{\partial\tilde{G}_n}{\partial r(t_n)}\Phi_0(t_0,t_n)
\end{pmatrix}
\begin{pmatrix} \Phi_\varepsilon(t_0,t_1) \\ \Phi_\varepsilon(t_0,t_2) \\ \vdots \\ \Phi_\varepsilon(t_0,t_n) \end{pmatrix} X_0^*
$$

(27)

Form the matrix formula function (27), it can be get that $S_{di} = S(t_{di}),(i=1,2,\cdots,n)$ is the model error value estimated by semi-linear model orbit determination, $\frac{\partial G_i}{r_i}$ is the differential coefficient of observation function to satellite state at t_i , Φ_0 is the state transfer matrix of known perturbation force. So only the $\Phi_\varepsilon(t_i,t_0)$ in matrix function (27)is an unknown quantity. $\Phi_\varepsilon(t_i,t_0)$ shows the influence of error perturbation force to satellite state, the function(27) cannot be solved because the matrix function contains $n\times 6\times m$ unknown parameters. There are two methods to solve the problem.

3.1.1 Fitting the sate transfer matrix by base functions

Because the $\Phi_\varepsilon(t_i,t_0)$ can be adapted to dynamic function, then it can be drawn that

$$\frac{d^2}{dt^2}\frac{\partial r_\varepsilon(t)}{\partial X_0} = \frac{\partial f_\varepsilon}{\partial r_\varepsilon}\frac{\partial r_\varepsilon(t)}{\partial X_0} + \frac{\partial f_\varepsilon}{\partial \dot{r}_\varepsilon}\frac{d}{dt}\frac{\partial r_\varepsilon(t)}{\partial X_0} + \frac{\partial f_\varepsilon}{\partial X_0} \tag{28}$$

Where, r_ε is the position vector effected by perturbation force error f_ε.

So the element in matrix $\dfrac{\partial r_\varepsilon(t)}{\partial X_0}$ is consecutive and slippery, and can be approached with

limited base functions. Let the base functions b $\displaystyle\sum_{i=1}^{\lambda} a_i B_i(t)$, and the λ is the estimated number of base function, a_i, B_i are coefficients of base functions and base functions.

$$S_{di} + \frac{\partial \tilde{G}_i}{\partial r(t_i)}\Phi_0(t_0,t_i)X_0^* = \frac{\partial \tilde{G}_i}{\partial r(t_i)}\Phi_0(t_0,t_i)\hat{\Phi}_\varepsilon X_0^* \tag{29}$$

$$\hat{\Phi}_\varepsilon = \begin{pmatrix} \sum_{j=1}^{\lambda} a_{j_{11}} B_j(t_i) & \sum_{j=1}^{\lambda} a_{j_{12}} B_j(t_i) & \sum_{j=1}^{\lambda} a_{j_{1m}} B_j(t_i) \\ \sum_{j=1}^{\lambda} a_{j_{21}} B_j(t_i) & \sum_{j=1}^{\lambda} a_{j_{22}} B_j(t_i) & \sum_{j=1}^{\lambda} a_{j_{2m}} B_j(t_i) \\ \sum_{j=1}^{\lambda} a_{j_{61}} B_j(t_i) & \sum_{j=1}^{\lambda} a_{j_{62}} B_j(t_i) & \sum_{j=1}^{\lambda} a_{j_{6m}} B_j(t_i) \end{pmatrix}_{6\times m}$$

There are $6 \times m \times \lambda$ parameters in the formula (29), so when $n \geq 6 \times m \times \lambda$, the form of state transfer of dynamic model error can be solved.

3.1.2 Method of orbit state

It is easy to be matrix singularity for to estimated parameters, if the state transfer matrix error is solved directly by formula (27). There is other method to solve the problem to get the orbit error ΔX based on the above formula (26).

$$S_d = \frac{\partial \tilde{G}}{\partial \Phi_0 X_0^*}\Delta X \tag{30}$$

Then, the same fitting method can be used to the orbit error ΔX, and get the right value with only $6 \times \lambda$ parameters.

3.2 Orbit determination function with uncertainty observation model error

The uncertainty observation model error mains the error of unable to model and modifying residual error with no form. With the more and more complicated observation, the system error will be more and more intricate. If no correct disposal method, the system observation error will mix into the observation data, and to guide the wrong direction.

There are two kinds of uncertainty observation error, one is residual error, such as atmosphere modify error and residual error of measure system error of radar. All of the form can be noted as

$$H = \tilde{G}(r,\dot{r},p_c,t) + S + \varepsilon = G(X_0,t_0,t) + S + \varepsilon \tag{31}$$

The other is occurred in elements of measurement data, it can be caused by any error of measurement parameters, such as position error of station.

$$H = \tilde{G}(r,\dot{r},p_c,t,S) + \varepsilon \tag{32}$$

In fact the above formula (32) can be noted as the formula (31), and then the uncertainty observation orbit determination function can be described as the following,

$$\begin{cases} \dot{X} = f(X,t) \\ H = G(X_0,t_0,t) + S + \varepsilon \end{cases}$$

$$Y = B\Delta X_0 + S + \varepsilon$$

3.3 Uniform format of orbit determination semi-linear model

When there are uncertainty model error both the observation model and the dynamic model, the orbit determination function can be drawn as,

$$\begin{cases} \dot{X} = f(X,t) + f_\varepsilon \\ H = G(X_0,t_0,t) + g_\varepsilon + \varepsilon \end{cases}$$

The satellite transfer matrix function is $\Phi = \Phi_0\Phi_\varepsilon$, then

$$Y = H - G\left(X_0^*,t_0,t\right) - g_\varepsilon = H - \tilde{G}\left(\Phi X_0^*,t\right) - g_\varepsilon = H - \tilde{G}\left(\Phi_0\Phi_\varepsilon X_0^*,t\right) - g_\varepsilon \tag{33}$$

For the uncertainty of matrix Φ_ε , the actual model of formula (33) can be expressed as

$$Y = H - \tilde{G}\left(\Phi_0 X_0^*,t\right) - S - g_\varepsilon \tag{34}$$

Where $S = \tilde{G}(\Phi_0\Phi_\varepsilon X_0^*,t) - \tilde{G}(\Phi_0 X_0^*,t)$ is the error caused by dynamic model error mixed in observation model. The uniform can be get, because the value S and g_ε are the same kind of uncertainty model error.

The uniform formula function of orbit determination with uncertainty model in dynamic model and observation model can be shown as,

$$Y = B\Delta X_0 + S + \varepsilon \tag{35}$$

Although the uncertainty error S is uniform, it is different in the character. For example, the form of dynamic model error is generally periodic with given frequency.

4. The solve method of uniform orbit determination with uncertainty model error based on semi-parametric model

4.1 Semi-parametric model

Let $b_1, b_2, \cdots, b_p$ and $t_1, t_2, \cdots, t_q$ be the variable array of L, and $b_1, b_2, \cdots, b_p$ are main part with linear character to L, and $t_1, t_2, \cdots, t_q$ can be explained as disturbed factors with non linear property, so the semi-parametric model will be noted as,

$$L_i = b_i^T x + s(t_i) + \varepsilon_i, \quad i = 1, 2, \cdots, n$$

Where $b_i = (b_{i1}, b_{i2}, \cdots, b_{id})^T$, $x = (x_1, x_2, \cdots, x_d)^T$, x is the parameters to be estimated, ε_i is the stochastic error.

4.2 Two stage parametric estimator

It should be noticed that nonparametric component $g(t)$ contains not only dynamic model error, but also measurement system error and so on, nonparametric component $g(t)$ cannot be described by parameters, so two stage estimator can be applied to solve $g(t)$.

Suppose $E(g(t)) = \alpha$, $v = g(t) - \alpha + \varepsilon(t)$, the error function can be noted as:

$$v(t) = B \cdot \Delta r + I \cdot \alpha - h \tag{36}$$

To construct the function based on Lagrange Method,

$$\Phi = v^T v + 2\lambda^T (B\Delta r + I\alpha - h - v) \tag{37}$$

Calculate the minimum value of formula (37), then the first stage estimation of Δr and α can be described as,

$$\Delta \hat{r}_1 = (B^T B)^{-1} B^T (h - I\hat{\alpha}_1) \tag{38}$$

$$\hat{\alpha}_1 = (I^T I)^{-1} I^T (h - B\Delta \hat{r}_1) \tag{39}$$

$$S(t) = h - B \cdot \Delta \hat{r}_1 \tag{40}$$

Formula (40)can be estimated with kernel estimator,

$$\hat{g}(t) = \frac{\sum_{i=1}^{n} S_i W_{in}(x)}{\sum_{i=1}^{n} W_{in}(x)}$$

Where, $W_{in}(x)$ is kernel weight function. The common kernel weight function is Nadaraya-Watson (41), and Gasser-Müller(42),

$$W_{in}(x) = \frac{K\left(\dfrac{t-t_i}{h}\right)}{\displaystyle\sum_{i=1}^{n} K\left(\dfrac{t-t_i}{h}\right)} \tag{41}$$

$$W_{in}(x) = \frac{1}{h}\int_{T_{i-1}}^{T_i} K\left(\frac{t-u}{h}\right)du \tag{42}$$

Where $T_0 = t_0, T_i = (t_i + t_{i+1}/2)$, h is the window to be set.

4.3 Stahel-Donoho kernel estimator based on data depth

For the influence of stochastic error, the nonparametric component contains not only system error but also stochastic error, so the information and reliability of different sampling measurement data are different, and so as the weight of observation data. Data depth can describe the degree of every data in the swatch. Many scholar put forward different data depth function based on different requirement. Though defines of data depth are different, the basic idea is the same, which can be shown that the values of data depth are big near the middle of data , on the contrary, the values are small far from the middle.

Define the data depth of ε with distribution F based on the basic idea [6],

$$D(\varepsilon, F) = \frac{1}{1 + O(S)^2} \tag{43}$$

Where, $O(S) = |\varepsilon - Med(F)| / MAD(F)$, $Med(F)$ is median of F, $MAD(F)$ is median of $|\varepsilon - Med(F)|$.

The depth D_i describes the degree of data S_i in the total data.

The common weight function are as following:

$$\omega(s) = \begin{cases} \dfrac{\exp\left(-k\left(1 - \dfrac{s}{med(D(Z,F))}\right)^2\right) - \exp(-k)}{1 - \exp(-k)}, & s < med(D(Z,F)) \\ 1, & s \geq med(D(Z,F)) \end{cases}$$

Where, k is constant parameter, and $med(D(Z,F))$ is median of depth set.

Then the nonparametric component estimate can be noted as:

$$\hat{g}(t) = \frac{\displaystyle\sum_{i=1}^{n} S_i \omega_i W_{in}(x)}{\displaystyle\sum_{i=1}^{n} \omega_i W_{in}(x)} \tag{44}$$

Then the nonparametric parts had been estimated with data depth weighted kernel method, put $\hat{g}(t)$ into formula (40), the second estimation of $\Delta r(t)$ will be calculated. The non-paramatric regression can be called Stahel-Donoho Kernel Estimator (SDKE).

4.4 Simulation experiments

4.4.1 Simulation conditions

The TLE of LEO satellite is as the following:

COSMOS2221
    1 22236U 92080A   08107.80786870 .00000150 00000-0 15532-4 0 1936
    2 22236 082.5088 327.4593 0016739 264.6279 095.3018 14.831301358342996

4.4.2 Simulation results and analysis

The results of orbit determination based on different methods are printed in figure 5 and figure 6. Numeric results are given in table 1.

method	Radial/m	Cross track/m	Along track/m
Traditional method	420.3	533.4	504.4
Stahel–Donoho Kernel Estimator	120.68	164.8	150.8

Table 1. Orbti determination results of different methods.

In figure1, random error and system error are included with max swing 30 arcsec. In traditional orbit determination method, the ME is treated as white noise, the results of orbit determination based on traditional method was shown in figure 2. It can be seen that the OC residual error still contained the ME(see in figure 4), while the OC residual error based on SDKE contained no system errors, in another word, OC residual based on SDKE is white noise after model error compensated, so the precision of SDKE was improved largely.

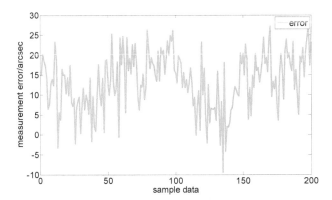

Fig. 4. Measurement data error in simulation experiment.

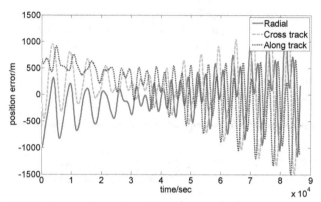

Fig. 5. Results of orbit determination based on traditional method.

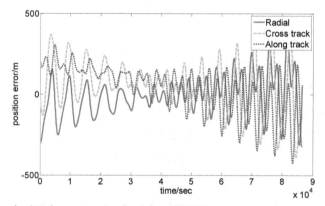

Fig. 6. Results of orbit determination based on SDKE.

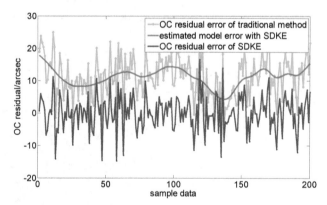

Fig. 7. OC residual of different method.

Model error is the main factor to badly pollute the precision of orbit determination, while semi-parametric is the effective way to compensate the model error. Stahel –Donoho Kernel

Estimator is with great superiority in robust and efficiency, especially its data depth weight matrix can restrain the gross error in the ME, which will happen usually in space based observation. SDKE method of orbit determination can highly improve the precision of orbit determination in space based surveillance system.

5. Reduced dynamic orbit determination based on spline method

The precision of orbit determination depends on the precision of measurement data and the precision of dynamic model. In modern times, the degree of measurement can be limited in millimeter degree, so the key method to improve the precision of orbit determination is to increase the exactness of dynamic model, or to compensate the model error. In the above chapters, some mathematic method are proposed to compensate the uncertainty model error, all those method is focus on the mathematics models, in this chapter, a new orbit determination method based on model error compensation is put forward to deal with directly the dynamic force.

5.1 Orbit determination based on experiential acceleration

In order to remove the error caused by dynamic force, many scholars applied the experiential acceleration method. The general form of experiential acceleration is,

$$f_{RTN} = \begin{pmatrix} C_R \cos u + S_R \sin u \\ C_T \cos u + S_T \sin u \\ C_N \cos u + S_N \sin u \end{pmatrix} \cdot \begin{pmatrix} e_R \\ e_T \\ e_N \end{pmatrix} \tag{45}$$

where u is earth latitude angle, e_R, e_T, e_N is unit vector of radial, track and cross of satellite.

The experiential acceleration form of (45)is proposed based on the frequency error with 1-cycle-per-revolution, and this model can absorb effectively the dynamic model error, and used in the orbit determination of satellite CHAMP. When the form of dynamic model changed, the above model may not be available. So the spline model is proposed to deal with the uncertainty dynamic model error.

Decompose the observation arc into some little arc,

$$\pi : a = u_0 < u_1 < \cdots < u_N = b, u_{i+1} - u_i = h \quad u_{-1} = u_0 - h, u_{N+1} = u_N + h$$

$$f_{RTN} = \begin{pmatrix} \sum_{j=-1}^{N+1} d_{Rj}\varphi_j(u) \\ \sum_{j=-1}^{N+1} d_{Tj}\varphi_j(u) \\ \sum_{j=-1}^{N+1} d_{Nj}\varphi_j(u) \end{pmatrix} \cdot \begin{pmatrix} e_R \\ e_T \\ e_N \end{pmatrix} \tag{46}$$

Where, u is earth latitude angle, $\varphi_j(u)$ is spline base function in arc, e_R, e_T, e_N is unit vector of radial, track and cross of satellite.

5.2 Precision of model based on dynamic model smoothing

It can be realized by dynamic model smoothing to evaluate the capability of experiential acceleration compensation method. Suppose the actual orbit data be observation data, calculate the orbit parameters by compensation orbit determination. In the following experiment, the actual orbit data is the CHAMP orbit data download by GFZ, and different method of orbit determination will be used to smooth the orbit in orbit to prove the experiential acceleration model. The condition of experiment simulation is shown in table 2.

Type of dynamic model smoothing	Dynamic model	Number of parameters to be estimated
D1	With no experiential acceleration, C_D, C_R with whole arc	8
D2	With no experiential acceleration C_D /1hour, C_R /6hour	20
D3	With traditional experiential acceleration C_D /1hour, C_R /6hour, C_R /n^*	92
D4	With spline experiential acceleration, C_D /1hour, C_R /6hour, C_R /1 hour	136

*n is orbit period.

Table 2. Type of dynamic model smoothing.

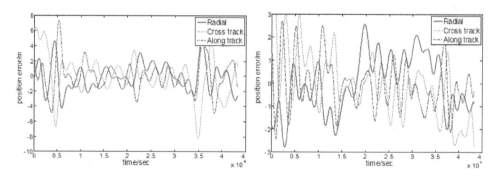

Fig. 8. Smoothing results of type D1 and D2.

From the above experiment, it can be drawn that the spline experiential acceleration model can effectively compensate the dynamic model error. In the type D1, the dynamic model with no experiential acceleration and with one group parameters cannot describe the actual orbit of CHAMP, so the smoothing error is high. In the type D2, although there is still no experiential model, the grouping parameters will absorb some error, the smoothing error is smaller than type D1. In the type D3, the smoothing error is smaller, and it is show that the traditional experiential acceleration model is effective. At the same time, the residual error means there are still some dynamic model error could not be explain by the traditional experiential acceleration, for the form of traditional experiential acceleration 1-cycle-per-

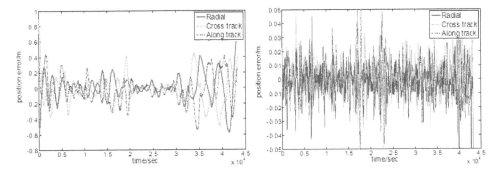

Fig. 9. Smoothing results of type D3 and D4.

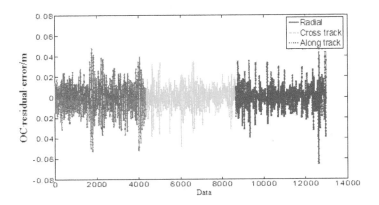

Fig. 10. The OC residual error of smoothing results of type D4.

revolution. At last, in the type D4, the spline model can explain dynamic error with any frequency, the error of smoothing is the most small in all the type experiments.

6. References

Tapley B D, Born G., Schutz B. (1986).*Orbit Determination Fundamental And Application.* Texas: The U niversity of Texas

Yunck T P, William G M , Thornton C L.(1985) .*GPS Based Satellite Tracking System For Precise Positioning.* IEEE Transactions on Geo science and Remote Sensing, Vol 23, pp 450~457.

Yunck T P, S C Wu, W I Bertiger, et al.(1994) .*First Assessment of GPS-based Reduced Dynamic Orbiter Determination on Topex/Poseidon.* Geophysics Research Letter (S0094-8276), Vol 21,pp 541-544.

Wu S C, T P Yunck, C L Thornton.(1991). *Reduced-Dynamic technique for Precise Orbit Determination of Low Earth Satellites.* Journal of Guidance, Control and Dynamics (S0371-5090), Vol 14,pp24-30.

Xue Liugen, Han Jianguo.(2001). *Asymptotic Properties of Two Stage Estimator in Semiparametric Regression Model.* Appl Math J. Chinese Univ ser A, Vol16, pp 87-94.

Lin Lu, Cui Xia. (2006).*Stahel –Donoho Kernel Estimator of model under nonparametric fixed design.* Science in China ser.A. 2006,vol.36,pp 1156~1172.

Part 2

Spacecraft Systems

5

Coordination Control of Distributed Spacecraft System

Min Hu[1], Guoqiang Zeng[2] and Hong Yao[1]
[1]*Academy of Equipment, Beijing,*
[2]*College of Aerospace and Material Engineering,*
National University of Defense Technology, Changsha,
China

1. Introduction

Spacecraft formation flying has received significant attention over the past decade, it has been a topic of interest because of its unique technical advantages and good application features. Several small, unconnected satellites operating in a coordinated way may achieve a better performance than a monolithic satellite, and possess advantages such as increased instrument resolution, reduced cost, reconfigurability, and overall system robustness, which can in turn enhance the scientific return (Zhang et al., 2008).Several ambitious distributed spacecraft missions are currently being put in operation or planned. The PRISMA satellite, which is an on-orbit technology demonstrator for autonomous formation flying and rendezvous, was launched on 15 June 2010 (Ardaens et al., 2011). The TanDEM-X satellite was launched on 21 June 2010 and orbited in close formation with the TerraSAR–X satellite on 15 October 2010. The twin satellites began a routine acquisition of the digital elevation model with flexible baselines on 12 December 2010 (Kahle et al., 2011). The F6 program of the Defense Advanced Research Projects Agency, the Terrestrial Planet Finder of the National Aeronautics and Space Administration, and the Darwin mission of the European Space Agency will all utilize the technology of formation flying.

The modelling of relative motion of distributed spacecraft has been extensively investigated in the past. The Hill-Clohessy-Wiltshire (HCW) equations are widely used. The equations describe the relative motion of two close formation flying satellites in near circular orbits about a spherical Earth, and no disturbances are included in the Hill equations. Using orbital elements to parameterize the relative motion is another important way (D'Amico & Montenbruck, 2006; Ardaens & D'Amico, 2009), which is extremely efficient and was successfully demonstrated during the swap of the GRACE satellites (Montenbruck et al., 2006). By a proper design of the relative orbit elements, a minimum distances in the cross-track plane is guaranteed and the collision hazard is minimized.

In recent years, a significant amount of work has been focused on formation relative orbit estimation. Liu considered the relative navigation for formation flying using an unscented Kalman filter (UKF) and showed that the error of the relative position and velocity estimation can be estimated in the centimeter and millimeter per second scales, respectively (Liu et al., 2008). The original Kalman filter is widely used in relative navigation; however, its inherent linearization process typically introduces significant biases in the estimation results. A particle filter (PF) achieves a recursive Bayesian estimation via a non-parametric Monte Carlo method and shows significant advantages in the nonlinear estimation problem (Rigatos, 2009). A way of generating the importance density function of a PF is essential to improve its performance. EKF and UKF are effective in generating the importance density function.Therefore, because of the strong non-linearity of the dynamics of satellites formation flying; the extended PF (EPF) is adopted to improve the precision of the relative orbit estimation for autonomous formation flying. Moreover, the nonlinear least squares method is applied to determine the relative orbit for the ground-in-the-loop control mode, which is more accurate and suit for the short-arc observation data of the ground station.

An accurate relative orbit control is also very important for the practical implementation of distributed spacecraft. A number of effective controllers are presented in recent literature, such as the linear quadratic regulator, the sliding mode control, and relative orbital elements. Scharf divided the formation flying control problem into five architectures: Leader/Follower, Multiple-Input Multiple-output, Virtual Structure, Cyclic, and Behavioral. We adopt the Leader/Follower approach for practical implementation (Scharf et al., 2004).

It is now known that finite-time stabilization of dynamical system usually demonstrate some nice features such as finite-time convergence to the equilibrium, high-precision performance, faster response as well as better disturbance rejection properties (Ding & Li, 2011). A number of effective methods to achieve the FTC are presented in recent literature (Wu et al., 2011), such as the time-optimal control, TSM control, adaptive control, homogeneous system approach and finite time stability approach. TSM control has been widely used in many applications. By designing a nonlinear switching manifold, the states reach the equilibrium in finite time and exhibit insensitive properties, such as robustness to parameter perturbations and external disturbances (Hu et al., 2008). Man proposed a robust control scheme for rigid robotic manipulators using the TSM technique (Man et al., 1994). However, the controller has a singularity problem. Feng presented a global non-singular TSM controller for a second-order nonlinear dynamic systems (Feng et al., 2002). On the one hand, TSM controllers converge to the equilibrium quickly once in the neighbourhood of the equilibrium, however, when the states are far away from the equilibrium, the system states converge slowly. On the other hand, the linear-hyperplane-based sliding mode controllers converge to the equilibrium quickly when the states are far away from the equilibrium, but they only guarantee asymptotic stability and convergence. All these controllers can not achieve global fast convergence performance in finite time. Therefore, the current study concentrates on the FTC technique to deal with this problem. Currently, the FTC approach has been applied in many fields, such as spacecraft attitude tracking control, consensus for multi-agent systems, robotic manipulators control and missile guidance law design. We will adopt the FTC approach for formation maintenance.

With an increasing number of projects in operation, a practical formation control has also become an area of concern. Relative orbital elements were demonstrated during the GRACE, PRISMA, and TanDEM-X missions. Therefore, the current study will concentrate on formation reconfiguration based on relative orbital elements. Ardaens and D'Amico proposed a dual-impulse method for the in-plane relative control and a single-impulse control for the cross-track motion (Ardaens & D'Amico, 2009); when the control period increases, the dual-impulse maneuver causes an additional along-track drift. Hence, the use of a dual-impulse maneuver for an extended control period of formation keeping may be restricted. The formation reconfiguration control is divided into fuel-optimal triple-impulse in-plane motion control and single-impulse cross-track motion control.

Safe trajectory planning methods are often employed in collision avoidance maneuver. By considering the minimum distances among the satellites as the constraints, safe trajectories can be generated using various planning algorithms, and collision avoidance can be realized by controlling the satellites along the planned trajectory. Tillerson and Richards introduced fuel-optimal trajectories for spacecraft using mixed-integer linear programming, which includes various avoidance constraints (Tillerson et al., 2002; Richards et al., 2002). The artificial potential function method for formation flying satellites has also received considerable attention in recent years (Nag et al., 2010; Bevilacqua et al., 2011). By constructing artificial fields, the goal position provides the attractive forces, whereas the collision avoidance constraints provide the repulsive forces, thereby enabling formation flying satellites to move into their target positions without colliding. Mueller used a robust linear programming technique for the collision avoidance manoeuvre of the PRISMA mission, enabling the satellites to rapidly exit the avoidance region through the application of a single impulse at a specified time (Mueller, 2009; Muelleret et al., 2010). Therefore, the linear programming algorithms are used for the collision avoidance manoeuvre of proximity operations.

The topics concerning simulation or experiment testbeds which focus on the verification of the new technologies of distributed spacecraft have been studied by many researchers in recent years. J. Leitner firstly developed a closed-loop hardware-in-the-loop simulation environment for GPS based formation flying (Leitner, 2001). The SPHERES testbed provided a verification environment for formation flying, rendezvous, docking and autonomy algorithms (Mark, 2002). Wang developed a real-time simulation framework for development and verification for formation flying satellites, which provides access of real sensor system via serial interface (Wang &Zhang, 2005). D'Amico presented an offline and hardware -in-the-loop validation of the GPS-based real-time navigation system for the PRISMA formation flying mission (D'Amico et al., 2008). D'Amico developed the TanDEM-X Autonomous Formation Flying (TAFF) system which is to support the design, implementation, testing and validation of real-time embedded GPS-based GNC system (D'Amico et al., 2009).

The organization of this chapter is as follows: In Section 2, the relative orbit dynamics are introduced, and the general formation description parameters are presented. In Section 3, the relative orbit estimation based on extended particle-filter and nonlinear least squares are presented, respectively. The different coordination control methods are proposed in Section 4. Section 5 presents the processor-in-the-loop distributed simulation system. Section 6 summarizes our conclusions.

2. Preliminaries

With respect to a near-circular reference orbit, and assuming the satellites are taken sufficiently close to each other, the relative motion given by several Keplerian elements differing can be treated to first order.

2.1 Coordinate systems

The relative motion dynamics has been discussed in many papers. We consider two neighbour satellites flying in Earth orbit. The inertial reference frame used is the J2000 frame. The origin of the coordinate system is the centre of the Earth; the X_I axis points toward the mean equinox of J2000.0, the Z_I axis points toward the mean north celestial pole of J2000.0, and the Y_I axis completes the right-handed system. The relative reference frame used is the Hill frame. The origin of the coordinate system is placed at the centre of mass of the master satellite; the x axis is aligned in the radial direction, the z axis is aligned with the angular momentum vector and the y axis completes the right-handed system (Fig. 1).

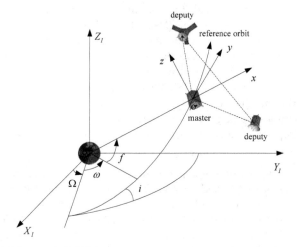

Fig. 1. J2000 inertial frame and Hill frame.

2.2 Relative orbit dynamics

2.2.1 Dynamics equations

With respect to the circular reference orbit, the relative motion can be described as the following equations:

$$\begin{cases} \ddot{x} - 2n\dot{y} - 3n^2x = 0 \\ \ddot{y} + 2n\dot{x} = 0 \\ \ddot{z} + n^2z = 0 \end{cases} \tag{1}$$

where $[x \quad y \quad z \quad \dot{x} \quad \dot{y} \quad \dot{z}]$ is the relative position and velocity in Hill's frame, n is the mean orbit rate.

The relative dynamics for the circular orbits can be expressed in a linear time-invariant (LTI) system in state-space.

$$\begin{aligned} \dot{\mathbf{x}}(t) &= A\mathbf{x}(t) + B\mathbf{u}(t) \\ \mathbf{y}(t) &= C\mathbf{x}(t) \end{aligned} \tag{2}$$

where x is the state vector in Hill's frame, u is the applied acceleration in Hill's frame and y is the output which is equal to the state vector(C is identity).

For a circular reference orbit, the A and B are independent of time:

$$A = \begin{bmatrix} 0 & 0 & 0 & 1 & 0 & 0 \\ 0 & 0 & 0 & 0 & 1 & 0 \\ 0 & 0 & 0 & 0 & 0 & 1 \\ 3n^2 & 0 & 0 & 0 & 2n & 0 \\ 0 & 0 & 0 & -2n & 0 & 0 \\ 0 & 0 & -n^2 & 0 & 0 & 0 \end{bmatrix} \quad B = \begin{bmatrix} 0 & 0 & 0 \\ 0 & 0 & 0 \\ 0 & 0 & 0 \\ 1 & 0 & 0 \\ 0 & 1 & 0 \\ 0 & 0 & 1 \end{bmatrix} . \tag{3}$$

2.2.2 Kinematics equations

The Keplerian orbital elements are a, e, i, Ω, ω, and u, which correspond to the semi-major axis, eccentricity, inclination, right ascension of the ascending node, argument of perigee, and mean argument of latitude ($u = \omega + M$, where M is the mean anomaly, and can be obtained from the true anomaly f), respectively. Spacecraft-1 is the master satellite, and Spacecraft-2 is the deputy satellite. For near-circular satellite orbits, the relative eccentricity vector can be defined as follows:

$$\Delta e = \begin{bmatrix} \Delta e_x \\ \Delta e_y \end{bmatrix} = \delta e \begin{bmatrix} \cos\theta \\ \sin\theta \end{bmatrix} = e_2 \begin{bmatrix} \cos\omega_2 \\ \sin\omega_2 \end{bmatrix} - e_1 \begin{bmatrix} \cos\omega_1 \\ \sin\omega_1 \end{bmatrix} \tag{4}$$

where δe represents the amplitude of Δe and θ defines the initial phase angle of the in-plane motion.

The inclination vector Δi can be defined using the law of sines and cosines for the spherical triangle:

$$\Delta i = \begin{bmatrix} \Delta i_x \\ \Delta i_y \end{bmatrix} = \delta i \begin{bmatrix} \cos\varphi \\ \sin\varphi \end{bmatrix} \approx \begin{bmatrix} \Delta i \\ \Delta\Omega \sin i_1 \end{bmatrix} \tag{5}$$

where $\Delta i = i_2 - i_1$, $\Delta\Omega = \Omega_2 - \Omega_1$, δi represents the amplitude of Δi, and φ defines the initial phase angle of the cross-track plane motion.

2.3 General formation configuration description parameters

For a near-circular reference orbit, the relative motion of the formation flying satellites can be described by the following equations (Hu et al., 2010):

$$\begin{cases} x = \Delta a - p\cos(u - \theta) \\ y = 2p\sin(u - \theta) + l \\ z = s\sin(u - \varphi) \end{cases} \tag{6}$$

where $\{p,s,\alpha,\theta,l\}$ are the five general formation configuration description parameters; $p = a\delta e$ represents the semi-minor axis of the relative in-plane ellipse; $s = a\delta i$ denotes the cross-track amplitude; $\alpha = \theta - \varphi$ defines the relative initial phase angle between the in-plane and cross-track plane motions; and θ is the initial phase angle of the in-plane motion. $\Delta u = u_2 - u_1$, $l = a(\Delta u + \Delta\Omega \cos i) - \dfrac{3}{2}(u - u_0)\Delta a$, u_0 is the initial mean argument of latitude of the deputy satellite, and l represents the along-track offset of the centre of the in-plane motion. An example trajectory is shown in Fig. 2.

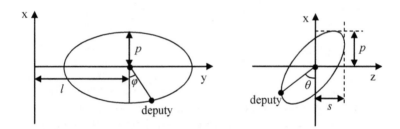

Fig. 2. Example of a relative motion in a near-circular reference orbit.

2.4 Passively safe formation configuration

The distance in the cross-track plane can be expressed as

$$r = \sqrt{x^2 + z^2} \tag{7}$$

By substituting equation (6) into equation (7), we obtain r :

$$r = \sqrt{p^2 \cos^2(u - \varphi) + s^2 \sin^2(u - \theta)} = \sqrt{\frac{p^2 + s^2 + p^2 \cos 2(u - \varphi) - s^2 \cos 2(u - \theta)}{2}} \tag{8}$$

where

$$[p^2 \cos 2(u - \varphi) - s^2 \cos 2(u - \theta)]^2 + [p^2 \sin 2(u - \varphi) - s^2 \sin 2(u - \theta)]^2 =$$
$$= p^4 + s^4 - 2p^2 s^2 \cos 2(\theta - \varphi) = p^4 + s^4 - 2p^2 s^2 \cos 2\alpha \tag{9}$$

so that

$$\left| p^2 \cos 2(u - \varphi) - s^2 \cos 2(u - \theta) \right| \le \sqrt{p^4 + s^4 - 2p^2 s^2 \cos 2\alpha} \tag{10}$$

Minimum distance $r_{\min}$ in the cross-track plane is

$$r_{\min} = \sqrt{\frac{p^2 + s^2 - \sqrt{p^4 + s^4 - 2p^2 s^2 \cos 2\alpha}}{2}} \tag{11}$$

Eq. (11) shows that $r_{min} = 0$ when $\alpha = \pi/2$ or $\alpha = 3\pi/2$, $r_{min} = \min(p,s)$ when $\alpha = 0$ or $\alpha = \pi$. Fig. 3 shows the minimum distances in the cross-track plane with different relative phase angles.

From Fig. 3(a), we can see that the radial and cross-track separations vanish at the same time, and that when the along-track distance is zero, the two satellites will collide. From Fig. 3(b), we can see that radial separation reaches its maximum when the cross-track separation vanishes, and the cross-track separation reaches its maximum when the radial separation vanishes. The safety of the formation flying satellites is guaranteed even in the presence of along-track uncertainty.

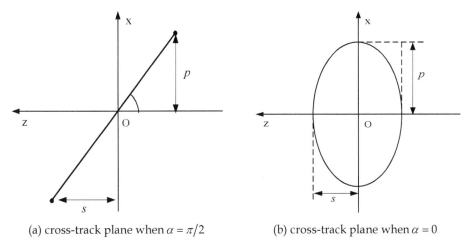

(a) cross-track plane when $\alpha = \pi/2$ (b) cross-track plane when $\alpha = 0$

Fig. 3. Cross-track plane with different relative phase angles.

3. Relative orbit estimation

3.1 Extended Kalman-particle filter

3.1.1 Measurement model

Formation flying satellites often operate in close proximity. Their relative measuring instruments include laser range finders and radio-frequency, infrared, and visible measurements. In the current work, we adopt the laser range finder and radio-frequency ranging equipment as the relative measurements. Thus, the high-precision relative distance, elevation, and azimuth angles can be obtained. The measurement geometry is shown in Fig. 4.

The relative range ρ, the azimuth angle A, and the elevation angle E can be calculated according to the following equations:

$$h(x) = \begin{bmatrix} \rho \\ A \\ E \end{bmatrix} = \begin{bmatrix} \sqrt{x^2 + y^2 + z^2} \\ arc\tan(x/y) \\ arc\sin(z/\sqrt{x^2 + y^2 + z^2}) \end{bmatrix} \tag{12}$$

where $h(x)$ is the measurement matrix, and x, y, and z are the coordinates of the deputy satellite in the body-fixed frame of the master satellite. As we know, the transformation matrix between the body-fixed frame and the Hill frame is a function of the attitude of the master satellite. In this paper, the attitude determination problem was not considered. Therefore, the relative measurements are defined with respect to the Hill frame.

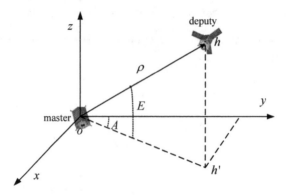

Fig. 4. Relative measurement geometry.

The state and measurement equations can be established as follows:

$$X_k = f(X_{k-1}, t_k) + W_k$$
$$Y_k = h(X_k, t_k) + V_k \tag{13}$$

where X_k is the relative state vector in the Hill frame at time t_k; Y_k is the relative measurements at time t_k, which can be obtained using Eq. (12); W_k is the zero mean value white Gaussian process noise with the covariance Q_k; and V_k is the zero mean value white Gaussian observation noise with the covariance R_k.

Five typical measurement errors, namely, the relative range and angle measurement error, the absolute position and velocity measurement error, and the attitude determination error, are considered.

3.1.2 EPF algorithm

The Kalman filter is the most common method of relative navigation. However, the PF shows better performance in a nonlinear relative state and measurement equations. The principle of PF is to implement the recursive Bayesian filter using Monte Carlo simulations, in which the choice of the importance density function is very important. We employ EKF to realize the importance sampling, which not only makes full use of the latest measurement information, but also avoids the particle exhaustion problem. The particle weights, which are closely associated with the observation, increase, whereas the other particle weights decrease.

The EPF algorithm is summarized as follows: the variable $p(x_0)$ is the prior probability density; $\hat{x}_{k/k-1}$ and $\hat{x}_{k/k}$ are the predicted and updated estimates of the states at time t_k, respectively; $P_{k/k-1}$ and $P_{k/k}$ are their error covariance matrices, respectively; $\Phi_{k,k-1}$ is the state transition matrix, K_k represents the Kalman gain matrix; and w_k^i represents the importance weight. The Jacobian matrix H_k is defined as follows:

$$H_k = \left. \frac{\partial h(X_k, t_k)}{\partial X_k} \right|_{\hat{x}_{k/k-1}} \tag{14}$$

We initialize the particles using:

$$x_0^i \sim p(x_0), w_0^i = 1/N \qquad i = 1, 2, \cdots, N$$

Importance sampling:

a.   The particles are updated using the following equations:

$$\hat{x}_{k/k-1} = \Phi_{k,k-1} x_{k-1/k-1} + Q_{k-1}$$
$$P_{k/k-1} = \Phi_{k,k-1} P_{k-1/k-1} \Phi^T_{k,k-1} + Q_{k-1}$$
$$K_k = P_{k/k-1} H^T_k [H_k P_{k/k-1} H^T_k + R_k]^{-1}$$
$$\hat{x}_{k/k} = \hat{x}_{k/k-1} + K_k [\overline{Y}_k - H_k \hat{\overline{x}}_{k/k-1}]$$
$$P_{k/k} = [I - K_k H_k] P_{k/k-1}$$

$$\hat{x}_k^i \sim q(x_k \mid x_{0:k-1}^i, y_{1:k}) = N(\overline{x}_k^i, P_k^i) \qquad i = 1, 2, \cdots, N$$

b.   The importance weights are calculated using the following equations:

$$w_k^i = \tilde{w}_{k-1}^i p(y_k \mid \hat{x}_k^i) p(\hat{x}_k^i \mid \hat{x}_{k-1}^i) / q(\hat{x}_k^i \mid x_{k-1}^i, y_{1:k})$$
$$\tilde{w}_k^i = w_k^i / \sum_{i=1}^{N} w_k^i \qquad\qquad i = 1, 2, \cdots, N$$

Re-sampling is conducted using

$$\left\{ \hat{x}_{k/k}^i, w_k^i \right\} \rightarrow \left\{ \hat{x}_{k/k}^i, 1/N_k \right\} \qquad i = 1, 2, \cdots, N$$

Thus, the state update is expressed as follows:

$$\hat{x}_k = \sum_{i=1}^{N} w_k^i x_k^i \qquad i = 1, 2, \cdots, N$$

3.2 Nonlinear least squares method

The nonlinear state equation and observation equation are as follows (Hu et al., 2010):

$$X_l = f(X_0, t_l) \tag{15}$$

$$Y_l = G(X_l, t_l) + V_l \tag{16}$$

where X_l is the state vector at the time t_l, which includes the J2 perturbations. Y_l is the observation vector at the time t_l. V_l is the observation noise with normal Gauss distribution.

Equation (16) can expanded at the approximation point X_0^* by using the Taylor series equation, the following equations can be derived by keeping the linear items:

$$Y_l - G\left(X_0^*, t_l\right) = A\left(X_0^*, t_l\right)\left(X_0 - X_0^*\right) + V_l \tag{17}$$

where

$$A\left(X_0^*, t_l\right) = \left[\left(\frac{\partial Y_l}{\partial X_l}\right)\left(\frac{\partial X_l}{\partial X_0}\right)\right]_{X_0 = X_0^*} \tag{18}$$

$$\left(\frac{\partial Y_l}{\partial X_l}\right) = \left(\frac{\partial G\left(X_l, t_l\right)}{\partial X_l}\right) = H(t) \tag{19}$$

$$\left(\frac{\partial X_l}{\partial X_0}\right) = \left(\frac{\partial f\left(X_0, t_l\right)}{\partial X_0}\right) = \Phi\left(t, t_0\right) \tag{20}$$

Let $y_l = Y_l - G\left(X_0^*, t_l\right)$ and $x_0 = X_0 - X_0^*$, we get the linear equation as follows:

$$y_l = H(t)\Phi\left(t, t_0\right)x_0 + V_l \tag{21}$$

where V_l is the residual error, $H(t)$ is the Jacobian matrix. The transition matrix $\Phi\left(t, t_0\right)$ can be calculated as follows:

$$\begin{cases} \dot{\Phi}\left(t, t_0\right) = F(t)\Phi\left(t, t_0\right) \\ F(t) = \left(\frac{\partial f}{\partial X}\right)_{X^*} \end{cases}$$

Therefore, the nonlinear model turns out to be the following form:

$$\begin{cases} X_l = \Phi_{l,0}X_0 \\ Y_l = A_l X_l + V_l \end{cases} \tag{22}$$

By using the least square method, the estimation value of epoch time can be derived by iteration:

$$\hat{x}_{0/k} = \left(\sum_{l=1}^{k} A_l^T A_l\right)^{-1}\left(\sum_{l=1}^{k} A_l^T y_l\right) \tag{23}$$

The optimal estimation should be calculated iteratively, and usually can converge by 3-5 steps.

3.3 Numerical simulations and results analysis

A numerical simulation is conducted to verify the effectiveness of the presented EPF algorithm. The simulation conditions are as follows: the mean orbital elements of the master and deputy satellites are as shown in Table 1, and Fig. 5 shows the three-dimensional formation configuration. The formation configuration parameters are p = 400 m, s = 350 m, α = 0°, θ = 90°, l = 0 m. The absolute position and velocity measurement precision are 10 m and 0.1 m/s, respectively; and the relative range and angle measurement precision are 0.1 m and 0.01°, respectively. The sampling interval is 1 s. Perturbations of Earth oblateness, atmospheric drag, solar radiation pressure, perturbation of the third-body of the sun and moon, and perturbation of the earth body tide are considered in the dynamics simulation. The fourth-order Runge–Kuta algorithm is employed for the numerical integration.

	a (m)	e	i (deg)	Ω (deg)	ω (deg)	M (deg)
master	6892937.0	0.001170	97.443823	100.0	90.0	0.0
deputy	6892937.0	0.001112	97.443823	99.997066	89.999620	0.0

Table 1. Mean orbital elements of the master and deputy satellites.

The absolute orbit of the master and deputy satellites can be generated using the Satellite Tool Kit based on the initial elements given in Table 1. The observation values can be simulated by the absolute orbit information and the measurement covariance using the Gaussian distribution random number series. The measurement sampling period is 1 s, and the simulation time is 3000 s.

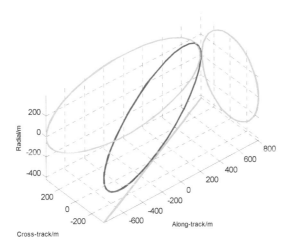

Fig. 5. Three-dimensional formation configuration.

The relative position and velocity estimation errors are shown in Figs. 6 and 7, respectively. The estimation curves are globally convergent, and the EPF algorithm achieved much faster convergence rate in the relative orbit estimation. The relative position estimation errors converge to 2×10^{-2} m within 500 s, and that of the relative velocity estimation are within 1×10^{-4} m/s.

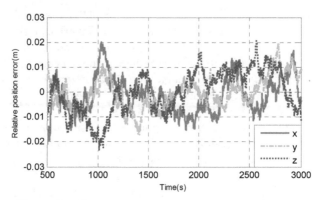

Fig. 6. Relative position estimation errors.

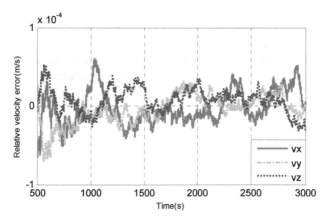

Fig. 7. Relative velocity estimation errors.

4. Relative orbit control

4.1 Coordinated control scheme

We consider an operational scenario with two formation flying satellites, and the deputy satellite performs the relative orbit correction maneuvers. Fig. 8 shows the schematic diagram of the formation flying guidance, navigation, and control (GNC) system.

The deputy satellite obtains the relative measurements and performs the relative orbit estimation to obtain the high-precision relative position and velocity. The formation control software generates control commands according to the current states and mission goals. Thrusters are used to control the geometry and phase angle of the formation, and the yaw angle maneuver commands are used to control the along-track drift. The ground station can monitor the formation flying system in autonomous mode and generate formation control commands in the ground-in-the-loop mode.

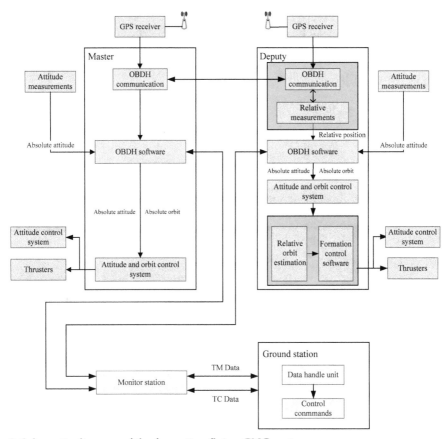

Fig. 8. Schematic diagram of the formation flying GNC system.

4.2 Finite time control for formation maintenance

4.2.1 Control objective

The finite-time control for distributed spacecraft is to design the controller u_m which guarantees that the trajectory tracking errors of the deputy satellite with respect to the master satellite converge to zero in finite time.

The trajectory tracking errors are defined as

$$e = \rho - \rho_d, \quad \dot{e} = \dot{\rho} - \dot{\rho}_d \tag{24}$$

where ρ_d, and $\dot{\rho}_d \in R^3$ are the desired relative position and velocity vectors, respectively.

4.2.2 Finite-time controller

In this section, a robust sliding mode controller is proposed to improve the transient performance and to guarantee the finite-time stability and convergence. The formation flying satellites are close to each other, thus, the disturbances acting on the satellites will be almost the same, and the total relative disturbances D can generally be treated as bounded forces. Suppose that $|D_i| \leq F_i$, $i = 1,2,3$, where F_i is a positive constant.

We propose the following controller

$$u_m = C(\omega_n)\dot{\rho} + N(\rho, \omega_n, r) + \ddot{\rho}_d - \alpha\dot{e} - \beta\gamma|e|^{\gamma-1}\dot{e} - k\,\text{sgn}(S) \tag{25}$$

where $\alpha, \beta > 0$, $k_i > 0$, $i = 1,2,3$, $0 < \gamma < 1$. S is given by

$$S = \dot{e} + \alpha e + \beta|e|^\gamma \text{sgn}(e) \tag{26}$$

Theorem 1. For the formation flying system, the controller (25) can achieve the control objective of trajectory tracking presented in Section 4.2.1.

Proof: Step 1: The system will reach the sliding mode $S = 0$ in finite time.

Consider Lyapunov function

$$V = \frac{1}{2}S^T S \tag{27}$$

Obtaining the time-derivative of V

$$\begin{aligned}
\dot{V} &= S^T\dot{S} = S^T(\ddot{e} + \alpha\dot{e} + \beta\gamma|e|^{\gamma-1}\dot{e}) \\
&= S^T(u_m - C(\omega_n)\dot{\rho} - N(\rho, \omega_n, r) - D - \ddot{\rho}_d + \alpha\dot{e} + \beta\gamma|e|^{\gamma-1}\dot{e}) \\
&= S^T[-D - k\,\text{sgn}(S)]
\end{aligned} \tag{28}$$

Let $k_i > F_i$ yields

$$\dot{V} \leq \sum_{i=1}^{3}\left(|s_i|F_i - k_i|s_i|\right) = -\sum_{i=1}^{3}|s_i|(k_i - F_i) \leq 0 \tag{29}$$

V is positive, and $\dot{V}$ is negative. Therefore, the sliding mode $S = 0$ is achieved in finite time.

Step 2: The system will converge to the equilibrium in finite time once under the condition of $S = 0$.

Once $S = 0$, the system is transformed as

$$\dot{e} = -\alpha e - \beta|e|^\gamma \text{sgn}(e) \tag{30}$$

$e = 0$ is the terminal sliding attractor of system (30). By integrating Eq. (30), we obtain the convergence time T:

$$T = \frac{1}{\alpha(1-\gamma)}\ln\frac{\alpha e_0^{1-\gamma} + \beta}{\beta} \tag{31}$$

where e_0 is the initial error state.

Therefore, once the system states reach the sliding mode manifold (26), the system will converge to the equilibrium in the time T. Combining step 1 and step 2 completes the proof of Theorem 1.

Remark 1. For linear controller, to increase the robustness of the closed-loop system, we can only modify the control gain; however, the control gain can not be too large considering the fuel consumption and system stability. For the FTC approach, we have an additional parameter γ to modify, which exhibits better disturbance rejection performance.

Remark 2. In order to reduce chattering due to high-frequency switching, the boundary layer approach is adopted to replace the signum function of (25) with a continuous saturation one

$$sat(S/\phi) = \begin{cases} S/\phi & |S| \leq \phi \\ sgn(S) & |S| > \phi \end{cases} \tag{32}$$

where ϕ denotes the thickness of the boundary layer. Therefore, the proposed controller (25) can be rewritten as follows:

$$u_m = C(\omega_n)\dot{\rho} + N(\rho, \omega_n, r) + \ddot{\rho}_d - \alpha \dot{e} - \beta\gamma|e|^{\gamma-1}\dot{e} - ksat(S/\phi) \tag{33}$$

However, when $|S| \leq \phi$, the controller (33) can only guarantee asymptotic convergence, although chattering phenomenon can be substantially alleviated. Therefore, a new saturation function is put forward.

$$fsat(S/\phi) = \begin{cases} |S|^\tau \, sgn(S)/\phi^\tau & |S| \leq \phi \\ sgn(S) & |S| > \phi \end{cases} \tag{34}$$

where $0 < \tau < 1$.
Then, we obtain the following controller:

$$u_m = C(\omega_n)\dot{\rho} + N(\rho, \omega_n, r) + \ddot{\rho}_d - \alpha \dot{e} - \beta\gamma|e|^{\gamma-1}\dot{e} - kfsat(S/\phi) \tag{35}$$

The theoretical proof of the finite time convergence inside the boundary layer is provided by Ding (Ding & Li, 2007). Hence, we can guarantee the finite time convergence by adopting the modified controller (35).

4.2.3 Numerical simulations and results analysis

In this scenario, formation keeping simulation is conducted to verify the effectiveness of the proposed controller (35). The initial orbital elements of the master satellite are as shown in Table 2.

a (m)	e	i (deg)	Ω (deg)	ω (deg)	M (deg)
6934386.0	0.001075	97.617093	0.0	0.0	0.0

Table 2. Initial orbital elements of the reference orbit.

The initial relative states in the Hill frame are as shown in Table 3.

x (m)	y (m)	z (m)	v_x (m/s)	v_y (m/s)	v_z (m/s)
-14.99910	0.32800	0.21871	0.00018	0.03285	0.02189

Table 3. Initial relative states in the Hill frame.

We design the formation with $\alpha = 0$, then, the projected trajectory in the cross-track plane is an ellipse, which guarantee the formation safety even in the presence of along-track uncertainty. The threshold of starting formation keeping control is set as 10% of the nominal formation geometry, namely, 50 m. The orbit propagator model includes perturbations of Earth oblateness, atmospheric drag, solar radiation, third-body of Sun and Moon and Earth body tides. The Earth's gravity field adopts EGM96 model, and the atmospheric density model adopts Jacchia70. The eighth-order Runge–Kuta algorithm is employed for the numerical integration. The simulation time is 20000 s.The controller parameters are given by $\alpha = [0.01, 0.01, 0.01]^T$, $\beta = [0.01, 0.01, 0.01]^T$, $\gamma = 0.6$, $k = [0.96, 0.96, 0.96]^T$, and $\phi = 1$.

Fig. 9(a) shows the three-dimensional formation configuration; Fig. 9(b) shows the variations of relative position error vs. time, Fig. 8(c) is the enlargement view of Fig. 9(b) and Fig. 9(d) shows the variations of sliding mode manifold vs. time.

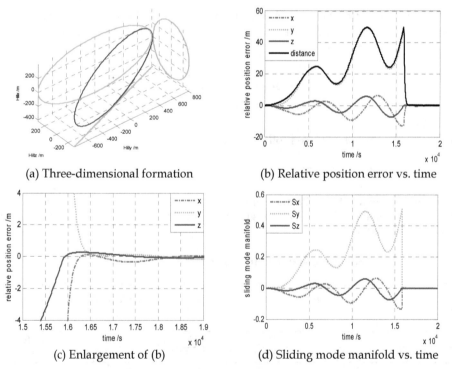

(a) Three-dimensional formation

(b) Relative position error vs. time

(c) Enlargement of (b)

(d) Sliding mode manifold vs. time

Fig. 9. Simulation results of formation keeping scenario.

As shown in Figs. 9(b) and 9(c), when the relative distance error reaches 50 m at the time t = 15888 s, high position tracking accuracy and fast convergence are achieved, which shows that the proposed controller (35) is effective and robust, since finite time convergence is still obtained in the presence of model uncertainties and environment perturbations.

4.3 Impulsive control for formation reconfiguration

4.3.1 Triple-impulse in-plane control

We assume that the nominal configuration parameters in the orbital plane are p_1 and θ_1, and the current configuration parameters in the orbital plane are p_2 and θ_2. According to Eq. (6), the relative position in the orbital plane can be described as

$$\begin{cases} x = -p_2 \cos(u - \theta_2) + p_1 \cos(u - \theta_1) \\ y = 2p_2 \sin(u - \theta_2) - 2p_1 \sin(u - \theta_1) \end{cases} \tag{36}$$

which is equal to

$$\begin{cases} x = -p_0 \cos(u - \theta_0) \\ y = 2p_0 \sin(u - \theta_0) \end{cases} \tag{37}$$

where

$$\begin{cases} p_0 = \sqrt{p_1^2 + p_2^2 - 2p_1 p_2 \cos(\theta_2 - \theta_1)} \\ \varphi_0 = arc\tan(p_2 \sin\theta_2 - p_1 \sin\theta_1, p_2 \cos\theta_2 - p_1 \cos\theta_1) \end{cases} \tag{38}$$

The problem of controlling the current configuration to achieve the nominal configuration is equivalent to the problem of setting p_0 to zero. According to Gauss variation equation, the variances in the relative orbital elements can be expressed by the along-track Δv_T:

$$\begin{cases} \Delta\Delta a = (2a / v)\Delta v_T \\ \Delta\Delta l = -(3t)\Delta v_T \\ \Delta\Delta e_x = (2 / v)\Delta v_T \cos u \\ \Delta\Delta e_y = (2 / v)\Delta v_T \sin u \end{cases} \tag{39}$$

where v is the orbital velocity.

The relative orbital element and the configuration parameters have the following relationship:

$$\begin{bmatrix} \Delta e_{x0} \\ \Delta e_{y0} \end{bmatrix} = \frac{p_0}{a} \begin{bmatrix} \cos\theta_0 \\ \sin\theta_0 \end{bmatrix} \tag{40}$$

Setting p_0 to zero is equivalent to setting Δe_{x0} and Δe_{y0} to zero. Therefore,

$$\begin{cases} (2 / V)\Delta v_T \cos u = -(p_0 / a)\cos\theta_0 \\ (2 / V)\Delta v_T \sin u = -(p_0 / a)\sin\theta_0 \end{cases} \Rightarrow \begin{cases} \Delta v_T = np_0 / 2 \\ u = \theta_0 + \pi \end{cases} \tag{41}$$

The dual-impulse method mentioned by D'Amico equate to (D'Amico & Montenbruck, 2006; Ardaens & D'Amico, 2009)

$$\begin{cases} \Delta v_1 = \Delta v_T / 2 \\ \Delta v_2 = -\Delta v_T / 2 \end{cases} \tag{42}$$

The first impulse will cause an additional along-track drift during the time span between the two impulses. The influence can be neglected if the control period is small; however, if the control period is large, the influence must be considered.

The conventional dual-impulse in-plane control method causes an additional along-track drift because of the time span between the two impulses. Hence, we implement the corrections three times. The maneuver sizes are Δv_1, Δv_2, and Δv_3, respectively, and the respective locations are u_1, u_2, and u_3. The triple-impulse locations must be equal to $\theta_0 + \pi$ or θ_0 and satisfy the following constraints:

$$\begin{cases} \Delta v_1 + \Delta v_2 + \Delta v_3 = 0 \\ |\Delta v_1| + |\Delta v_2| + |\Delta v_3| = \Delta v_T \end{cases} \tag{43}$$

We let $u_1 = \theta_0$, $u_2 = u_1 + (2k+1)\pi$, and $u_3 = u_2 + (2k+1)\pi$. Thus,

$$\Delta v_2 = -2\Delta v_1 == -2\Delta v_3 \tag{44}$$

We obtain the maneuver commands when $u_1 = \varphi_0 + \pi$, as expressed by

$$\begin{cases} \Delta v_1 = \Delta v_T / 4 \\ \Delta v_2 = -\Delta v_T / 2 \\ \Delta v_3 = \Delta v_T / 4 \end{cases} \tag{45}$$

and another solution when $u_1 = \varphi_0$, as expressed by the following equations:

$$\begin{cases} \Delta v_1 = -\Delta v_T / 4 \\ \Delta v_2 = \Delta v_T / 2 \\ \Delta v_3 = -\Delta v_T / 4 \end{cases} \tag{46}$$

The along-track drift caused by the first impulse will be compensated by the subsequent two impulses. The maneuver sizes and locations can be easily calculated according to the initial and nominal formation parameters. Eq. (41) shows that the total Δv needed for formation control can be calculated once the initial and nominal formation parameters are provided, which is helpful in formation-flying mission design and analysis.

4.3.2 Single-impulse out-of-plane control

The relative inclination vector of the initial and target formation configurations is $\Delta \Delta i$, the argument is φ_0, and the single burn can be provided by Gauss variation equation. Thus,

$$\begin{cases} \Delta v_N \cos u / v = \Delta i \cos \varphi_0 = \Delta \Delta i_x \\ \Delta v_N \sin u / v = \Delta i \sin \varphi_0 = \Delta \Delta i_y \end{cases} \tag{47}$$

and

$$\begin{cases} \Delta v_N = v \| \Delta \Delta i \| \\ u = \varphi_0 = arc \tan(\Delta \Delta i_y, \Delta \Delta i_x) \end{cases} \tag{48}$$

4.3.3 Numerical simulations and results analysis

In this scenario, formation reconfiguration simulation is conducted to verify the effectiveness of the proposed method. The initial orbital elements of the formation flying satellites are as shown in Table 4.

	a (m)	e	i (deg)	Ω (deg)	ω (deg)	M (deg)
master	6 892 937.0	0.00117	97.4438	90	0	0
deputy	6 892 937.0	0.00116	97.44698	89.9973	357.888	2.112

Table 4. Initial orbital elements of the reference orbit.

The formation is reconfigured from the initial configuration $\{ p = 300 \text{ m}, \quad s = 500 \text{ m}, \theta = 100° , \quad \varphi = 40° \}$ to the target configuration $\{ p = 500 \text{ m}, \quad s = 300 \text{ m}, \quad \theta = 90° , \quad \varphi = 60° \}$.

Fig. 10(a) shows the reconfiguration of the relative eccentricity vector, Fig. 10(b) shows the reconfiguration of the relative inclination vector.

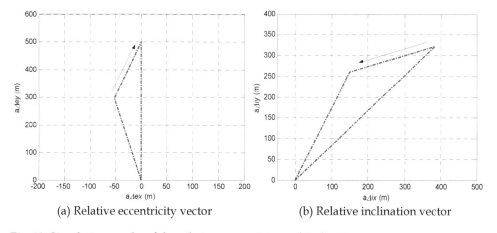

(a) Relative eccentricity vector   (b) Relative inclination vector

Fig. 10. Simulation results of the relative eccentricity and inclination vector.

As shown in Figs. 11 and 12, we can see that formation was successfully reconfigured to the target configurations.

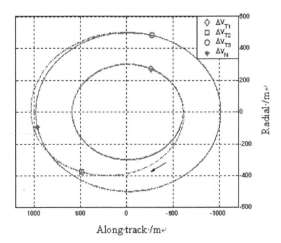

Fig. 11. In-plane motion.

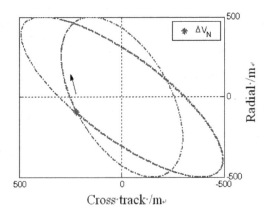

Fig. 12. Cross-track motion.

4.4 Linear programming method for collision avoidance maneuver

4.4.1 Linear programming method

The dynamic system mentioned in Section 2.2.1 can be discretized using zero-order hold as follows (Paluszek et al., 2008):

$$x_{k+1} = Ax_k + Bu_k$$
$$y_k = x_k$$
(49)

where $k = 0, \cdots, N-1$, and the time-step is Δt.

The problem of optimal collision avoidance manoeuvre can be described as follows. Given the initial and the terminal states, equation (21) is minimized by a sequence of u_k and manoeuvre time T :

$$\min \frac{1}{2} \sum_{k=0}^{N-1} \|u_k\|_2^2 \tag{50}$$

with the constraints

$$\begin{cases} x_{k+1} = Ax_k + Bu_k, k = 0,\cdots,N-1 \\ \left| x_{N-1} - x^* \right| \le \varepsilon \\ Lb \le u_k \le Ub \end{cases} \tag{51}$$

where ε is the small error vector of the terminal state, Lb and Ub are the boundaries of the thrust.

The problem mentioned above can be converted into a standard linear programming problem:

$$x_1 = Ax_0 + Bu_0$$

$$\begin{aligned} x_2 &= Ax_1 + Bu_1 \\ &= A\left(Ax_0 + Bu_0\right) + Bu_1 \\ &= A^2 x_0 + ABu_0 + Bu_1 \end{aligned}$$

$$\cdots\cdots$$

$$x_N = A^N x_0 + \begin{bmatrix} A^{N-1}B & A^{N-2}B & \cdots & AB \end{bmatrix} \times \begin{bmatrix} u_0 \\ u_1 \\ \vdots \\ u_{N-1} \end{bmatrix} \tag{52}$$

Let $B_p = \begin{bmatrix} A^{N-1}B & A^{N-2}B & \cdots & AB \end{bmatrix}$ and $u_p = \begin{bmatrix} u_0 & u_1 & \cdots & u_{N-1} \end{bmatrix}^T$,

so that

$$x_N = A^N x_0 + B_p u_p \tag{53}$$

The terminal constraint can then be written as

$$-\varepsilon \le x_N - x^* \le +\varepsilon \tag{54}$$

Let $\tilde{A} = \begin{bmatrix} -B_p \\ B_p \end{bmatrix}$ and $\tilde{b} = \begin{bmatrix} \varepsilon + A^N x_0 - x^* \\ \varepsilon - A^N x_0 + x^* \end{bmatrix}$. We obtain

$$\tilde{A}u_p \le \tilde{b} \tag{55}$$

The problem of optimal collision avoidance manoeuvre can be written as

$$\min \frac{1}{2} \sum_{k=0}^{N-1} \|u_k\|_2^2 \tag{56}$$

$$\text{s.t.} \begin{cases} \tilde{A}u_p \leq \tilde{b} \\ Lb \leq u_k \leq Ub \end{cases}$$

4.4.2 Numerical simulations and results analysis

Scenario 1

We take the TanDEM-X formation as an example. When the relative measurement sensors fail, the formation satellites cannot obtain the relative states, which rapidly increases the collision probability. To minimize the collision hazard, we can manoeuvre the chaser satellite from the formation with $\alpha = 90$ ° to a safe formation with $\alpha = 0$ °. The safe configuration parameters are {$p = 400$ m, $s = 300$ m, $\alpha = 0$ °, $\theta = 0$ °, $l = 0$ m}, and the terminal state error vector is [1 m, 1 m, 1 m, 0.1 m/s, 0.1 m/s, and 0.1 m/s]. When the initial and terminal configurations are given, the control sequences can be calculated while minimizing total delta-v by the proposed linear programming method. The method is flexible and independent of the time window. The maneuver time is 600 s.

Fig. 13 shows the control input for the maneuver, Fig. 14 indicates the three-dimensional collision avoidance trajectory, and Fig. 15 displays the projected trajectory in the cross-track plane.

Total delta-v is 0.646 m/s. The safe trajectory is reached within a short period. Fig. 15 shows that the trajectory reached has a minimum separation of 300 m. The two cases above illustrate that shorter maneuver time gives rise to a larger total delta-v, and that a collision avoidance strategy can be formulated by considering time urgency and residual propellant mass.

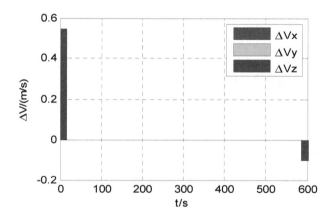

Fig. 13. Impulsive control input.

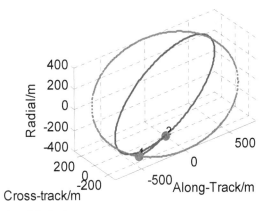

Fig. 14. Three-dimensional trajectory.

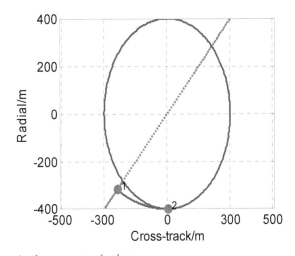

Fig. 15. Trajectory in the cross-track plane.

Scenario 2

The TanDEM-X formation is taken as an example. The nominal configuration is passively safe with $\alpha = 0$ °. When the 40 mN cold gas thrusters are open for a certain period given some uncertainties, the collision hazard increases. After failure is eliminated, the chaser satellite should be controlled so that it immediately returns to safe orbit.

The initial configuration parameters are $\{p = 300 \text{ m}, s = 400 \text{ m}, \alpha = 0 °, \theta = 23 °, l = 0 \text{ m}\}$, and the safe configuration parameters are $\{p = 507.2 \text{ m}, s = 400 \text{ m}, \alpha = 0 °, \theta = 37.3 °, l = 0 \text{ m}\}$. We assume that the chaser satellite burns only in the along-track direction; thus, the cross-track motion amplitude remains unchanged. The collision avoidance region is defined as a circle with a 200 m radius. The optimal maneuver trajectory is shown in Fig. 16.

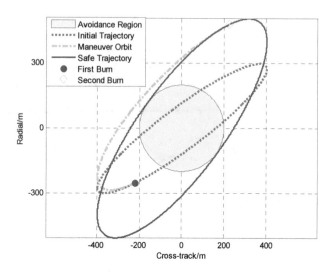

Fig. 16. Optimal maneuver that enables reaching the safe ellipse.

As seen in Fig. 16, the trajectory intersects with the collision avoidance region after the thrusters malfunction is eliminated. The proposed optimal collision avoidance manoeuvre is used to steer the chaser satellite toward the safe trajectory within a minimum distance of 200 m. Total delta-v is 0.126 m/s. The manoeuvre requires only two burns; hence, it is simple, effective, and suitable for on-board implementation.

5. Processor-in-the-loop simulation system for distributed spacecraft

5.1 System architecture

In order to simulate the control architecture of distributed spacecraft, the distributed system architecture is selected. The main elements in the platform are the formation control embedded computers, which builds a VxWorks environment in a PowerPC8245 board and runs the GNC flight software. The dynamic simulation computers exchange data with the formation control embedded computers via CAN bus. The formation control embedded computer receive the high precision orbit, attitude and measurement data provided by the corresponding dynamic simulation computer real-time, and produce a series of time-tagged maneuver commands to add to the dynamic simulation environment, which forms the close-loop processor-in-the-loop simulation of the GNC system. The formation control embedded computers not only communicate with each other through wireless to emulate the communication among distributed spacecraft, but also communicate with the ground station to emulate the ground-in-the-loop communication. One workstation sets the simulation parameters and displays the simulation scenarios by a plasma displayer. One industrial control computer generates impulse to guarantee synchronization among different subsystems. Fig. 17 shows the system architecture diagram of the distributed simulation system (Hu et al., 2010).

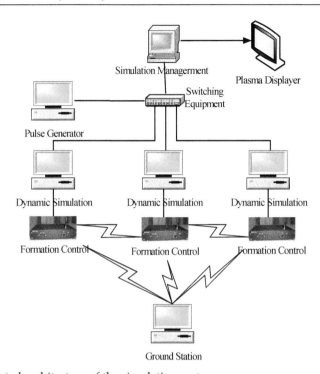

Fig. 17. Distributed architecture of the simulation system.

5.2 System implementation

The dynamic simulation computers are the backbone of the close-loop simulation platform. The software is written in C language and compute orbit, attitude, sensor models and actuators of distributed spacecraft system. The simulation computers are synchronized with the pulse generator and the real-time simulation time step can be set as 10 milliseconds. It provides the epoch time, ECI states of each spacecraft, relative states to the master satellite and attitude data to the formation control computers via CAN bus. The typical error models of the motion data as Guassian noise are also added to evaluate the control performance and the fuel consumption. The adopted dynamic models for orbit propagation include the Earth's gravity field (such as EGM96、JGM3、JGM2 or GEMT1 model), atmospheric drag(such as Harris-Priester or Jacchia70 atmospheric density model), solar radiation pressure, gravity of Sun and Moon and solid Earth tides. The dynamic simulation software also includes the attitude dynamic models based on quaternions to simulate six degree-of-free motions of each spacecraft.

The dynamic simulation computers can receive the maneuver commands from the formation control computers via CAN bus. The maneuver commands include the start control time, the execution time and the delta-V of the desired impulsive maneuver. The net force error and the direction error of thrust are added to emulate the natural environment. The effect of the maneuver is then reflected to the motion data sent to the formation control computers.

The formation control embedded computers receive the absolute and relative states with typical errors from the dynamic simulation computers. The Extended Kalman Filter is used to determine the relative orbit real-time for the autonomous formation flying, and the non-linear least squares estimation is used to determine the relative orbit for the ground-in-the-loop control mode. The formation initialization, formation keeping, formation reconfiguration and collision avoidance maneuver control algorithms are realized.

The ground station is run on a workstation and developed by Visual C++ 6.0, it can produces the control commands in the ground-in-the-loop control, which is sent to the OBDH modules in the formation control embedded computers.

The simulation manager is developed by Visual C++ 6.0, it has a friendship user interface, and enabled the user to select the simulation parameters such as the control model (autonomous mode or ground-in-the-loop mode), the mission scenario (formation initialization, formation keeping, formation reconfiguration or collision avoidance maneuver), the simulation time and the time step etc. It also receives the position and velocity from the dynamic simulation computers and drive the STK VO 3D window through STK's Connect Module.

5.3 Numerical simulations and results analysis

This scenario demonstrates the autonomous formation keeping experiment.

Fig. 18 shows the relative navigation error in RTN frame. The statistical performance of relative position is 3cm respectively and the relative velocity is 0.2mm/s respectively.

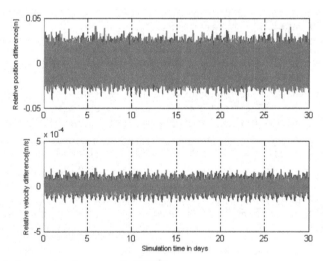

Fig. 18. The relative navigation error in RTN frame.

Fig. 19 shows the key results of formation keeping scenario. The simulation time is 30 days, the in-plane control period is 7 days and the cross-track control period is 28 days. The 1st plot shows the change of the relative semi-major axis(Δa),the 2st plot shows the change of

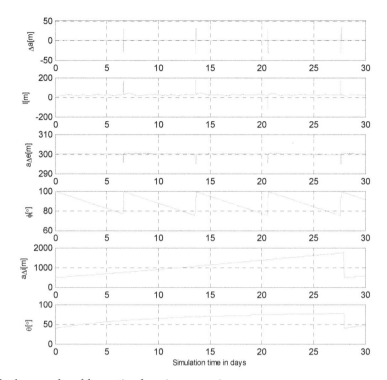

Fig. 19. The key results of formation keeping scenario.

the along-track drift(l), the 3st plot shows the change of the in-plane geometry($a\delta e$), the 4st plot shows the change of the in-plane phase angle(φ), the 5st plot shows the change of the cross-track geometry($a\delta i$), the 6st plot shows the change of the cross-track phase angle(θ).The relative semi-major axis and the relative eccentricity vector are controlled by three in-plane impulse maneuvers in the along-track direction separated by half an orbital period interval. The relative inclination vector is controlled by out-of-plane maneuvers only.

The relative semi-major axis and the long-track drift are affected by the execution of the three in-plane impulse maneuvers. The relative eccentricity vector and the relative inclination vector are properly moved from one perturbation side to the desired side in order to compensate their natural drift caused by J2.

Through the formation keeping test and the formation reconfiguration test, the functionalities and the performance of the process-in-the-loop simulation testbed are validated.

6. Conclusions

This chapter investigates several key technologies of distributed spacecraft, such as the high precision relative orbit estimation, the formation maintenance and reconfiguration strategies, the collision avoidance maneuver and the distributed simulation system.

Simulation results show that the relative position estimation errors are within 2×10^{-2} m, and that of the relative velocity estimation are within 1×10^{-4} m/s.

A robust sliding mode controller is designed to achieve formation maintenance in the presence of model uncertainties and external disturbances. The proposed controller can guarantee the convergence of tracking errors in finite time rather than in the asymptotic sense. By constructing a particular Lyapunov function, the closed-loop system is proved to be globally stable and convergent. Numerical simulations are finally presented to show the effectiveness of the developed controller. The full analytical fuel-optimal triple-impulse solutions for formation reconfiguration are then derived. The triple-impulse strategy is simple and effective. The linear programming method is suitable for collision avoidance maneuver, in which the initial and terminal states are provided.

A real-time testing system for the realistic demonstration of the GNC system for the distributed spacecraft in LEO is presented. The system allows elaborate validations of formation flying functionalities and performance for the full operation phases. The test results of autonomous formation keeping and formation reconfiguration provide good evidence to support performance and quality of the coordination control algorithms.

The key aim of this chapter is to introduce the important aspects of the distributed spacecraft, and pave the way for future distributed spacecraft.

7. References

Ardaens J S & D'Amico S. (2009). Spaceborne Autonomous Relative Control System for Dual Satellite Formations, *Journal of Guidance, Control, and Dynamics*, Vol.32, No.6, pp. 1859–1870

Ardaens J S, D'Amico S & Montenbruck O. (2011). Final Commissioning of the PRISMA GPS Navigation System, *22nd International Symposium on Spaceflight Dynamics*, Sao Jose dos Campos, Brazil, 28 February-4 March, pp.1-16

Bevilacqua R, Lehmann T & Romano M. (2011). Development and Experimentation of LQR/APF Guidance and Control for Autonomous Proximity Maneuvers of Multiple Spacecraft, *Acta Astronautica*, Vol.68, pp.1260-1275

D'Amico S , Florio S D, Ardaens J S & Yamamoto T. (2008). Offline and Hardware-in-the-loop Validation of the GPS- based Real-Time Navigation System for the PRISMA Formation Flying Mission, *The 3rd International Symposium on Formation Flying, Missions and Technologies*, Noordwijk, The Netherlands

D'Amico S, Florio S D, Larsson R & Nylund M. (2009). Autonomous Formation Keeping and Reconfiguration for Remote Sensing Spacecraft, *The 21st International Symposium on Space Flight Dynamics*, Toulouse, France, September

D'Amico S & Montenbruck O. (2006). Proximity Operations of Formation-Flying Spacecraft Using an Eccentricity/Inclination Vector Separation, *Journal of Guidance, Control and Dynamics*, vol. 29, no. 3, pp. 554-563

Ding S H & Li S H.(2007). Finite Time Tracking Control of Spacecraft Attitude, *Acta Aeronautica et Astronautica Sinica*,Vol.28 No.3, pp. 628-633

Ding S H & Li S H. (2011). A Survey for Finite-time Control Problems, *Control and Decision*,Vol.26, No.2, pp. 161-169

Feng Y, Yu X H & Man Z H.(2002). Non-singular Terminal Sliding Mode Control of Rigid Manipulators, *Automatica*,Vol.38, No.9, pp.2159-2167

Hu M, Zeng G Q & Song J L.(2010). Collision Avoidance Control for Formation Flying Satellites, *AIAA Guidance, Navigation, and Control Conference*, 2-5 August 2010, Toronto, Ontario Canada, AIAA 2010-7714

Hu M, Zeng G Q & Song J L.(2010). Navigation and Coordination Control System for Formation Flying Satellites. *2010 International Conference on Computer Application and System Modeling*. October 21-24, Taiyuan, China

Hu M, Zeng G Q & Yao H.(2010). Processor-in-the-loop Demonstration of Coordination Control Algorithms for Distributed Spacecraft, *Proceedings of the 2010 IEEE International Conference on Information and Automation*, 20-23 June , Harbin, China, pp. 1008–1011

Hu Q L, Ma G F & Xie L H. (2008). Robust and Adaptive Variable Structure Output Feedback Control of Uncertain Systems with Input Nonlinearity, *Automatica*,Vol.44, No.2, pp. 552-559

Kahle R, Schlepp B, Meissner F, Kirschner M & Kiehling R. (2011). TerraSAR-X/TanDEM-X Formation Acquisition: Analysis and Flight Results, *21st AAS/AIAA Space Flight Mechanics Meeting*, New Orleans, Louisiana,13-17 February,AAS 11-245

Leitner J.(2001). A Hardware-in-the-Loop Testbed for Spacecraft Formation Flying Applications, *IEEE Aerospace Conference*, Big Sky, MT

Liu J F, Rong S Y & Cui N G.(2008). The Determination of Relative Orbit for Formation Flying Subject to J2, *Aircraft Engineering and Aerospace Technology: An International Journal*, Vol.80, No.5, pp. 549-552

Man Z H, Paplinski A P & Wu H R.(1994). A Robust MIMO Terminal Sliding Mode Control Scheme for Rigid Robotic Manipulators, *IEEE Transactions on Automatic Control*,Vol.39, No.12, pp.2464-2469

Mark O H.(2002). *A Multi-Vehicle Testbed and Interface Framework for the Development and Verification of Separated Spacecraft Control Algorithms*, Massachusetts Institute of Technology

Montenbruck O, Kirschner M, D'Amico S & Bettadpur S. (2006). E/I Vector Separation for Safe Switching of the GRACE Formation, *Aerospace Science and Technology*, Vol. 10, pp. 628–635

Mueller J B. (2009). Onboard Planning of Collision Avoidance Maneuvers Using Robust Optimization, *AIAA Infotech@Aerospace Conference*,Seattle, Washington, AIAA 2009-2051

Mueller J B, Griesemer P R & Thomas S. (2010). Avoidance Maneuver Planning Incorporating Station-keeping Constraints and Automatic Relaxation, *AIAA Infotech@ Aerospace Conference*, Atlanta, Georgia, AIAA 2010-3525

Nag S, Summerer L & Weck O. (2010). Comparison of Autonomous and Distributed Collision Avoidance Maneuvers for Fractionated Spacecraft,*6th International Workshop on Satellite Constellation and Formation Flying*, Taipei, Taiwan, 1-3 November

Paluszek M, Thomas S, Mueller J & Bhatta P. (2008). *Spacecraft Attitude and Orbit Control*, Princeton Satellite System, Inc., Princeton, NJ

Richards A, Schouwenaars T, How J & Feron E. (2002). Spacecraft Trajectory Planning with Avoidance Constraints Using Mixed-Integer Linear Programming, *Journal of Guidance, Control, and Dynamics*, Vol.25 No.4,pp. 755-764

Rigatos G G.(2009). Particle Filtering for State Estimation in Nonlinear Industrial Systems, *IEEE Transaction on Instrumentation and Measurement* ,Vol.58, No.11, pp. 3885-3900

Scharf D P, Hadaegh F Y & Ploen S R. (2004). A Survey of Spacecraft Formation Flying Guidance and Control (Part 2): Control.*Proceeding of the 2004 American Control Conference*,Boston, Massachusetts June 30 -July 2. pp.2976-2985

Tillerson M, Inalhan G. & How J. (2002). Coordination and Control of Distributed Spacecraft Systems Using Convex Optimization Techniques, *International Journal of Robust and Nonlinear Control*, Vol.12 No.2, pp. 207-242

Wang Z K & Zhang Y L. (2005). A Real-Time Simulation Framework for Development and Verification of Distributed Satellite Control Algorithms, *Asia Simulation Conference/ the 6th International Conference on System Simulation and Scientific Computing*, Beijing, October

Wu S N, Radice G & Gao Y S. (2011). Quaternion-based Finite Time Control for Spacecraft Attitude Tracking, *Acta Astronautica*,Vol.69, pp.48-58

Zhang Y L., Zeng G Q, Wang Z K & Hao J G. (2008). *Theory and Application of Distributed Satellite*, Science Press, Beijing, pp. 1-2

6

Modularity and Reliability in Low Cost AOCSs

Leonardo M. Reyneri[1], Danilo Roascio[1], Claudio Passerone[1],
Stefano Iannone[1], Juan Carlos de los Rios[1], Giorgio Capovilla[1],
Antonio Martínez-Álvarez[2] and Jairo Alberto Hurtado[1]
[1]Department of Electronics and Telecommunications, Politecnico di Torino, Torino,
[2]Department of Computer Technology, University of Alicante, Alicante,
[1]Italy
[2]Spain

1. Introduction

The use of off the shelf electronic components is now common in university and low cost satellites. The advantage against space qualified parts consists in a significant cost reduction, a wider range of components selection, better second sourcing capabilities and an effective reuse of existing technologies, devices, circuits and systems from other engineering domains.

Commercial Off The Shelf Components (**COTS**) are sometimes subject to reliability requirements which are often tougher than those applied to space devices, as they have to be used in markets (e.g., automotive) where safety concerns and the huge number of systems manufactured set demanding constraints on the components. Yet, the drawbacks of COTS components and other low cost design methods, mostly in space missions, remain the higher sensitivity to radiation-induced effects and the reduced system level tolerance to faults.

Compensation of these weak points is possible with the development of appropriate design techniques and their proper application throughout the lifecycle of the system, from system level design down to manufacturing. The use of COTS components has thus enormous capabilities and benefits in, but not limited to, small satellite missions.

Low cost spacecraft design not only refers to COTS devices but to several other aspect of the design of a spacecraft.

A low-cost approach to spacecraft design will have a huge impact on the development of space technology in the future, provided that system-level approaches are applied to the design in order to contemporarily reduce cost and increase reliability.

This chapter will analyze several aspects in the low cost and high reliability design of a specific spacecraft subsystem, namely an Attitude and Orbit Control System (**AOCS**), developed at Politecnico di Torino as part of the AraMiS modular architecture for small satellites. We will show how an appropriate mixture of innovative techniques can produce a high reliability and high performance, low cost AOCS.

The topics will be covered at both system and subsystem level, with references to commercially available devices (sensors, actuators, drivers and microcontrollers) and following the object-oriented modeling used for both system management, subsystem development and programming.

This chapter will therefore present several technical solutions (at different design levels) which the authors have developed and applied to the development of the AraMiS built-in AOCS system. In particular, after analyzing the **Requirements on small satellites AOCS**, the following items will be discussed in detail: i) **Attitude and Orbit Codetermination**, a mean to substitute the high-cost space-grade GPS with an appropriate analysis of the images acquired by our solar and horizon sensor; ii) **Fine-grain Modularity**, a mean to cut down design, development, testing and assembly costs, while maintaining a high level of design flexibility; iii) **Sensor Fusion and Reconfigurability**, a system-level mean to increase reliability of low-cost sensors and actuators by processing the data from the large network of sensor intrinsic in the architecture of the AraMiS spacecraft; iv) **Latchup Protection**, achieved by a hybrid anti-latchup protector developed by the authors; v) **Open hardware/software plug and play architecture,** allowing a simple and straightforward integration of several sensors and actuators into the SW architecture of the spacecraft, while minimizing the risk of errors and guaranteeing a high level of reliability, also thanks to the pervasive use of the proposed vi) **Hardening techniques against radiation of SW code and HW interfaces HW/SW**; vii) the application of an innovative **micropropulsor** found in literature is considered for future developments; viii) the aspects of **Sensor and Actuator Calibration** will be analyzed and it will be described how these will be integrated into the HW/SW architecture of AraMiS.

2. System overview

The proposed AOCS has been conceived as a part of the AraMiS modular architecture for small satellites developed at Politecnico di Torino (Reyneri et al., 2010). The basic idea behind AraMiS is the concept of **tile**, that is, a standardized building block to be used to build small spacecrafts according to the specific requirements. As many tiles as required can be assembled to build spacecrafts of virtually any size and shape, as shown in figure 1.

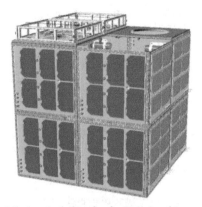

Fig. 1. Mechanical sketch of a 2x2x2 cubic configuration using 22 tiles (left). Photograph of the mockup of ah hexagonal telescope made using 12 tiles.

Each tile contains a number of attitude and housekeeping sensors and actuators, such that a spacecraft made of a number of tiles incorporates all the standard subsystems needed, among which the AOCS which is being described in this chapter. Note that, although the AOCS will be described as a whole, in practice its subsystems are distributed among all the tiles, that is, throughout the spacecraft. Several functions are therefore replicated multiple times, offering a relevant degree of redundancy evaluated in detail later in the chapter.

Figure 2 shows a simplified overview of the proposed AOCS (internally called 1B2 Attitude and Orbit Subsystem), seen in its entirety. It is made of five major blocks.

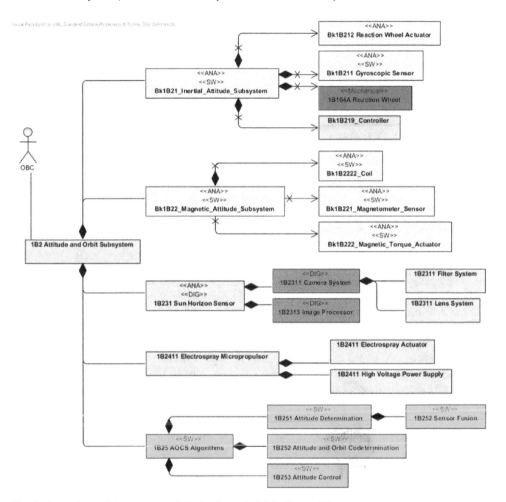

Fig. 2. Overview of the proposed Attitude and Orbit Control System.

- A Bk1B21_Inertial_Attitude_Subsystem, aimed at attitude determination and control with inertial techniques. It is composed of a Bk1B212 Reaction Wheel Actuator (namely, a brushless micromotor to operate a small 1B164A Reaction Wheel), a Bk1B211 Gyroscopic Sensor to measure satellite spin, plus a software Bk1B219_Controller to interpret user and attitude commands from OBC and drive the other subsystems accordingly.

- A Bk1B22_Magnetic_Attitude_Subsystem, aimed at attitude determination and control using magnetic techniques. It is composed of a Bk1B222_Magnetic_Torque_ Actuator (namely, a power driver driving an appropriate magnetic Bk1B2222_Coil), a Bk1B221_Magnetometer_Sensor to measure satellite attitude with respect to earth magnetic field; an embedded software controller (not shown) interprets user and attitude commands from OBC and drive the other subsystems accordingly.
- An optical 1B231 Sun Horizon Sensor to measure satellite attitude with respect to sun and earth horizon. It is made basically of a 1B2311 Camera System (incorporating a 1B2311 Filter System plus a 1B2311 Lens System) and a 1B2313 Image Processor which interpret images taken from the camera. It returns (to the OBC) the satellite attitude, whenever either sun or earth is visible.
- An 1B2411 Electrospray Micropropulsor, not yet implemented as its possible usage is currently under analysis. An appropriate micropropulsor has been taken from literature as a driver example to evaluate its applications to the proposed small satellite.
- A set of appropriate 1B25 AOCS Algorithms to gather and fuse magnetic, attitude and optical sensor measurements and to drive magnetic and inertial attitude actuators.

3. Applications and requirements of small satellites AOCS

The goal of the AOCS is to determine the position of the satellite along the orbit and its attitude, i.e. where pre-defined axes of the satellite are pointing to. A detailed mathematical treatment of the AOCS problem will be considered in the next sections, while here the focus will be on its requirements and use.

The ability to determine the satellite position and attitude is of paramount importance in most space missions, as there are only a few and special cases where these capabilities are not needed. Satellite applications can be divided in a number of broad classes, among which the most common are: i) Telecommunication, ii) Earth observation, iii) Science and iv) Navigation. In each of these cases the AOCS plays a role to achieve the mission goals, but requirements with respect to several metrics can be extremely different, even within missions belonging to the same class. The obvious consequence is that methods and technologies vary considerably among various cases, which makes the choice of an AOCS architecture a difficult task when designing a satellite.

Typical applications of an AOCS subsystem can be summarized as follows.

- **Despinning** of a satellite after deployment. The launcher generally releases the satellite with an unknown spin, which must be compensated prior to starting the mission. An initial coarse attitude determination is sufficient, while orbit control can often be neglected, as the focus is on stabilizing satellite attitude. Despinning may take a long time (e.g., several orbits), but is applied only once during the life of the satellite.
  Despinning can be useful in other situations, as well. Light pressure or atmospheric drag can impose unknown spins to the satellite, so this procedure might be occasionally applied to recover from them. However, the magnitude of the correction is typically much smaller with respect to the initial deployment case.
- **Pointing** of the satellite towards a specific direction. This task has a number of uses depending on the application. In telecommunication missions, it is needed to accurately direct the satellite antennas lobe to the ground stations to improve the radio signal

strength. In Earth observation, instruments such as cameras shall be pointed towards those locations that are of interest for the application. Science mission often need to know the attitude of the satellite, as well.

Pointing can be used in different ways. Some satellites need to always point to the Nadir, regardless of their position in the orbit. Other satellites need to constantly track a given point on the Earth surface, and thus require a constant attitude correction as the satellite moves along the orbit. In other cases, inertial pointing is necessary, i.e. the ability to always point towards the same direction with respect to a fixed reference frame.

The applications of pointing are too many to list them all here, and strongly depend on the satellite mission. In many cases, pointing is used for different goals, even within the same mission. For instance, many satellites will also use pointing to steer the solar panel arrays in order to increase the generated electrical power, or to avoid forbidden attitudes that may endanger the spacecraft.

- **Positioning** is also needed in many different cases. When coarse orbit determination is sufficient, the orbit position can be computed using a mathematical model starting from ephemerides, which can be periodically updated from ground. However, if the satellite requires to autonomously determine its position, other techniques based on several kinds of sensors are used.

  The ability to know the satellite position is extremely important in navigation missions, as these satellites are the base for many other applications that require accurate position determination, on the Earth or in orbit. However, positioning is also important in Earth observation and telecommunication satellites, especially when coupled with pointing: a typical example is that of pointing a specific location on Earth, which requires both abilities to work together to reach the final goal.

- **Deorbiting**, namely the capability to move to a low enough orbit (at the end of mission) to guarantee the spacecraft destruction impacting Earth atmosphere. This is desirable (and soon it may become compulsory) to keep orbit pollution by space debris within reasonable limits.

- **Formation Flight and Docking**, where sets of satellites cooperate to reach a common goal and where their position must be carefully controlled. The AOCS must provide the capability to slightly correct the orbit, to keep the relative distance among a set of spacecrafts within given bounds, or to control the distance among them accurately to perform a safe docking.

All the applications listed above, and many others as well, use the AOCS subsystem with different requirements, and several metrics shall be considered in order to evaluate its performance. One of the most important is the **accuracy**, both of pointing and of positioning. Antennas pointing, especially for small satellites in LEO orbit which don't have sophisticated radio systems, does not usually require a fine accuracy, as the transmission lobe is generally broad enough to allow for large errors. Typical values are in the order of several degrees. Similarly, a small camera with a short focal length objective can image a wide area on the Earth surface, thus not needing high accuracy in order to frame a specific target within the picture. Longer focal lengths coupled with small image sensors impose more stringent requirements on the pointing accuracy, as other applications, especially in science missions, also do.

Another important aspect of pointing is its **stability**, i.e. the ability to keep the satellite pointing towards a specific direction for a given amount of time. Consider again a small camera with a short focal length objective: while the absolute pointing requirements might not be very tight due to the large area swath, a single pixel will only cover a very small angle, which should not change during the acquisition of the frame. For example, the angle covered by a 5μm pixel coupled with a 50mm objective is around 20 arcsec, corresponding to approximately 60m on ground as seen from a 600km high orbit. To avoid smearing of the picture due to movements of the satellite, a very high precision in pointing should be guaranteed for the entire period of time needed to take a shot. While this can be a short time (1/100s or less) for pictures in visible light taken during the day, it can be significantly longer for night pictures, in other bands of the electromagnetic spectrum, or when imaging stars or other celestial objects in astronomical related missions.

Performance of the AOCS system is also related to the **pointing time**, namely the time needed to complete a given pointing or positioning command. Higher accuracy usually requires a longer time, but this has a negative impact on the number of activities that the satellite can carry out during the lifespan of the mission. Large changes in pointing or positioning also require a longer time, due to more complex maneuvers, so a careful scheduling of the satellite tasks can minimize these kinds of overheads.

Finally, **cost**, **size** and **mass** should also be considered when designing an AOCS subsystem. These metrics deal more with the technologies and the kind of sensors and actuators that are involved in the satellite. Small satellites often require several trade-offs that may sacrifice accuracy and performance to the sake of decreasing costs; larger satellites can accommodate better sensors and actuators, thus achieving better metrics. Modularity can significantly increase the efficiency and availability of the AOCS, allowing the satellite to get good accuracy and performance at moderate costs, as will be detailed in the rest of this chapter.

4. Attitude and orbit codetermination

A satellite in the free space, corresponds to a model of an object with six degrees of freedom, rotation in three dimensions and translation in a three dimensional space.

The attitude of a body, in our case a satellite, is its angular position or orientation with respect to a defined frame of reference. Attitude in a rigid body can be represented by Euler angles, heading (rotation about Z-axis, ψ), elevation (rotation about Y-axis, θ) and bank (rotation about X-axis, ϕ). A diagram of this is shown in Figure 3.

In a similar way, the attitude of a satellite can be represented by a rotation matrix, defined by Euler angles, corresponding to a series of positive right hand rotation. If a 1-2-3 Euler rotation is used, the rotation matrix will be:

$$Q = \begin{bmatrix} C\theta C\psi & C\theta S\psi & S\theta \\ S\phi S\theta C\psi + C\phi S\psi & -S\phi S\theta S\psi + C\phi C\psi & -S\phi C\theta \\ -C\phi S\theta C\psi + S\phi S\psi & C\phi S\theta S\psi + S\phi C\psi & C\phi C\theta \end{bmatrix} \qquad (1)$$

where ϕ, θ, and ψ are the Euler angles, while $C\theta = \cos\theta$, $S\Psi = \sin\Psi$.

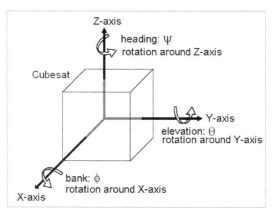

Fig. 3. Satellite coordinate system and Euler angles.

A part of the AOCS is the Attitude Determination and Control Subsystem (ADCS). It is a set of sensors, actuators and software that must be able to determine, change or keep the attitude of the satellite within a predefined range of values.

Attitude determination can be achieved through sensors such as gyroscopes, Sun, Earth and horizon sensors, orbital gyrocompasses, star trackers and magnetometers. A body sensor, such as Earth, Moon and Sun sensors, is a device that senses the direction to the respective object and returns the unit vector to the corresponding body.

A **Sun and Horizon sensor** can be based on solar cells, photodiodes, CCD cameras or CMOS cameras; these devices can determine the incidence angle (unit vector orientation) of the sun with respect to the normal to their plane. However, this information only gives a cone of possible sun positions and another sensor (e.g., a magnetometer) must be used to solve the ambiguity.

When a Sun sensor is used, the terrestrial albedo can generate errors especially in low Earth orbit (LEO), but it can be corrected using appropriate processing.

The second vector that will be used to estimate the attitude will be taken from a **Magnetometer**. It measures direction and magnitude of Earth's magnetic field and the measurement is compared to a simulated model or with a previously available map of the expected magnetic field. To avoid errors in the measurement of the magnetic field, all other (active and passive) parts inside the satellite must have a minimal residual magnetic field.

4.1 Attitude determination

To determine the attitude, a set of three sensor-measured vectors (with a non-null cross product between them) is used with another set of three corresponding reference vectors (based on models, simulations or data stored on memory) to find a rotation matrix Q, as shown in (1), that satisfies:

$$m = Q \cdot r \tag{2}$$

where m = measurement vector and r = reference vector. The first vectors of the sets are obtained with:

$$\hat{a}_1 = \frac{m_1}{|m_1|}, \quad \hat{b}_1 = \frac{r_1}{|r_1|} \tag{3}$$

The second vector in each set (a_2, b_2), is given by the second pair (vector measurement and the corresponding reference, m_2 and r_2, respectively). The third vectors are the result of a cross product of the first two vectors in each set:

$$\hat{a}_3 = \hat{a}_1 \times \hat{a}_2, \quad \hat{b}_3 = \hat{b}_1 \times \hat{b}_2 \tag{4}$$

So, like these vectors, the sets satisfy:

$$A = Q \cdot B \tag{5}$$

Where A and B are the two matrices 3x3 made with the set of three vectors. Finally, the attitude matrix Q, is determined by:

$$Q = A \cdot B^{-1} \tag{6}$$

In some cases there may be singularities, for instance when the first two vectors are parallel to each other.

In this case, the two main vectors are given by the Sun sensor and a magnetometer; with these vectors the analysis can be completed. However, since the Sun sensor will not work properly in the eclipse phase of the orbit (near to 40 minutes in LEO), a third vector can be used. To have a set of three measurement vectors, the sensor suite is complemented with a gyroscope which senses satellite spin without any external reference object. MEMS gyroscopes (Micro Electro-Mechanical Systems) have a reduced size making them suitable to be used in AraMiS.

All measurements from sensors are processed by the microcontroller. Once the reading is done, the microcontroller applies the algorithms to estimate the attitude of the satellite.

4.2 Optical orbit determination

Another processing that must be done is to estimate the orbit of the satellite. Twenty years ago, tracking and processing in the ground stations was used for orbit determination. Later, some satellites (LEO in particular) computed their own positions using GPS receivers. It reduced the work that ground stations should do to collect and transfer data.

Now the idea is to have another option to estimate the orbit without the use of GPS while still reducing the work of the ground segment, and this could be feasible using image sensors. The satellite, capturing an image of the Earth, can use this information to estimate its own orbit.

The processing is based in several parts: image capture of the Earth, knowledge of the satellite attitude, shine-dark zones detection on Earth's surface, Sun position.

First step in the process is to detect when the image shows a day-night transition or vice versa. In each rotation the satellite could detect two kinds of transitions (day-night or night-day). See Figure 4.

Fig. 4. Night-day transition (left). Inclination angle of the transition (right).

Once the transition is detected, the inclination angle of this transition in the image must be calculated, as shown in Figure 4. This inclination angle on the image must be corrected with the attitude angles of the satellite. Now, taken this angle as a reference, we can calculate the inclination angle of the orbit satellite with respect to the shine-dark angle.

From the image some reference points are taken and they are compared in the successive frames to estimate their movement in the sequence. With at least two different images, the trajectory of the points can be estimated as shown in Figure 5.

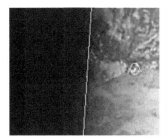
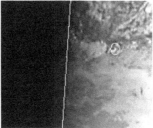
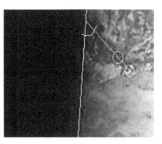

Fig. 5. Sequence of images tracing a point (left and center). The angle between the two lines (right).

The angle between the shine-dark line (calculated previously) and the points trajectory can then be estimated, as shown in Figure 5.

Finally, having the angle between the border in the image and the relative movement of the points on Earth, applying the satellite attitude and Earth's axial tilt correction factors it's possible to estimate the orbit angle of the satellite. This angle will have some error, but after several orbits this error will decrease.

With this process, the orbit inclination angle can be estimated. Trajectory prediction, instead, is still in development.

5. Fine-grain modularity

In electronics, a useful and alternative design methodology to reduce system complexity, among others, is the modularity; whereby, through the usage of standardized proper interfaces, it is possible to split the main functions of a complex system into some independent components (modules). Thanks to modularity, our system also becomes flexible because it can be configured in several ways just plugging/unplugging few modules.

The main goal of this project is to integrate in an innovative way, for the development of nano and micro satellites, mechanical structure, harness, signal routing, and other basic and common functions like solar panels, into standard off-the-shelf modular panels. This new approach is expected to allow the development of new space missions with a mass, cost and time budget significantly lower than present missions.

The outcome will be a standardized modular platform, equipped with power management and attitude control subsystems, integrated harness and data routing capabilities, for low-cost nano and micro satellites.

One of the keys for achieving the goal is the creation of loosely coupled component designs by specifying standardized component interfaces that define functional, spatial, and other relationships between components. Once specified, this will allow all of the developers to design their own components in an independent way. In the AraMiS modular architecture, major bus functions are split over a number of properly placed and identical modules, so to simplify design, maintenance, manufacturing, testing and integration.

The modules interconnect and dynamically exchange data and power in a distributed and self configuring architecture, which is flexible because standardized interfaces between components are specified. Product variations can then take place substituting modular components without affecting the rest of the system. This design strategy offers, additionally, a high degree of built in redundancy and different configurations providing a larger number of system variations adaptable to different missions, spacecraft sizes and shapes.

AraMiS is characterized by its design reuse, which allows to effectively reduce as much as possible design and non-recurrent fabrication costs (e.g., qualification and testing costs), while reducing as well time-to-launch.

The basic architecture of AraMiS is based on one or more small intelligent modules (**tiles**), located on the outer surface of the satellite. The inner part of the satellite is mostly left empty to suit several kind of user-defined payload, which is the only part to be designed and manufactured ad-hoc for each mission.

Each tile is designed, manufactured and tested in relatively large quantities. There is an increased design effort to compensate for the lower reliability of COTS components, therefore achieving reasonable system reliability at a reduced cost.

Thanks to the features of compatibility, design reuse, integration and expandability, while keeping the low-cost and COTS approach of CubeSats, the AraMiS architecture extends its modularity in two directions:

1. Possibility to reconfigure the system according to the needs of the payload, to target different satellite shapes and sizes (from 5 to 100 kg and even more) based on one of the following design options: Smallest cubic shape; larger cubic (or prismatic) shape; small hexagonal / octagonal satellites.
2. Functional modularity, achieved using smart tiles (or Panel Bodies) which have, at the same time, thermo-mechanical, harness, power distribution, signal processing and communication functionalities.

Power and data handling capabilities are embedded in each tile (*Power Management tile*), which incorporates solar panels on the external side and basic power and data routing capabilities on the internal side, and can host a number of small sensors, actuators or payloads (up to 16 for each tile). An AraMiS tile is illustrated in Figure 6. Each tile offers a power and data standardized interface with mechanical support for small subsystems.

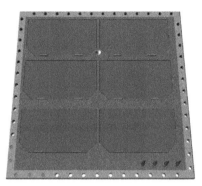
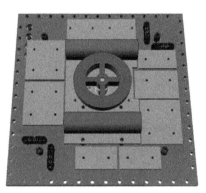

Fig. 6. Mechanical drawing of external (left) and internal (right) side of an AraMiS tile.

Figure 6 also shows about twelve standard interconnection points for small and light-weighted subsystems, among which a reaction wheel (centre) and a small power storage subsystems (batteries).

Figure 7 shows a photograph of the outer side of a tile, bearing solar cells, and a detail of the honeycomb structure of one of the possible AraMiS configurations:

Each subsystem is housed in small daughter boards which can be connected with spring-loaded connectors, like the one shown in Figure 8. Connections are electrically and mechanically modular, so they are suitable for a large range of systems, from those with just 1 or 2 analog channels up to larger systems with up to 8 analog channels, 16 digital I/O and possibly a CPU with I2C and RS232 communication.

Fig. 7. Photograph of the outer side of a tile (left). Detail of honeycomb structure (right).

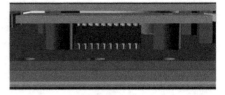

Fig. 8. Detail of a spring-loaded connector (left). Detail of modules interconnection (right).

Figure 9 shows examples of a single connector board and a double connector board respectively. The bottom side of the small modules is also shown, where the pads for spring-loaded connectors are clearly visible.

Fig. 9. Top view of single module (left); double module (center); bottom view (right).

The basic idea is that the outer tiles must also possess structural properties, i.e. that they may become an important part of mechanical subsystems. This results in reducing both weight and costs and the simplification of processes and assembly and testing times.

Therefore, the basic Power Management tile is composed of solar cells, control electronics with a dual CPU and A/D converters for handling sensors, storage, signal processing, housekeeping functions, rechargeable batteries and the Al-alloy panel which holds everything together and encloses the satellite. All of these modules are connected with a power and data distribution bus to share power and to exchange system information.

The power generation system is consequently extremely modular, since it is composed by the Power Management tile replicated as many times as needed to get the desired power. All these tiles work in parallel and offer a good level of redundancy, making the system able to tolerate several faults and allowing graceful performance degradation.

6. Sensor fusion and reconfigurability

An AraMiS satellite is built using several replicated modules connected together and capable of sharing information. The module in charge of attitude and orbit control is located on the power management tile and can send the measurements, acquired by its sensors and stored in a housekeeping vector, only to the on-board computer (OBC).

Having several tiles with the same structure and equipment, means that also the housekeeping vectors contain the same information. This solution offers a high degree of redundancy but also a more accurate and less error prone readout; in order to exploit the characteristics of the AraMiS architecture a sensor fusion algorithm was developed.

The sensor fusion algorithm is included in a module of the OBC which allows system level management. Its main task is to check for the status of the tile, that includes also the AOCS, in order to detect any anomalies; if some anomalous conditions are identified the software starts the self configuration.

At regular intervals, the OBC reads from each tile in the satellite the housekeeping and the configuration vectors. From the housekeeping vector analysis, the OBC obtains a view of the general status of the satellite, while the configuration vector shows the mechanical status. Then the sensor fusion algorithm analyzes these vectors in order to detect anomalies or non-working sensors. The word "analyze" is a bit vague, so from now on the discussion will be about the operations which supervise the system and reconfigure the parameters.

6.1 Fault detection

A way to detect faulty condition and anomalies is to do some calculation on the housekeeping vectors; in fact, comparing the telemetry information of each tile can be a very effective method to find errors, especially if the dimensions of the satellite are small. For each type of information there will be an appropriate operation to detect faulty conditions and the choice of what to do depends on the characteristics of the information.

The intrinsic subsystem redundancy of the AraMiS architecture can be effectively exploited as a fault detection mechanism. Subsystems compatible both in the measured quantity and in its reference points can be critically compared to detect anomalies and faults. Comparison on indirect measurements (e.g., on attitude parameters obtained from sun and magnetic sensors) is to be considered risky mainly due to the possible ambiguities in the comparison.

As a fault detection example, the magnetic field measured by two parallel tiles should be similar in absolute values. With the magnetic field, the arithmetic mean of the values can be an effective detector for anomalies. Also, comparing the newly acquired information and the previous one helps the detection of anomalies in slowly varying measurements like temperature.

Also, it's possible to use a range detection method. There are some kinds of information that can't exceed a certain range, like the minimum battery voltage. If the measured value exceeds the threshold it's easy to detect an anomaly.

6.2 Self reconfiguration

When a fault is detected, for example if a sensor stops working, the system needs to check if the parameters are still valid or become obsolete with respect to the new configuration of the system. The critical parameters to update are all those related to vector information, such as magnetic field or spin. In these cases, the parameter is a pseudo matrix so the calculation is more complex; in the following example is considered the calculation of the magnetic field, but the same process can be used for other fields.

The computation of the magnetic field vector with respect to the centre of the satellite is a very important step in order to determine the maneuvers to control attitude. All values read from magnetometers are referred to the coordinate system of their tile, which means that, in order to compute the magnetic vector in the satellite's coordinate system, the algorithm needs to apply a roto-translation to all measures finding the solution of a linear system. For each tile are available two magnetometers, which measure the component x and y with respect to the coordinate system of the tile.

This operation is done using the director cosine matrix of each tile, which represents the relative position of the tile's coordinate system with respect to the coordinate system of the entire satellite. Using the first two rows of the director cosine matrix of each tile, the algorithm builds a $2T * 3$ matrix A, where T indicate the number of the tiles with working sensors. This matrix represents the roto-translation needed to switch from the tile coordinate system to the one in the centre of the satellite. In order to find the magnetic field vector from the measures of the magnetometer, matrix A needs to be pseudo inverted (the normal inversion works only on square matrix).

One of the methods that can be used for the calculation of a pseudo inverse matrix is the Singular Values Decomposition (SVD). Using this method the starting matrix A can be decomposed in the matrices U, V^* and Σ, which represent, respectively, the left-singular vectors, the right-singular vectors and the singular values of A. With U, V^* and Σ, the pseudo-inverse of A is equal to

$$A^+ = V\Sigma^+ U^*$$

The only problem related to this method is that it's a numerical approximation one, so it needs to use floating point variables, which grant a greater dynamic range, in order to return acceptable values. But considering that the self reconfiguring algorithm is, hopefully, used only few times (or never used at all) the floating point computation is not a major concern.

7. Latchup protection

Single event latchup (SEL) in silicon devices is the most damaging condition to be considered during the design of space-based electronic systems. Latchup in CMOS gates is a condition where a parasitic SCR thyristor connecting the supply rails is activated by a spurious event. This creates a low-resistance path which draws an excess current, possibly leading to the destruction of the affected device (Sexton, 2003). This condition is self-sustaining.

The latchup condition can be triggered in several ways, e.g. from power supply and input signal glitches or from interaction of the substrate with ionizing particles, heavy ions, etc. (Voldman, 2007). Commonly found in the space environment, ionizing particles and heavy ions pose then a serious threat to CMOS electronic devices since the particles cannot be easily stopped from interacting with the silicon substrate. The use of commercial, non rad-hard, components in space requires then proper mitigating measures to avoid damages and recover proper operation of the affected systems.

The hold voltage for self-sustainment of thyristor activation is on the order of 1 V. To extinguish the latchup, the supply voltage has to be reduced below this threshold, i.e., by means of a simple power cycle of the affected ICs. At system level, a trade-off needs to be established since the protection of every CMOS device cannot satisfy the low-complexity requirements. System partitioning is needed to isolate vulnerable sections that will be protected by a limited number of adjustable protection modules.

The latchup is detected when an excess current draw is measured in a circuit section and the power supply of multiple devices will be disconnected at once. Before power supply re-activation however, stored energy within the section circuitry (e.g., in bypass capacitors) needs to be depleted through an explicit short circuiting of the interested supply rails. This ensures that discharge transients complete and latchups are extinguished before the timed re-activation of the power supply.

During the system partitioning, also the power supply requirements of the protected devices have to be taken into account. On one hand, impulsive loads may be erroneously detected as latchup conditions, causing periodic and rather systematic reboots in the protected section. On the other hand, some components may be rather sensitive to the high current draw of a real latchup and may need a quicker/more sensitive setting on the current consumption threshold. Also, a sensitive component should not be grouped in a high power consumption section since its latchup condition may become difficult to detect. The different devices should then be grouped in sections based on the expected power consumption profile. Each section will be overseen by a single protection module and its intervention parameters will have to be tuned accordingly.

The central part of the latchup protection subsystem can then be isolated in a replicated latchup protection module. The module needs to be both radiation tolerant (to be immune from the problems affecting the protected circuits), flexible (to adapt to different power consumption profiles) and simple (in terms of basic functionalities, circuit complexity, occupied board space and cost). We have then decided to develop the **1B127** (1B127 Datasheet, 2009), an appropriate ad-hoc anti-latchup and over current protection device manufactured using Neohm's hybrid technology.

For the protection circuit to be radiation tolerant and not being affected by SELs and total dose induced phenomena, a proper component selection is needed. The electronic components will have to be readily available and based on a latchup-free technology. The obvious choice is a bipolar process that may also be dielectrically isolated for improved SEL tolerance. The use of dielectric trenches and oxide isolation layers, allows for a complete isolation of the single transistors, avoiding resistive paths and parasitic junctions and capacitances (National Semiconductor, 2000). Available at the NSRE Components Database (IEEE Radiation Effects Data Workshop, 2010) is also a list of radiation tolerances of selected devices that may provide valuable information. The accurate selection of commercial devices will contain costs while providing a design stage guarantee over the expected radiation tolerance of the subsystem. This early guarantee will have to be complemented by real world radiation exposure to reach the target tolerance levels.

The flexibility in adapting the protection circuit to the needs of the different sections is centered on a limited number of parameters. The *turn-off* time controls the time needed for the protection to trigger, allowing control of the circuit tolerance to spikes and inrush currents. The *turn-on* time controls the time needed for the output of the module to reach (almost linearly due to an intrinsic output current limitation) the nominal value after supply re-activation; the slow ramp limits the inrush current of the protected load avoiding instability. The *low-pass filtering* of the current reading allows masking high frequency noise that may trigger the protection and contributes, with the *turn-off* time, to protection stability.

An excessive level of flexibility however may hinder circuit adoption within the system or module re-use in future applications. The need of external components adds to the used board space, requires additional design and manufacturing time, and increases cost. The module is then designed to work with a minimal configuration needing, essentially, only an external sense resistor. The parameters are internally tuned to achieve optimal performances in average subsystem conditions.

The module block diagram is shown in figure 10. A high-side current sense is implemented with the use of a specialized *current sense* (or *shunt*) *amplifier*. This solution avoids low-side sense resistors that would introduce interruptions and extraneous resistances on ground planes. The specialized amplifier helps avoiding common mode rejection issues typical of differential amplifiers used in high-side applications. A precision voltage reference allows for predictable triggering of the protection over an extended temperature and supply voltage range. The comparison result is then fed into a monostable circuit controlling the delay before load re-activation and a slew rate controller that introduces the necessary asymmetry to differentiate *turn-off* and *turn-on* times. Two MOS transistors are finally used (exclusively) to connect the load rails to the supply or to short circuit them to extinguish latchups during protection activation. Additionally, the protection module outputs the current reading for telemetry purposes and allows for the use of external higher power transistors to switch high current loads.

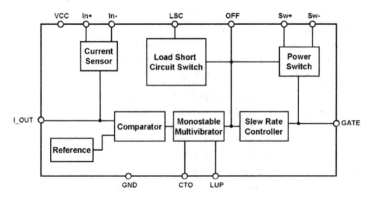

Fig. 10. Internal block diagram of the latchup protection module.

The intrinsic flexibility of this approach also allows for more advanced approaches like the one depicted in figure 11, where two modules are used in parallel to provide some degree of module fault tolerance.

8. Open hardware/software plug and play architecture

The modularity at the hardware level which has been described previously requires an open software architecture, in order to really be effective.

We originally aimed at having a kind of **plug and play** approach, similar to that available in many other terrestrial applications and also in the Space Plug and Play (SPA, 2011) but we soon pointed out that this was posing excessive constraints on the capabilities of the supporting CPU and its tolerance to radiations. It simply could not be afforded by a low cost, high reliability approach. Furthermore, a real-time, full plug and play solution was not required, as the flexibility of plug and play was only needed during the integration phase and not later.

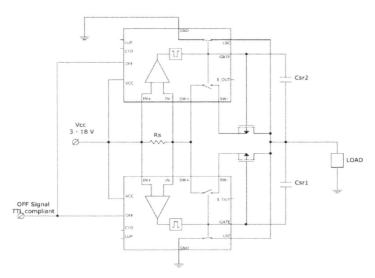

Fig. 11. Advanced protection stage with redundancy of protection modules and external transistors. A simple system can use just one device plus an external resistor.

We therefore decided to develop an alternative approach, nearly as flexible as our original aim, but which requires significantly lower CPU capabilities and power and which could offer better tolerance to radiation. This solution can be run on the MSP430 family of processors, a standard COTS from Texas Instruments.

We found that this capability was straightforward if a proper design approach was used. We therefore developed an appropriate system-level approach based on the extensive use of UML and all its capabilities, strongly supported by the Visual Paradigm VP Suite tool.

With the proposed approach, every subsystem is composed of: i) a HW part (namely, each of the small module boards described in section 5 and shown in figure 9); ii) a corresponding SW support, properly hardened using the techniques which are discussed in section 9; iii) an appropriate interface to the sensor fusion and reconfiguration algorithms described in section 6 and the AOCS algorithms which are shown in figure 2.

An **UML class** is associated with each of our subsystems, as shown in figure 2 (overview), while figure 12 shows a detailed (although simplified) view of one of those subsystems. From that, we can see a few important items:

- The internal hierarchical architecture of the module: the "root" class (the left box) is composed of three major blocks (the three boxes at the right); in turn, each block is made of smaller and simpler blocks (not shown);
- The electromechanical parameters of the module, which are described by appropriate **class attributes**, properly documented (documentation not shown). For instance, attribute **float SENS_MAGNETIC_FIELD_RAW** indicates the sensitivity (in V/T) of the pin measuring magnetic field. Its value is given by a formula **1.0 / (sensor. SENS_MAGNETIC * adc.SENS_ADC)**, and therefore it is given in term of the subsystem parameters (including for instance also the sensitivity of the ADC which is inherited by

the hosting CPU). Any change of its subsystem (or, similarly, any element of its subsystems) or any change in the ADC is automatically inherited by the overall system and, from here, to the sensor fusion algorithm, which requires this value to properly interpret values received from the module. This inheritance is obtained at compile time, that is, during HW/SW system integration: when the integrator chooses to assemble a given module, he will bring the whole UML class inside the UML model of the spacecraft under construction and, from here, the correct SW is automatically generated, both for the tile CPUs and for the OBC supporting the sensor fusion and AOCS algorithms.

- Some template parameters (included in the dashed box above the root class), which allow to configure, for instance, in which physical slot in each tile the module is plugged (see figure 6) and in which positions of the overall housekeeping, status and control vectors the relevant data of the module shall be stored by the tile CPU.

This is a form of UML-based Object Oriented Design (OOD) approach which allows building either a spacecraft or, in this case, one of its subsystems according to specific mission requirements, with very limited risks, effort and design time.

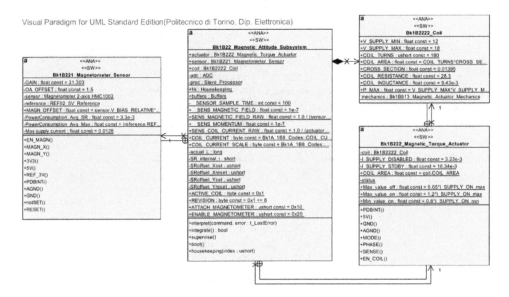

Fig. 12. A detailed (although not complete) UML view of the architecture and the parameters of our 1B22 Magnetic Attitude Subsystem.

9. Hardening techniques against radiation of SW code and HW interfaces

This section presents a general methodology depicting how to address the overall protection of code to mitigate transient faults induced by radiations. The code is supposed to be written in either C/C++ or assembly language. COTS devices running this code are supposed to suffer from all possible Single Event Effects (SEE): single event upsets (SEU), both single (SBU) and multiple (MBU), single event functional interrupts (SEFI) and possibly others. They are also supposed to: survive a certain desired total ionization dose

(TID) present in its operational environment (given orbit and mission lifetime) and be either latchup-free or protected from disruptive radiation-induced effects as shown in section 7.

Having in mind the mitigation of the harmful effect of soft errors, we have determined a number of use cases to design an effective radiation hardening of a COTS processor-based system. Although no hardware redundancy can be applied within the processor, preventing us from applying a Triple Modular Redundancy (TMR) over a set of critical registers from the register file, we can think of a rich set of software-side fault-tolerant strategies. Moreover, exploring the software hardening design space reveals not only the importance to take into account the type and physical location of code/data, but also the need to take into consideration the hardening of the interface between processor and outer elements such as ports, DACs, etc.

We describe the following software hardening targets and the selected hardening action:

9.1 Data storage

With independence of its physical location, data can be labeled as having an infinite, medium or short lifetime:

- **Infinite lifetime**, if lifetime of data is much higher than most other variables in the system. In this case, the proposed hardening method is performing periodic refreshing (compatible with system time constraints) using software DMR or TMR (Double/Triple Module Redundancy) when applicable.
- On the other hand, having a **medium** or **short lifetime**, that is, a lifetime comparable or shorter than storage time of most variables in the system, led us to propose DMR or TMR carrying out its action (fault detection or correction) every time data is accessed.

9.2 Data driven routines

This use case consists of programs without state automata, whose execution flow does not depend on past events and system states. A data-driven routine is supposed to enter, get input data (either from the calling program, input devices or data storage), execute in a predefined time and with a predefined algorithm, outputs results (either to the calling program, the output devices or data storage) and either exit or repeat execution endlessly.

A data-driven routine can either be **flat** (that is, without internal calls to other routines) or **hierarchical** (that is, with internal calls to subroutines). This characterization implies two different hardening processes:

- **Flat code**: A data-driven flat code is defined as a software routine which either has no input and output data (so all data are and remain inside the routine itself, as for example a delay function), or input and output data come from or go to the calling routine, but they need not be hardened. The code can either have or not have calls to internal routines or accesses to external devices which either requires no parameter passing or parameter passing need not be secured. We have evaluated two possible hardening strategies. At low level (i.e., at assembler level) we propose using the Software Hardening Environment (SHE) as described in (Antonio Martínez-Álvarez et al., 2012). SHE is a proven compiler-directed soft error mitigation system especially designed to harden a certain software code against radiation-induced faults. SHE makes a special emphasis on flexibility and selectivity, that is, different hardening techniques can be carried out on different subsets microprocessor resources. At a higher level, for example, hardening a C++ code, we propose using a library with hardened-types. This library must declare dual hardened types for every C++ native type such as char, int, and so on. So we can choose at C++ level either using char or Hchar, the hardened version for char. Each H-type ensures redundancy (based on TMR) by overloading every possible operation on it.

- **Hierarchical code**: a software routine which requires securing the exchange of data from one element to another. All data transfers must be hardened. We have focused our attention to the process of hardening parameter passing.

  Different hardening sub-cases can be described depending of the nature of sender/receptor:

- **Hardware to hardware**: data is passing from a hardware device to another hardware device, usually via one or more wires (bus). If the bus handles critical information, bus-TMR can be applied; otherwise a higher level software solution (like using a hardened data packet exchange protocol) is a good solution where time overhead doesn't exceed design constraints.

- **Hardware to software**: data is passing from a hardware device to a software routine. It usually implies a read operation by the code from a microprocessor port or an internal device register. It does not include interrupts.

- **Software to hardware**: data is passing from a software routine to a hardware device. It usually implies a write operation by the code to a microprocessor port or an internal device register.

- **Software to software**: hardens parameter passing between two software routines. This case encloses hardening of different lifetime variables, data memory and register file.

The last three sub cases are supported by the SHE tool so we propose its use to ensure predictable fault coverage.

9.3 Program flow

The hardening against Control Flow Errors (CFE) is a mandatory hardening task. Without the possibility of a hardware hardening of special involved flags nor registers (e.g. carry, zero, overflow flags presented in conditional jump/calls), we take into consideration two possible alternatives. The first one consists of the classic use of a watchdog (see 9.6 harden code memory). The later one consists of **control-flow checking by software signatures** (Oh et al., 2002b). This technique is based on adding compiled-time calculated signatures on every node from Control-Flow Graph (CFG). These signatures will be checked during execution of code. The best choice depends on the time-constraints of the system (e.g. keep time overhead less or equal than a certain limit)

9.4 Control driven programs

We find here programs with state automata, whose execution flow usually depend on past events and a number of system states. The most sensible parts within the hardening of these programs have to do with the hardening of parameter passing as described in 9.2 (**Hierarchical Code**) and the hardening of the state storage, which is also described in 9.1 (**Data Storage**).

9.5 Precompiled libraries

Although we have supposed the software to be given as a C/C++ or assembler code (allowing us to perform code transformations), it is worth assessing other possibilities. A binary code (statically linked or not) usually makes a number of subroutines calls that have to be under control. In the assumption that there is no operating system, the usual library calls are related to the compiler execution runtime (e.g. GNU-GCC **libgcc** support library), the classical C/C++ runtime libraries (e.g. GNU-GCC **libc/libstd++** support libraries) or the user libraries. Our primary choice to ensure a correct hardening of a precompiled library is writing in C/C++/assembler the subset of used calls and proceeding with its hardening as described in 9.2.

9.6 System configuration

In this use case, we are interested in hardening the processor configuration registers (clock configuration registers, peripheral configuration registers and interrupt configuration registers) and code memory. Our research group has already developed an FPGA-based **smart watchdog** able to assist the hardening of the next three cases of system configuration of a COTS system:

- **Harden code memory**. This code storage may contains either one or more software programs (for CPUs), as well as one or more fuse maps (or configuration files for FPGAs). Code memory has to be protected more than any other storage in the system, as any corruption forever affects the whole system functionality. A SEU in code memory always causes a SEFI. There are different types of code memory, and consequently three ways to harden them. In particular:
  - **Radiation-tolerant ROMs**, which can never be affected by SEUs. For instance, true ROMs and PROMs and several types of FLASH memories. In this subcase, no hardening tasks are needed.
  - **Radiation-tolerant ROM copied into radiation-sensitive RAM**, to speed up its execution. The master copy of the program is stored in a radiation-tolerant storage but the working copy is subject to SEU. Detecting this situation always requires an external agent (like our smart watchdog), since the code alone is not able to detect it in all cases. To correct this situation a reboot action is usually enough, as reboot reloads code RAM from code ROM
  - Rewritable radiation-tolerant ROM: there are situations where the code resides in a radiation-tolerant FLASH but the CPU has the capability to write onto it. Even if the program does not foresee rewriting the FLASH, a SEU-induced error might unexpectedly rewrite and corrupt the code memory. As in the previous situation, detecting this situation always requires and external agent (like our smart watchdog), while its correction requires and external agent able to reload the FLASH from a backup copy from an external radiation-tolerant ROM through an appropriate bootloader interface.
- **Harden static configuration**. Most systems configure their peripherals or I/O systems by means of appropriate configuration words. For instance, baud rate and modulation for UARTS; counting mode and period for timers/counters; acquisition modes for ADCs and DACs; pin direction for I/O ports. It requires an external agent like our smart watchdog.

- **Harden Dynamic configuration.** Hardening those processor or peripheral configuration registers which are occasionally modified during program execution. Depending on the modification rate, it can be necessary an external agent (smart watchdog) or a periodic refresh similar to the one described in 9.1.

9.7 Interrupt Service Routines (ISRs)

Hardening of interruptions masks is covered by hardening of system configuration (see 9.6) whereas the code itself is supposed to be a data driven routine which are covered in section 9.2. In relation to the expected calling rate, we can distinguish between periodic (non reentrant) and occasionally called ISRs, which is also covered in section 9.6 (Harden static/dymanic configuration).

10. Micropropulsor

The most appropriate micropropulsor for a nano or micro satellite as AraMiS turns out to be an electrospray electric propulsor, such as that described by (Lozano and Courtney, 2010).

The propellant, which is an ionic liquid, is confined in a tank of porous metal, and by capillarity forces it can be driven into a pseudo-conical structure called emitter. Here the ions are accelerated by a strong electric field established among the emitter itself and a pair of electrodes, called extractor electrode and accelerator electrode. The global potential difference varies in the range of 1 – 2 kV. The thrust produced by a single emitter is too low to meet mission requirements, and therefore the emitters are grouped into arrays. Each propulsor houses a couple of arrays that emits ions of opposite species, to preserve the electrical neutrality of the satellite; in addition, the polarity of emitted ions alternates in time, at a frequency of about 1 Hz, to avoid accumulation of ions on the outer surface of the propulsor.

A set of achievable maneuvers are discussed below, considering propulsor performance similar to those reported by (Lozano and Courtney, 2010): a specific impulse of 3500 s and a thrust of 100 µN. A cubic configuration is adopted, with one tile per face and two propulsors per tile; to remain in the category of nano satellites, the mass varies between 1 kg and 10 kg. Two orbital maneuvers in LEO circular orbits (altitude between 120 km and 1000 km) are considered here: a deorbiting and a change of inclination (both with unchanged remaining orbital parameters), plus an attitude maneuver: a complete rotation about a body axis, in open loop. For each maneuver, propellant consumption and maneuver time are calculated. Other maneuvers have been considered although not reported here.

10.1 Orbital maneuvers

The used model involves ideal Keplerian orbits, i.e., it considers only the gravitational forces acting between the satellite and the Earth. The model refers to the so-called Edelbaum solution (Edelbaum, 1961), which provides the minimum velocity change ΔV to apply to the satellite to modify its altitude, and/or to provoke a change i in its orbital inclination, between the initial orbit, marked by subscript o, and the final orbit. The orbital radius r (and therefore the altitude) is strictly related, in case of ideal circular orbits, to the orbital speed V by the relation $V = \sqrt{\frac{GM_T}{r}}$, where G is the gravitational constant and M_T the Earth mass. From Edelbaum solution follows:

$$\Delta V = \sqrt{V_0^2 - 2VV_0 \cos\left(\frac{\pi}{2}i\right) + V^2} \tag{7}$$

the consumption of fuel Δm is obtained from ΔV through Tsiolkovsky's equation:

$$\Delta m = m_f \left(e^{\frac{\Delta V}{c}} - 1 \right) \tag{8}$$

where c is the speed of the expelled propellant and m_f the mass of the satellite. The maneuver time Δt is obtained thanks to the second law of dynamics:

$$\Delta t = \Delta V \frac{m_f}{T} \tag{9}$$

where T is the overall thrust of two propulsors. The model therefore does not take into account the following phenomena:

- Gravitational interaction with other bodies;
- Other forces of different nature, such as aerodynamic drag;
- Attitude maneuvers to keep the thrust vector in the desired optimum direction, as the engines have no chance of movement when installed on an AraMiS tile.

The results for the deorbiting maneuver are collected in figures 13 and 14 (left graphs) and also apply in the case of an orbital climb. An increase in the final mass of the satellite is detrimental to maneuver characteristics, as it would imply that more mass has to be accelerated. This maneuver is regarded as practically feasible thanks to the results of the model.

The change of inclination (figures 13 and 14, bottom) is limited to 2 rad ≈ 114.59°, due to limitations of the Edelbaum solution; in a precautionary way a minimum altitude of 120 km is used, as it implies the maximum ΔV, and therefore the worst characteristics. This maneuver would be much more challenging than the previous for the propulsion subsystem.

10.2 Attitude maneuvers

The open-loop rotation takes place in three phases:

1. *Angular acceleration.* A pair of engines, on opposite tiles, generates a pure moment on the satellite, up to a maximum angular speed;
2. *Drift.* The propulsors are turned off, and the satellite moves with uniform circular motion at the maximum angular velocity;
3. *Angular deceleration.* A torque is generated in the opposite direction, to allow the stopping of the satellite motion. Maneuver characteristics are the same of phase 1.

Knowing the geometric and inertial features of the satellite, the characteristics of the maneuver are calculated with simple considerations on the kinematics of circular motion. In this model only propulsive forces are considered. The results are described in figures 13 and 14 (right graphs). The graphs are restricted to a maximum angular speed of 0.19 rad/s, dictated by AraMiS features and by the considered thrust. As previously explained, an increase in mass worsens maneuver characteristics. The maneuver appears to be much more sustainable by the propulsive subsystem than the previous, even from a time extension point of view.

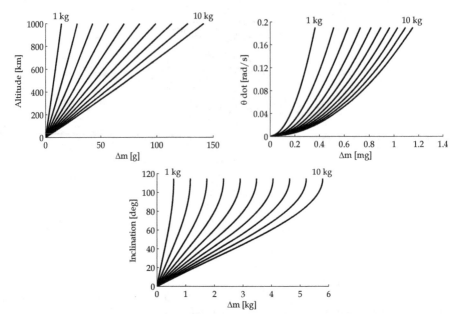

Fig. 13. Fuel consumption of chosen micropropulsor for three different attitude and orbit maneuvers: deorbiting (left); spin (right); orbital plane inclination (bottom). All plots are for satellite masses from 1 kg to 10 kg.

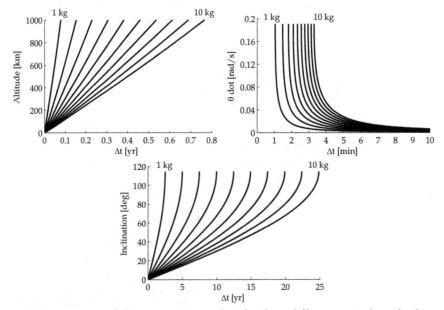

Fig. 14. Maneuver time of chosen micropropulsor for three different attitude and orbit maneuvers: deorbiting (left); spin (right); orbital plane inclination (bottom). All plots are for satellite masses from 1 kg to 10 kg.

Propellant consumption for deorbiting is of the order of tens of grams; therefore compatible with small satellites characteristics and maneuver time, even in the worst case of 10 kg of final mass, is less than one year, while fuel consumption and maneuver time for large changes of orbital inclination are too high for small spacecrafts and typical missions. Formation flight is still under consideration, while attitude maneuvers require little propellant, unless used frequently, therefore they can be occasionally used in support of the magnetic and inertial approaches.

11. Conclusion

This chapter has described a few aspects of the AOCS subsystem for small satellites which has been developed at Politecnico di Torino as a part of its AraMiS modular architecture for small satellites.

Several innovative aspects of the development of a low cost, high performance system have been considered in details, more than the complete description of the system, as these are the most technologically challenging aspects of the development and can find many applications in other low cost spaceborne system. In particular, the AOCS subsystem itself and all the innovative technological aspects described above can be used in other similar applications, like inside CubeSats or other small satellites.

12. References

Reyneri, L.M.; Sansoe, C.; Passerone, C.; Speretta, S.; Tranchero, M.; Borri, M.; Del Corso, D. (2010), Modular Design Solutions For Modular Satellite Architectures, *Book Chapter in Aerospace Technologies Advancements*, InTech Editor, 2010, pp. 165-188.

Hurtado, J.; Reyneri, L.; Saenz-Otero, A. (2011) CubeSat & NanoSat image simulator for a low cost model position and attitude sensor, *Proceedings of IEEE LARC & CCAC 2011*. October 2011, Bogota, Colombia.

Wertz, J.; Larson, W. (1999), Attitude Determination and Control, *Section 11.1 of Space Mission Analysis and Design*, 3rd edition, Space Technology Library, editor.

Sexton, F.W. (2003) Destructive single-event effects in semiconductor devices and ICs. *IEEE Transactions on Nuclear Science*, vol.50, no.3, pp. 603- 621, DOI 10.1109/TNS.2003.813137, June 2003.

Voldman, S.H. (2007) Latchup. John Wiley & Sons.

National Semiconductor (2000). National Semiconductor Develops New Complementary Bipolar Process. <http://portal.national.com/nationaledge/aug01/860.html> (retrieved November 2nd, 2011).

IEEE. Radiation Effects Data Workshop (REDW) Record, 2010. <http://nsrec.com/redw/> (retrieved November 2nd, 2011).

De Los Rios, J.C.; Roascio, D.; Reyneri, L.M.; Sansoè, C.; Passerone, C.; Del Corso, D.; Bruno, M.; Hernandez, A.; Vallan, A (2011). ARAMIS: A fine-grained modular architecture for reconfigurable space missions. *Proceedings of 1st Conference on University Satellite Missions*. Rome, Italy.

Speretta, S.; Reyneri, L.M.; Sansoè, C.; Tranchero, M.; Passerone, C.; Del Corso, D. (2007). Modular Architecture for Satellites. *Proceedings of 58th International Astronautical Congress.* Hyderabad, India.

Del Corso, D.; Passerone, C.; Reyneri, L.M.; Sansoè, C.; Borri, M.; Speretta, S.; Tranchero, M. (2007). Architecture of a Small Low-Cost Satellite. *Proceedings of 10th Euromicro Conference on Digital System Design*, 428–431.

Schilling, M.A (2000). Towards a general modular systems theory and its application to inter-firm product modularity. Academy of Management Review, Vol 25: 312–334.

Baldwin, C. Y. & Clark, K. B (2000). Design rules, Volume 1: The power of modularity, Cambridge, MA: MIT Press.

Carliss Young Baldwin, Kim B Clark (2000). Design Rules: The power of modularity. *MIT Press*, 63-64.

Ron Sanchez, Joseph T. Mahoney (1996). Modularity, Flexibility, and Knowledge Management in Product and Organization Design. *Strategic Management Journal*, Vol 17: 63-76.

Orton, J. & Weick, K (1990). Loosely coupled systems: A reconceptualization. Academy of Management Review, 15: 203–223.

Loose Coupling. [Online]. Available:
http://searchnetworking.techtarget.com/definition/loose-coupling

Lozano, P. & Courtney, D. (2010), On the Development of High Specific Impulse Electric Propulsion Thrusters for Small Satellites, Madeira, Portugal, June 2010

Edelbaum, T.N. (1961), Propulsion Requirements for Controllable Satellites, *ARS Journal*, August 1961

(Oh et al., 2002b) Oh, N.; Shirvani, P.; and McCluskey, E. (2002b). Control-flow checking by software signatures. *IEEE Transactions on Reliability*, 51(1):111-122.

Cuenca-Asensi, S.; Restrepo-Calle, F.; Palomo, F.R.; Guzmán-Miranda, H.; Aguirre, M.; (2011) Compiler-Directed Soft Error Mitigation for Embedded Systems, *IEEE Transactions on Dependable and Secure Computing*, 2011.

(1B127 Datasheet, 2009), 1B127 Datasheet, Neohm Components

(SPA, 2011), Space Plug-and-Play Architecture (SPA) Standard, System Capabilities, AIAA G-133-10-201X, Draft for Public Review, American Institute of Aeronautics and Astronautics 1801 Alexander Bell Drive, Reston, VA 20191.

7

Adaptive Fuzzy Sliding-Mode Attitude Controller Design for Spacecrafts with Thrusters

Fu-Kuang Yeh
Department of Computer Science and Information Engineering,
Chung Chou University of Science and Technology, Changhua, Taiwan,
R.O.C.

1. Introduction

Space exploration has often been used as a proxy competition for geopolitical rivalries. The early era of space exploration was driven by a "space race" among might countries. Nowadays, many advanced nations have entered the space arena and competed with one another for the space technology development as they recognize the importance of space technologies for national strength. In general, physical exploration of space is conducted both by human spaceflights and by robotic spacecrafts. To promote the scientific standard of space vehicle performance, stability, and control, it requires analysis of the six degrees of freedom of the vehicle's flight. One is the translational motion in three dimensional axes; the other is the orientation about the vehicle's center of mass in these axes, known as pitch, roll and yaw. In this paper, the main goal is to propose an adaptive fuzzy sliding-mode controller for spacecrafts with thrusters to follow the predetermined trajectory in outer space by use of employing the spacecraft attitude control. By using fuzzy inference mechanism, the upper bounds of the lumped uncertainty can be estimated, and the adaptive theory with center adaption of membership functions is designed to estimate optimal upper bounds of the lumped uncertainty, respectively. For the above reasons, we know that the attitude control of spacecrafts for the space exploration is essential to successfully develop the space activities [1].

In order to deal with the nonlinear spacecraft attitude dynamics, we employ the quaternion representation [2-4] to model the equation of rotational motion and the time derivative of quaternion, so that the nonlinear attitude control is applicable. Using the quaternion representation, the global control effect can be fulfilled and the singularity problem, which will be faced with the discontinuity by the three- dimensional Euler's representation [5], can be avoided.

To cope with non-ideal factors surrounding the spacecraft under attitude control and to enhance the robustness property of the attitude tracking system, the sliding mode control has been employed by Hu et al. [6], which integrate both the command input shaping and sliding mode output feedback control techniques to investigate the vibration problem for a flexible spacecraft during attitude maneuvering, and by Yeh [7], which proposes two nonlinear attitude controllers for spacecrafts with thrusters to follow the predetermined trajectory to estimate parameters and eliminate disturbances. Servidia and Pena [8] present the attitude stabilization of a spacecraft using thrusters, considering from a practical point of view. Hu [9] proposes a dual-stage control system design scheme for rotational maneuvers and vibration stabilization of a flexible spacecraft in the presence of the parameter uncertainty and external disturbances as well as the control input saturation to actively suppress certain flexible modes. Xia et al. [10] present the adaptive law and the extended state observer for a spacecraft model that is nonlinear in dynamics with the inertia uncertainty and external disturbances to converge to the reference attitude states. Despite the popularity of such control technique, it is however well known that the chattering problem is worthy of more attention for the sake of practical deployment. Taking into consideration of the aforementioned reason, a guide to sliding mode control for practical implementation has been proposed by Young et al. [11]. Eker [12] proposes a second-order sliding mode control for uncertain plants using the equivalent control approach to improve the performance of control systems, in which second-order plant parameters are experimentally determined using input-output measured data.

A spacecraft equipped with thrusters can effectively control its acceleration direction [8, 13], which in turn implies that the maneuverability/controllability of the spacecraft can be greatly enhanced during the stage when the spacecraft is flying in the outer space; whereas Janhunen et al. [14] propose a space propulsion concept of electric solar wind sail using the natural solar wind dynamic pressure for producing spacecraft thrust.

Estimation theory is used to deal with estimating values of parameters based on measured/empirical data that have a disturbance component. A parameter estimation approach called adaptive control has been developed by Slotine [15, 16] to achieve accurate attitude tracking of a rigid spacecraft with large loads of unknown mass. Zou et al. [17] investigate the robust adaptive output feedback controller based on Chebyshev neural networks (CNN) for an uncertain spacecraft to counteract CNN approximation errors and external disturbances. Huang et al. [18] propose a robust adaptive PID-type controller incorporating a fuzzy logic system and a sliding-mode control action for compensating parameter uncertainties and the robust tracking performance. An adaptive fuzzy theory is generally employed to approximate unstructured uncertainties and dynamic disturbances, such as Tong et al. [19] discuss an adaptive fuzzy output feedback control approach for nonlinear systems to estimate unmeasured states; Islam and Liu [20] propose a robust adaptive fuzzy control system for the trajectory tracking control problem of robotic systems to approximate the certainty equivalent-based optimal controller and to cope with uncertainties.

In this paper, we investigate the adaptive fuzzy sliding-mode control for spacecrafts with thrusters, employing the fuzzy sliding-mode controller to estimate upper bounds of the lumped uncertainty, and the adaptive fuzzy sliding-mode controller with center adaption of membership functions to estimate optimal bounds of the lumped uncertainty, respectively. This paper is organized as follows. In Section 2, preliminaries for deriving three-degree-of-freedom attitude models of a spacecraft equipped with thrusters. In Section 3, we respectively propose the sliding-mode, the fuzzy sliding-mode and the adaptive fuzzy sliding-mode attitude controllers aiming for tracking the predetermined trajectory in outer space. For tracking realization, three simulation results incorporating the so-called quaternion-based attitude control are developed in Section 4. To demonstrate the superior property of the proposed attitude controllers, three numerical simulations are provided in that Section. Finally, conclusions are drawn in Section 5.

2. Equations of rotational motion for spacecrafts

Assume the spacecraft is a rigid body; therefore, the Euler's equation of rotational motion is adopted with the following general form as

$$J\dot{\Omega} = -\dot{J}\Omega - \Omega \times (J\Omega) + T_b + D ,\qquad (1)$$

where all the variables are defined in and please referring to the Nomenclature.

Assume that the movable nozzle is located at the center of the spacecraft tail, and the distance between the movable nozzle center and the spacecraft's center of gravity is ℓ. Furthermore, we also assume that the spacecraft is equipped with a number of thrusters on the surface near the center of gravity that will produce a pure rolling moment whose direction is aligned with the vehicle axis, X_b, referring to Fig. 1. Thus, the vector L_b, defined as the relative displacement from the spacecraft's center of gravity to the center of the movable nozzle, satisfies $|L_b| = \ell$. Note that J is the moment of inertia matrix for a spacecraft with respect to the body coordinate frame, and hence is a 3×3 symmetric matrix. After referring to Fig. 1 and Fig. 2, the torque exerted on the spacecraft can be expressed in the body coordinate frame as

$$T_b = L_b \times F_{Tb} + M_b = \ell n \begin{bmatrix} m_{bx} / \ell n \\ \sin d_p \\ -\cos d_p \sin d_y \end{bmatrix} ,\qquad (2)$$

where n is the magnitude of the movable nozzle thrust, d_p and d_y are the pitch angle and yaw angle, respectively, of the movable nozzle, $M_b = \begin{bmatrix} m_{bx} & 0 & 0 \end{bmatrix}^T$ is the aforementioned variable moment in the axial direction of the spacecraft, and F_{Tb} is the force produced by the movable nozzle in the body coordinate frame.

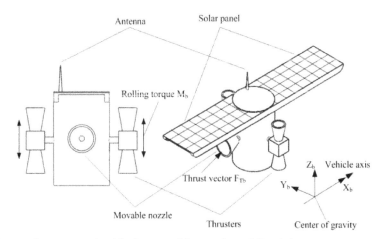

Fig. 1. Scheme of a spacecraft with the movable nozzle and fixed thrusters.

Let the rotation matrix B_b denote the transformation from the body coordinate frame to the inertial coordinate frame. Thus, the force exerted on the spacecraft observed in the inertial coordinate system is as follows:

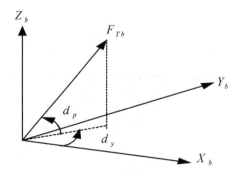

Fig. 2. Two angles of the movable nozzle in the body coordinate frame.

$$F_M = B_b F_{Mb} .\qquad(3)$$

From Eqs. (1) and (2), the rotational motion model of a spacecraft can then be derived as

$$J\dot{\Omega} = -\dot{J}\Omega - \Omega \times (J\Omega) + \ell n \begin{bmatrix} m_{bx} / \ell n \\ \sin d_p \\ -\cos d_p \sin d_y \end{bmatrix} + D ,\qquad(4)$$

where $D = \begin{bmatrix} d_1 & d_2 & d_3 \end{bmatrix}^T$ is a disturbance vector in the body coordinate frame.

Generally speaking, the attitude of a rigid body may be described in various ways, and "quaternion" is one of the means. According to Euler's rotation theory [21], there exist a unit vector U and an angle ϕ such that U is perpendicular to the rotation plane with respect to a rotation angle ϕ. Thus, for any quaternion, it can be defined as four parameters $Q = \begin{bmatrix} q_1 & q_2 & q_3 & q_4 \end{bmatrix}^T = \begin{bmatrix} \bar{Q}^T & q_4 \end{bmatrix}^T$ involving U and ϕ, i.e.,

$$\bar{Q} = \begin{bmatrix} q_1 \\ q_2 \\ q_3 \end{bmatrix} = U \sin(\phi/2),\qquad(5)$$

$$q_4 = \cos(\phi/2).$$

In theory, it can be verified that the time derivative of a quaternion is a function of the corresponding angular velocity and the quaternion itself [2-4], i.e.,

$$\dot{\bar{Q}} = \frac{1}{2}\langle \bar{Q} \times \rangle \Omega + \frac{1}{2} q_4 \Omega,$$

$$\dot{q}_4 = -\frac{1}{2}\Omega^T \bar{Q}\qquad(6)$$

From Eqs.(1) and (6), the dynamic model of a spacecraft, treated as a rigid body, can be derived by differentiation of the angular velocity and the associated error quaternion as a function of the corresponding angular velocity and its error as well as the error quaternion itself, i.e.,

$$\dot{\bar{Q}}_e = \frac{1}{2}\langle \bar{Q}_e \times \rangle \Omega_e + \frac{1}{2} q_{e4} \Omega_e$$

$$\dot{q}_{e4} = -\frac{1}{2}\Omega_e^T \bar{Q}_e \qquad (7)$$

$$J\dot{\Omega} = -\dot{J}\Omega - \Omega \times (J\Omega) + T_b + D$$

where $\Omega_e = \Omega - \Omega_d$ is the error between the angular velocity at the present attitude and the desired attitude, and T_b is the torque exerted on the spacecraft due to the movable nozzle and the rolling moment. Whereas the error quaternion $Q_e = \begin{bmatrix} q_{e1} & q_{e2} & q_{e3} & q_{e4} \end{bmatrix}^T = \begin{bmatrix} \bar{Q}_e^T & q_{e4} \end{bmatrix}^T$, $\bar{Q}_e^T = \begin{bmatrix} q_{e1} & q_{e2} & q_{e3} \end{bmatrix}$, is defined as the required rotation from the initial quaternion $Q = \begin{bmatrix} q_1 & q_2 & q_3 & q_4 \end{bmatrix}^T$ to the desired quaternion $Q_d = \begin{bmatrix} q_{d1} & q_{d2} & q_{d3} & q_{d4} \end{bmatrix}^T$, and can be derived in matrix form [2-4] as

$$Q_e = \begin{bmatrix} q_{e1} \\ q_{e2} \\ q_{e3} \\ q_{e4} \end{bmatrix} = \begin{bmatrix} q_{d4} & q_{d3} & -q_{d2} & -q_{d1} \\ -q_{d3} & q_{d4} & q_{d1} & -q_{d2} \\ q_{d2} & -q_{d1} & q_{d4} & -q_{d3} \\ q_{d1} & q_{d2} & q_{d3} & q_{d4} \end{bmatrix} \begin{bmatrix} q_1 \\ q_2 \\ q_3 \\ q_4 \end{bmatrix}, \qquad (8)$$

and $\langle \bar{Q}_e \times \rangle = \begin{bmatrix} 0 & -q_{e3} & q_{e2} \\ q_{e3} & 0 & -q_{e1} \\ -q_{e2} & q_{e1} & 0 \end{bmatrix}$ is a skew-symmetric matrix.

3. Nonlinear attitude controller design

3.1 Sliding-mode attitude controller design

Considering the presence of model uncertainties, parameter variations, and disturbances, we recognized that the sliding mode control is an effectively robust controller for various applications; in this paper, we consider a sliding-mode control to eliminate all variation influences of a spacecraft during the whole flying course for the practical controller design. We first design a sliding-mode attitude controller, which can compensate for the adverse effect owing to spacecraft variations.

The principal procedure to verify the stability and robustness of the sliding-mode attitude tracking problem consists of the sliding and reaching conditions, and that will be given in detail as follows.

Step 1: Choose the sliding surface such that the sliding condition will be satisfied and hence the origin of the error dynamic is exponentially stable.

From the sliding mode theory, once the reaching condition is satisfied, the system is eventually forced to stay on the sliding surface, i.e.,

$$S = P\overline{Q}_e + \Omega_e = 0, \tag{9}$$

where P is a positive definite diagonal matrix. The system dynamics are then constrained by the following differential equations, referring to Eq. (7), as

$$\dot{\overline{Q}}_e = -\frac{1}{2}\left\langle \overline{Q}_e \times \right\rangle P\overline{Q}_e - \frac{1}{2}q_{e4}P\overline{Q}_e,$$
$$\dot{q}_{e4} = \frac{1}{2}\overline{Q}_e^T P\overline{Q}_e. \tag{10}$$

Define another Lyapunov function $V_e(\overline{Q}_e) = \frac{1}{2}\overline{Q}_e^T \overline{Q}_e$, then

$$\dot{V}_e(\overline{Q}_e) = \overline{Q}_e^T \dot{\overline{Q}}_e$$
$$= -\frac{1}{2}\overline{Q}_e^T \left(\left\langle \overline{Q}_e \times \right\rangle P\overline{Q}_e + q_{e4}P\overline{Q}_e\right), \tag{11}$$
$$= -\frac{1}{2}q_{e4}\overline{Q}_e^T P\overline{Q}_e$$

where $\overline{Q}_e^T \left\langle \overline{Q}_e \times \right\rangle = 0$.

Recalling that the quaternion definition q_{e4} in general has two possible values different only in sign and the sign can be arbitrarily chosen to meet the design convenience. For the sake of design and analysis, q_{e4} is selected as $q_{e4}\big|_{t=0} = c > 0$, and because $\dot{q}_{e4} = \frac{1}{2}\overline{Q}_e^T P\overline{Q}_e \geq 0$, hence we can conclude that q_{e4} is a positive and growing variable, i.e., $q_{e4}(t) \geq c$. By quaternion definition, a quaternion always satisfies the so-called unit-norm property. That is, $q_1^2 + q_2^2 + q_3^2 + q_4^2 = 1$, which implies that $1 \geq q_{e4}$, and hence $c \leq q_{e4}(t) \leq 1$, $\forall t \geq 0$, so that the following relation will hold as

$$-\frac{c}{2}\overline{Q}_e^T P\overline{Q}_e \geq \dot{V}_e(\overline{Q}_e) \geq -\frac{1}{2}\overline{Q}_e^T P\overline{Q}_e. \tag{12}$$

By Lyapunov stability theory, it can be proved that $\overline{Q}_e$ will be driven to zero when the system is constrained on the sliding mode dynamics and so will the error angular velocity Ω_e. For the above reason, the system origin $\left(\overline{Q}_e, \Omega_e\right) = \left(0_{3\times1}, 0_{3\times1}\right)$ of the ideal system can be verified to be exponentially stable.

Step 2: Design the controller such that the reaching condition is satisfied.

Let us define the sliding surface as above equation (9) shown as $S = P\overline{Q}_e + \Omega_e$, where $P = diag[p_1 \quad p_2 \quad p_3]$ is a 3×3 positive definite diagonal matrix. Here, we make an assumption that J is a symmetric and positive definite matrix, and let the Lyapunov function candidate be set as

$$V_s = \frac{1}{2} S^T J S, \tag{13}$$

where $V_s = 0$ only when $S = 0$. Then, the time derivative of V_s can be derived as

$$\dot{V}_s = S^T J \dot{S} + \frac{1}{2} S^T \dot{J} S. \tag{14}$$

Substituting Eq. (7) and the time derivative of Eq. (9) into the above equation (14), we have

$$
\begin{aligned}
\dot{V}_s &= S^T J \dot{S} + \frac{1}{2} S^T \dot{J} S \\
&= S^T \left[J\dot{\Omega} - J\dot{\Omega}_d + J P \dot{\bar{Q}}_e + \frac{1}{2} \dot{J} S \right] \\
&= S^T \left[-\dot{J}\Omega - \Omega \times (J\Omega) + T_b + D - J\dot{\Omega}_d + JP \left(\frac{1}{2} \langle \bar{Q}_e \times \rangle \Omega_e + \frac{1}{2} q_{e4} \Omega_e \right) + \frac{1}{2} \dot{J} S \right]
\end{aligned}
\tag{15}
$$

Let the control torque input T_b be proposed as

$$T_b = -K_s S + \dot{J}_0 \Omega - \frac{1}{2} \dot{J}_0 S - J_0 P \left(\frac{1}{2} \langle \bar{Q}_e \times \rangle \Omega_e + \frac{1}{2} q_{e4} \Omega_e \right) + \Omega \times (J_0 \Omega) + J_0 \dot{\Omega}_d + \Lambda_s, \tag{16}$$

where $K_s = diag[k_{s1} \ k_{s2} \ k_{s3}]$ is a positive definite diagonal matrix, and $\Lambda_s =$ $[\lambda_{s1} \ \lambda_{s2} \ \lambda_{s3}]^T$, $\lambda_{si} = -c_{si}(Q, \Omega, Q_d, \dot{Q}_d, \ddot{Q}_d) \cdot sgn(s_i)$, with $sgn(s_i) = \begin{cases} 1 & s_i > 0 \\ 0 & s_i = 0, \ i = 1,2,3, \\ -1 & s_i < 0 \end{cases}$

and $S = [s_1 \ s_2 \ s_3]^T$ is a sliding surface. Let the external disturbance D and the induced 2-norm of $\Delta \dot{J}$ and ΔJ are all bounded, where $J = J_0 + \Delta J$, $\dot{J} = \dot{J}_0 + \Delta \dot{J}$. If the inequality condition shown below can be guaranteed

$$c_{si}(Q, \Omega, Q_d, \dot{Q}_d, \ddot{Q}_d) > \delta_i^{max}(Q, \Omega, Q_d, \dot{Q}_d, \ddot{Q}_d) \geq |\delta_i|, \ i = 1,2,3, \tag{17}$$

where

$$
\begin{aligned}
\bar{\delta} &= [\delta_1 \ \delta_2 \ \delta_3]^T \\
&= -\Delta \dot{J}\Omega - \Omega \times (\Delta J \Omega) + D - \Delta J \dot{\Omega}_d + \Delta J P \left(\frac{1}{2} \langle \bar{Q}_e \times \rangle \Omega_e + \frac{1}{2} q_{e4} \Omega_e \right) + \frac{1}{2} \Delta J S'
\end{aligned}
\tag{18}
$$

where bounding functions δ_i, $i = 1,2,3$ are obviously functions of Q, Ω, Q_d, $\dot{Q}_d$ and $\ddot{Q}_d$, then the exponential stability and robustness of the proposed controller for attitude tracking can be achieved.

It is evident that Eq. (15) becomes

$$\dot{V}_s = -S^T K_s S - \sum_{i=1}^{3} |s_i| \left[c_{si} - \delta_i \, \mathrm{sgn}(s_i) \right]$$

$$\leq -S^T K_s S - \sum_{i=1}^{3} |s_i| \left[c_{si} - \delta_i^{\max} \right] \quad , \qquad (19)$$

$$\leq -\sigma_{\min}(K_s) |S|^2 < 0$$

for $S \neq 0$, where $\sigma_{\min}(K_s)$ is the minimum eigenvalue of K_s, where K_s is a positive definite diagonal matrix as above mentioned. Therefore, the reaching and sliding conditions of the sliding mode $S = 0$ are guaranteed. As a result, the exponential stability and robustness of the sliding mode attitude controller can be achieved.

Remark 1: However, due to the existence of non-ideality in the practical implementation of the sign function $\mathrm{sgn}(s_i)$, the control law T_b in (16) always suffers from the chattering problem. To alleviate such undesirable phenomenon, the sign function can be simply replaced by the saturation function. The system is now no longer forced to stay on the sliding surface but is constrained within the boundary layer $|s_i| \leq \varepsilon$, where ε is a small positive value. The cost of such substitution is a reduction in the accuracy of the desired performance.

To alleviate the chattering phenomenon, the saturation function may be employed to the control input of the sliding mode attitude control system. Consequently, the term $\Lambda_s = \begin{bmatrix} \lambda_{s1} & \lambda_{s2} & \lambda_{s3} \end{bmatrix}^T$ in Eq. (16) can be replaced by

$$\lambda_{si} = -c_{si}(Q, \Omega, Q_d, \dot{Q}_d, \ddot{Q}_d) \cdot Sat(s_i, \varepsilon), \qquad (20)$$

where $Sat(s_i, \varepsilon) = \begin{cases} 1 & s_i > \varepsilon \\ \dfrac{s_i}{\varepsilon} & |s_i| \leq \varepsilon \\ -1 & s_i < -\varepsilon \end{cases}$, $i = 1, 2, 3$.

3.2 Fuzzy sliding-mode attitude controller design

Upper bounds of the lumped uncertainty, which includes the external disturbances and internal perturbations, for the sliding-mode control need to be decided before the controller is using. In general, upper bounds of the lumped uncertainty are difficult to be obtained in advanced by computing, but always by the method of try and error. In this section, a fuzzy sliding-mode controller is proposed, in which a fuzzy inference mechanism is used to estimate upper bounds of the lumped uncertainty. We recognized that the prior expert knowledge of the fuzzy inference mechanism, which can be used to estimate upper bounds of the lumped uncertainty, is available effectively.

Values of c_{s1}, c_{s2}, and c_{s3} in Eq. (17) can be estimated by the fuzzy inference mechanism. The control block diagram of the fuzzy sliding-mode controller is depicted as in Fig. 3. Based on fuzzy set theory, associated fuzzy sets involved in fuzzy control rules are defined and listed as follows:

PB: positive big; PM: positive medium; PS: positive small; ZE: zero;
NS: negative small; NM: negative medium; NB: negative big.

Here universes of discourse for inputs s_i, $\dot{s}_i$, $i = 1,2,3$ and outputs $\hat{c}_{si}$, $i = 1,2,3$ are assigned
to be $\begin{bmatrix} -50, & 50 \end{bmatrix}$, $\begin{bmatrix} -6000, & 6000 \end{bmatrix}$, and $\begin{bmatrix} -50, & 50 \end{bmatrix}$, respectively. Membership functions for the
fuzzy sets corresponding to switching surfaces s_i, $i = 1,2,3$, its derivative $\dot{s}_i$, $i = 1,2,3$ and
upper bounds of the lumped uncertainty $\hat{c}_{si}$, $i = 1,2,3$, are defined in Fig. 4.

Because the seven fuzzy subsets NB, NM, NS, ZE, PS, PM, and PB are used to divide every
element of sliding surfaces s_i, $i = 1,2,3$ and its derivative $\dot{s}_i$, $i = 1,2,3$, respectively, the
fuzzy inference mechanism contains 49 rules. The resulting fuzzy inference rules are given
as the following Table 1.

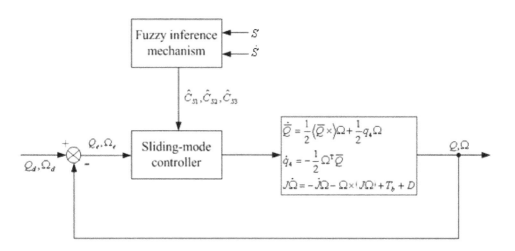

Fig. 3. Control block diagram of the fuzzy sliding-mode controller.

$\dot{s}_i$ \ s_i	NB	NM	NS	ZE	PS	PM	PB
PB	ZE	PS	PS	PM	PM	PB	PB
PM	NS	ZE	PS	PS	PM	PM	PB
PS	NS	NS	ZE	PS	PS	PM	PM
ZE	NM	NS	NS	ZE	PS	PS	PM
NS	NM	NM	NS	NS	ZE	PS	PS
NM	NB	NM	NM	NS	NS	ZE	PS
NB	NB	NB	NM	NM	NS	NS	ZE

Table 1. Rule base with 49 rules.

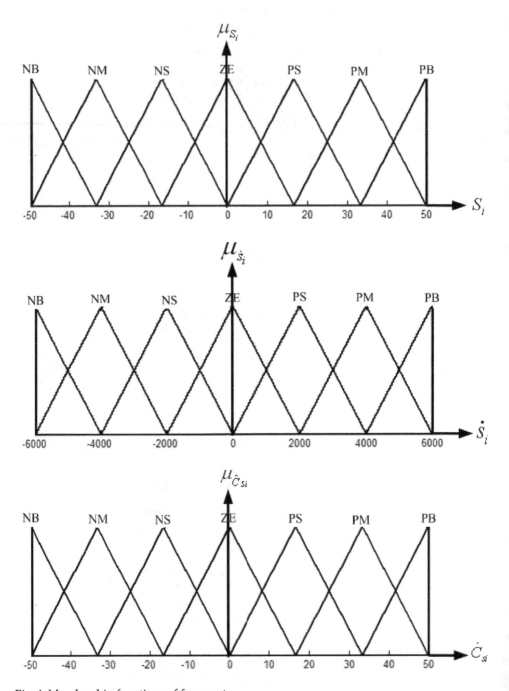

Fig. 4. Membership functions of fuzzy sets.

The fuzzy outputs $\hat{c}_{si}$, $i = 1,2,3$, can be calculated by the centre of area defuzzification as

$$\hat{c}_{si} = \left| \frac{\sum_{j=1}^{49} \omega_{ij} c_{ij}}{\sum_{j=1}^{49} \omega_{ij}} \right| = \frac{\begin{bmatrix} c_{i1} & \cdots & c_{i49} \end{bmatrix} \begin{bmatrix} \omega_{i1} \\ \vdots \\ \omega_{i49} \end{bmatrix}}{\sum_{j=1}^{49} \omega_{ij}} = \left| C_i^T W_i \right|, \quad i = 1,2,3, \tag{21}$$

where $C_i^T = \begin{bmatrix} c_{i1} & \cdots & c_{i49} \end{bmatrix}$ is an adjustable parameter vector, c_{i1} through c_{i49} are the centre

of the membership functions of $\hat{c}_{si}$, $W_i^T = \dfrac{\begin{bmatrix} \omega_{i1} & \cdots & \omega_{i49} \end{bmatrix}}{\sum_{j=1}^{49} \omega_{ij}}$ is a firing strength vector. Here,

the absolute value of $C_i^T W_i$ in Eq. (21) is used to satisfy the requirement of estimating upper bounds, so that upper bounds of the lumped uncertainty are greater than or equal to zero, that is, $\hat{c}_{si} \geq 0$, $i = 1,2,3$.

The following Lemma is introduced to enable the fuzzy sliding-mode controller such that the reaching condition of the switching surface is satisfied. By Lyapunov stability theory, the fuzzy sliding-mode attitude controller of the spacecraft can be proved as an exponentially stable system. The sufficient conditions for successful stability effect are stated in the following lemma.

Lemma 1 Fuzzy sliding-mode attitude controller: Let the dynamic model of a spacecraft corresponding the angular velocity and the quaternion be given by Eqs. (1) and (6), and if the control torque input T_b is proposed as

$$T_b = -K_s S + \dot{J}_0 \Omega - \frac{1}{2} \dot{J}_0 S - J_0 P \left(\frac{1}{2} \langle \bar{Q}_e \times \rangle \Omega_e + \frac{1}{2} q_{e4} \Omega_e \right) + \Omega \times (J_0 \Omega) + J_0 \dot{\Omega}_d + \hat{\Lambda}_s, \tag{22}$$

where $K_s = diag \begin{bmatrix} k_{s1} & k_{s2} & k_{s3} \end{bmatrix}$ is a positive definite diagonal matrix, and $\hat{\Lambda}_s =$

$\begin{bmatrix} \hat{\lambda}_{s1} & \hat{\lambda}_{s2} & \hat{\lambda}_{s3} \end{bmatrix}^T$, $\hat{\lambda}_{si} = -\hat{c}_{si}(C_i, W_i) \cdot \mathrm{sgn}(s_i)$, with $\mathrm{sgn}(s_i) = \begin{cases} 1 & s_i > 0 \\ 0 & s_i = 0, \\ -1 & s_i < 0 \end{cases}$ $i = 1,2,3$, and

$S = \begin{bmatrix} s_1 & s_2 & s_3 \end{bmatrix}^T$ is a sliding surface that are defined as Eq. (9). Let the external disturbance D and the induced 2-norm of $\Delta \dot{J}$ and ΔJ are all bounded, where $J = J_0 + \Delta J$, $\dot{J} = \dot{J}_0 + \Delta \dot{J}$. And values of $\hat{c}_{s1}$, $\hat{c}_{s2}$, and $\hat{c}_{s3}$ are chosen to be positive big enough and are computed by Eq. (21). Then the exponential stability and robustness of the fuzzy sliding-mode attitude control system can be achieved.

Proof: See the Appendix 1. From the proof in the Appendix 1, the exponential stability and robustness of the fuzzy sliding-mode attitude controller can be guaranteed, so that the spacecraft attitude tracking system can be achieved completely.

3.3 Adaptive fuzzy sliding-mode attitude controller design

Since the sliding-mode attitude controller for a spacecraft requires estimating upper bounds (constant values) of the lumped uncertainty, so that the lumped uncertainty during the entire flying course can be always eliminated. If upper bounds are not chosen appropriately, the high-gain problem will be suffered. That is, the control input torque is over-large definitely and the implementation cost is increased as well. Optimal upper bounds c_{si}, $i = 1,2,3$, cannot be obtained exactly by pure sliding mode control owing to the unknown of uncertainties. Therefore, the adaptive fuzzy control algorithm based on the principle of sliding mode control is developed to estimate optimal upper bounds of the lumped uncertainty and to achieve the minimum control torque.

Assume there exists estimated upper bounds $\hat{c}_{si}$, $i = 1,2,3$ which can achieve minimum control torque and satisfy the sliding-mode condition of the switching surface, errors between estimated and optimal upper bounds are shown as

$$\hat{c}_{si} - c_{si} = \tilde{c}_{si} \, , \, i = 1,2,3 \, , \tag{23}$$

where error upper bounds $\tilde{c}_{si}$, $i = 1,2,3$ are small values. Here, estimated upper bounds can be computed as Eq. (21).

Assume there exists estimated upper-bound vector $\hat{C}_s = \begin{bmatrix} \hat{c}_{s1} & \hat{c}_{s2} & \hat{c}_{s3} \end{bmatrix}^T$, which can achieve minimum control torque and satisfy the sliding-mode condition of the switching surface, the error vector between the estimated and optimal upper-bound vectors are shown as

$$\hat{C}_s - C_s = \tilde{C}_s \, , \tag{24}$$

The following theorem is introduced to enable the adaptive fuzzy sliding-mode controller such that optimal upper bounds can be obtained. By Lyapunov stability theory, the adaptive fuzzy sliding-mode attitude controller of a spacecraft can be proved as an exponentially stable system. The sufficient conditions for the successful stability effect are stated in the following theorem.

Theorem 1 Adaptive fuzzy sliding-mode attitude controller: Let the dynamic model of a spacecraft corresponding the angular velocity and the quaternion be given by Eqs. (1) and (6), and if the control torque input T_b is proposed as

$$T_b = -K_s S + \dot{J}_0 \Omega - \frac{1}{2} \dot{J}_0 S - J_0 P \left(\frac{1}{2} \langle \bar{Q}_e \times \rangle \Omega_e + \frac{1}{2} q_{e4} \Omega_e \right) + \Omega \times (J_0 \Omega) + J_0 \dot{\Omega}_d + \bar{\Lambda}_s \, , \tag{25}$$

where $K_s = diag\begin{bmatrix} k_{s1} & k_{s2} & k_{s3} \end{bmatrix}$ is a positive definite diagonal matrix, $\bar{\Lambda}_s = \begin{bmatrix} \bar{\lambda}_{s1} & \bar{\lambda}_{s2} & \bar{\lambda}_{s3} \end{bmatrix}^T$, $\bar{\lambda}_{si} = -c_{si}(\int s_i dt) \cdot \text{sgn}(s_i)$, and $S = \begin{bmatrix} s_1 & s_2 & s_3 \end{bmatrix}^T$ is a sliding surface. Let the external disturbance D and the induced 2-norm of $\Delta \dot{J}$ and ΔJ are all bounded, where $J = J_0 + \Delta J$, $\dot{J} = \dot{J}_0 + \Delta \dot{J}$, whereas $\hat{c}_{si}$, $i = 1,2,3$ is estimated upper bounds that are assumed to be optimal, and the bounding function is shown as Eq. (18).

For adapting upper bounds of the lumped uncertainty, the adaptation laws can be let as

$$\dot{c}_{si} = -\Gamma_i |s_i|, \; i = 1,2,3,$$ (26)

where the optimal upper-bound vector is defined as $C_s = \begin{bmatrix} c_{s1} & c_{s2} & c_{s3} \end{bmatrix}^T$ and $\Gamma = diag(\Gamma_1, \Gamma_2, \Gamma_3)$ is a positive definite diagonal matrix. Then the exponential stability and robustness of the adaptively fuzzy sliding-mode attitude control system can be achieved.

Proof: To achieve the exponential stability and convergence of the spacecraft attitude tracking system, let us define the sliding surface as above equation (9) shown as $S = P\bar{Q}_e + \Omega_e$, where $P = diag[p_1 \quad p_2 \quad p_3]$ is a 3×3 positive definite diagonal matrix. Here, we also make an assumption that J is a symmetric and positive definite matrix, now we choose the Lyapunov-like function as

$$V = \frac{1}{2} S^T J S + \frac{1}{2} \tilde{C}_s^T \Gamma^{-1} \tilde{C}_s.$$ (27)

where $S = \begin{bmatrix} s_1 & s_2 & s_3 \end{bmatrix}^T$ is the sliding surface, $\tilde{C}_s = \begin{bmatrix} \tilde{c}_{s1} & \tilde{c}_{s2} & \tilde{c}_{s3} \end{bmatrix}^T$ is the error upper-bound vector between the optimal upper-bound vector C_s and the estimated upper-bound vector $\hat{C}_s$, which elements are computed by Eq. (21), and $\Gamma^{-1} = diag(\Gamma_1^{-1}, \Gamma_2^{-1}, \Gamma_3^{-1})$ is a positive definite diagonal matrix. Then, the time derivative of V can be derived as

$$\dot{V} = S^T J \dot{S} + \frac{1}{2} S^T \dot{J} S + \tilde{C}_s^T \Gamma^{-1} \dot{\tilde{C}}_s.$$ (28)

Substituting Eqs. (1) and (6) and the time derivative of Eq. (9) into the above equation (28), we have

$$\dot{V} = S^T \left[J\dot{\Omega} - J\dot{\Omega}_d + JP\dot{\bar{Q}}_e + \frac{1}{2}\dot{J}S \right] + \tilde{C}_s^T \Gamma^{-1} \dot{\tilde{C}}_s$$
$$= S^T \left[-J\Omega - \Omega \times (J\Omega) + T_b + D - J\dot{\Omega}_d + JP\left(\frac{1}{2}\langle \bar{Q}_e \times \rangle \Omega_e + \frac{1}{2}q_{e4}\Omega_e \right) + \frac{1}{2}\dot{J}S \right] + \tilde{C}_s^T \Gamma^{-1} \dot{\tilde{C}}_s$$ (29)

where $\dot{\tilde{C}}_s = \dot{\tilde{C}}_s$, $\tilde{C}_s = C_s - \hat{C}_s$, here assume $\hat{C}_s$ is a slow-varying vector in which those elements are computed by Eq. (21), so that the first-time derivative of $\hat{C}_s$ can be neglected.

Let the control torque input T_b be proposed as Eq. (25), in which $\bar{\Lambda}_s = \begin{bmatrix} \bar{\lambda}_{s1} & \bar{\lambda}_{s2} & \bar{\lambda}_{s3} \end{bmatrix}^T$,

$\bar{\lambda}_{si} = -c_{si} \left(\int s_i dt \right) \cdot \mathrm{sgn}(s_i)$, with $\mathrm{sgn}(s_i) = \begin{cases} 1 & s_i > 0 \\ 0 & s_i = 0, \; i = 1,2,3, \text{ where the values of } c_{s1}, c_{s2}, \\ -1 & s_i < 0 \end{cases}$

and c_{s3} are optimal upper bounds. For adapting upper bounds of the lumped uncertainty, adaptation laws can be given as Eq. (26), and by the verification of fuzzy sliding-mode control in the Lemma 1, the inequality condition as Eq. (A.2) can be obtained. Bounding functions δ_i,

$i = 1,2,3$ can also be defined as Eq. (18) and estimated upper bounds $\hat{c}_{si}$, $i = 1,2,3$ can be computed by Eq. (21).

It is evident that Eq. (29) becomes

$$\dot{V} = -S^T K_s S - \sum_{i=1}^{3} |s_i| \left[c_{si} - \delta_i \, \text{sgn}(s_i) \right] - \sum_{i=1}^{3} |s_i| \left[\hat{c}_{si} - c_{si} \right]$$

$$\leq -S^T K_s S - \sum_{i=1}^{3} |s_i| \left[\hat{c}_{si} - \delta_i^{\max} \right] \tag{30}$$

$$\leq -\sigma_{\min}(K_s) |S|^2 < 0$$

for $S \neq 0$, where $\sigma_{\min}(K_s)$ is the minimum eigenvalue of K_s, where K_s is a positive definite diagonal matrix as above mentioned. Therefore, the reaching and sliding conditions of the sliding mode $S = 0$ are guaranteed. As a result, the exponential stability and robustness of the adaptive fuzzy sliding mode attitude controller can be achieved.

Therefore, the global stability of the attitude tracking system is guaranteed by the proposed adaptive fuzzy sliding-mode controller. In the following, let us in detail verify the convergence of the states $\overline{Q}_e$ and Ω_e and of the parameters $\tilde{C}_s$.

First, the function V in Eq. (27) is a Lyapunov-like function, in our case simply a positive continuous function of time. Expression (30) shows that output errors converge to the sliding surface [16], so that $S = 0_{3\times1}$. And because that a quaternion always satisfies the so-called unit-norm property, we have $\overline{Q}_e$, Ω_e are bounded. Let us now detail the proof itself. Since $\dot{V}$ is negative or zero and V is lower bounded, and because S, $\overline{Q}_e$, Ω_e are bounded.

So, in turn, from Eq. (30), we have $\ddot{V} = -2S^T K_s \dot{S} - \sum_{i=1}^{3} \left[\hat{c}_{si} - \delta_i \, \text{sgn}(s_i) \right] \dot{s}_i \, \text{sgn}(s_i)$ exists and is

bounded as a result that $\dot{S}$ is bounded, and referring to Eq. (5), we have that $\overline{Q}_e$, $\dot{\overline{Q}}_e$, and $\ddot{\overline{Q}}_e$ are all bounded, referring to reference [22], we have that $\dot{\Omega}_e$ is a function of $\ddot{\overline{Q}}_e$, such that $\dot{\Omega}_e$ is bounded. Therefore we can concluded that $\dot{V}$ is uniformly continuous on $t \in [0, \infty)$. Consequently by Barbalat's lemma, $\dot{V}$ tends to zero as $t \to \infty$. This implies from Eq. (30) that $S \to 0$ as $t \to \infty$.

Now, given that S converges to zero only exponentially, the actual differential equation that governs $\overline{Q}_e$ by some exponentially decaying term can be given as

$$S = P\overline{Q}_e + \Omega_e = \varepsilon(t), \tag{31}$$

where $\varepsilon(t) = e^{-k_a t} S_0$ is an exponentially decaying function of time t, whereas $S_0 = P\overline{Q}_e + \Omega_e \big|_{t=0}$.

From Eqs. (7) and (31), the dynamics of $\overline{Q}_e$ are

$$\dot{\bar{Q}}_e = \frac{1}{2}\langle\bar{Q}_e\rangle\Omega_e + \frac{1}{2}q_{e4}\Omega_e$$

$$= \frac{1}{2}\left[\langle\bar{Q}_e\rangle\left(-P\bar{Q}_e + \varepsilon\right) + q_{e4}\left(-P\bar{Q}_e + \varepsilon\right)\right]$$

$$= -\frac{1}{2}\left(\langle\bar{Q}_e\times\rangle P\bar{Q}_e + q_{e4}P\bar{Q}_e\right) + \frac{1}{2}\left(\langle\bar{Q}_e\times\rangle + q_{e4}I_{3\times3}\right)\varepsilon \tag{32}$$

$$= f_1(\bar{Q}_e) + f_2(\bar{Q}_e)$$

where $f_1(\bar{Q}_e) = -\frac{1}{2}\left(\langle\bar{Q}_e\times\rangle P\bar{Q}_e + q_{e4}P\bar{Q}_e\right)$ and $f_2(\bar{Q}_e) = \frac{1}{2}\left(\langle\bar{Q}_e\times\rangle + q_{e4}I_{3\times3}\right)\varepsilon$. As indicated by

Eq. (7), $\dot{\bar{Q}}_e = f_1(\bar{Q}_e)$ is an exponentially stable system, implying that the stability and convergence of Eq. (32) are governed by $f_2(\bar{Q}_e)$. If $f_2(\bar{Q}_e) \to 0$ exponentially as $t \to 0$, then the system is reduced to system Eq. (7), which is an exponentially stable system. From Eq. (31), ε is an exponentially decaying function, and because $\left|\frac{1}{2}\left(\langle\bar{Q}_e\times\rangle + q_{e4}I_{3\times3}\right)\right|$ is bounded, $\left|\frac{1}{2}\left(\langle\bar{Q}_e\times\rangle + q_{e4}I_{3\times3}\right)\right| \le \sigma$ for some constant $\sigma > 0$, $\left|f_2(\bar{Q}_e)\right| \le \sigma\varepsilon$ also becomes an exponentially decaying function, which now truly ensures the stability and convergence of Eq. (32).

To verify the convergence of the error parameter vector $\tilde{C}_s$, by definition in Lemma 1, due to estimated values $\hat{c}_{si} = \left|C_i^T W_i\right|$, $i = 1,2,3$ being positive values, the estimated parameter vector $\hat{C}_s$ is a slow-varying vector, referring to Fig. 4. From Eq. (30), S is an exponentially decaying function vector of time t, we can obtain that the optimal vector C_s is a bounded vector by integrating Eq. (26). Finally, the bounded convergence of the error parameter vector $\tilde{C}_s = C_s - \hat{C}_s$ in Eq. (24) can be confirmed.

Remark 2: Similar to alleviate the chattering phenomenon, the saturation function may also be employed to the control input of the adaptive fuzzy sliding-mode control system. Consequently, the control torque T_b in Eq. (25) can be re-expressed as

$$T_b = -K_s S + \dot{J}_0\Omega - \frac{1}{2}\dot{J}_0 S - J_0 P\left(\frac{1}{2}\langle\bar{Q}_e\times\rangle\Omega_e + \frac{1}{2}q_{e4}\Omega_e\right),$$
$$+\Omega\times\left(J_0\Omega\right) + J_0\dot{\Omega}_d - C_s Sat(S,\varepsilon) \tag{33}$$

where $C_s Sat(S,\varepsilon) = \left[c_{s1}Sat(s_1,\varepsilon) \quad c_{s2}Sat(s_2,\varepsilon) \quad c_{s3}Sat(s_3,\varepsilon)\right]^T$, whereas

$$Sat(s_i,\varepsilon) = \begin{cases} 1 & s_i > \varepsilon \\ \dfrac{s_i}{\varepsilon} & |s_i| \le \varepsilon, \ i = 1,2,3. \\ -1 & s_i < -\varepsilon \end{cases}$$

Now the system will not be forced to strictly stay on the sliding surface but constrained within the boundary layer $|s_i| \le \varepsilon$.

4. Simulations

To validate the proposed attitude tracking problem, we present three nonlinear controllers respectively consisting of the sliding-mode attitude controller design in Section 3.1, the fuzzy sliding-mode attitude controller design in Section 3.2 and the adaptive fuzzy sliding-mode attitude controller design in Section 3.3 for a rigid spacecraft, so as to demonstrate the performance and effectiveness of the respective controller designs.

For simulation the initial conditions of quaternion and angular velocity are set as $Q(0) = [0.866 \ -0.212 \ 0.283 \ 0.354]^T$ and $\Omega(0) = [0 \ 0 \ 0]^T$, respectively. Furthermore, the desired values of the quaternion $Q_d(t) = [0.985 \ 0.174\cos(0.2t) \ 0.174\sin(0.2t) \ 0]^T$ with period 10π seconds, the angular velocity $\Omega_d(t) = 0_{3\times1}$, and its time derivative $\dot{\Omega}_d(t) = 0_{3\times1}$ are also be given. Of course, the setting elements of the desired quaternion must satisfy the unit-norm property of quaternion, i.e., $Q_d^T Q_d = 1$.

The above initial conditions of $\Omega_d(0) = 0_{3\times1}$ and $\dot{\Omega}_d(0) = 0_{3\times1}$ will have two principle advantages. One is to simplify the simulation procedure but not to lose the practical implementation; the other is to smoothly stabilize the spacecraft's flying in the outer space. On the other hand, the nominal part and the uncertain part of inertia matrix of a spacecraft is set as

$$J_0 = \begin{bmatrix} a_1 & a_4 & a_5 \\ a_4 & a_2 & a_6 \\ a_5 & a_6 & a_3 \end{bmatrix} = \begin{bmatrix} 950 & 10 & 5 \\ 10 & 600 & 30 \\ 5 & 30 & 360 \end{bmatrix}, \ \Delta J = \begin{bmatrix} 95 & 1 & 0.5 \\ 1 & 60 & 3 \\ 0.5 & 3 & 36 \end{bmatrix}.$$

And the variation and the variation uncertain part of the inertial matrix are set as

$$\dot{J}_0 = \begin{bmatrix} -0.95 & -0.01 & -0.005 \\ -0.01 & -0.6 & -0.03 \\ -0.005 & -0.03 & -0.36 \end{bmatrix}, \ \Delta\dot{J} = \begin{bmatrix} -0.095 & -0.001 & -0.0005 \\ -0.001 & -0.06 & -0.003 \\ -0.0005 & -0.003 & -0.036 \end{bmatrix}.$$

Further, we also consider the disturbance vector $D = [d_1 \ d_2 \ d_3]^T$, where the disturbances d_1, d_2, and d_3, which are containing real white Gaussian noises of power 1 dBW, given in Eq. (1), are shown in Fig. 3(e). In our simulation, some positive definite diagonal matrices of P (see Eq. (9)) for the sliding surface and K_s (see Eq. (16)) for the control torque input will also be shown as follows.

$$P = \begin{bmatrix} 120 & 0 & 0 \\ 0 & 120 & 0 \\ 0 & 0 & 120 \end{bmatrix}, \ K_s = \begin{bmatrix} 1400 & 0 & 0 \\ 0 & 1400 & 0 \\ 0 & 0 & 1400 \end{bmatrix},$$

where parameters c_{s1}, c_{s2}, c_{s3} of the sliding-mode controller are appropriately set as $c_{s1} = c_{s2} = c_{s3} = 10$ (see Eq. (17)) to satisfy the better stability requirements of the overall system, such that external disturbances can be eliminated completely.

4.1 Simulation results of the sliding-mode attitude controller

The appealing effect of the sliding-mode attitude control presented in quaternion form is given in Fig. 5(a), which shows the present attitude and the desired one simultaneously. The solid line in each sub-figure denotes the current quaternion of the spacecraft, where the dashed line denotes the desired attitude in Fig. 5(a), we can see that the current and desired attitudes are coincident with each other, that is, the attitude tracking effect is fulfilled after about 3 seconds. This is to show feasible of the conclusion from Eq. (19) and to show well results of attitude tracking of the spacecraft. Here, the robustness and effectiveness can simultaneously be demonstrated by results shown in Fig. 3 for the attitude control system. The sliding surface and the torque input of the sliding-mode controller with varied desired attitude are shown in Figs. 5(b) and 5(c), respectively to demonstrate the practical effect. Variations of the moment of inertia matrix J and disturbances D, which are used in the proposed three nonlinear controllers, are shown in Figs. 3(d) and 3(e), respectively.

4.2 Simulation results of the fuzzy sliding-mode attitude controller

In this section, the revealing response of the fuzzy sliding-mode attitude control is given in Fig. 6(a), which also shows the present attitude and the desired one simultaneously. The solid line in each sub-figure denotes the current quaternion of the spacecraft, where the dashed line denotes the desired attitude in Fig. 6(a), we can see that the current and desired attitudes are coincident with each other. From Fig. 6(a), the attitude tracking effect is fulfilled almost totally after 3 seconds. This is to show feasible of the conclusion from Eq. (A.1) and to show well the results of attitude tracking of the spacecraft. The sliding surface and the torque input of the fuzzy sliding-mode controller with varied desired attitude are

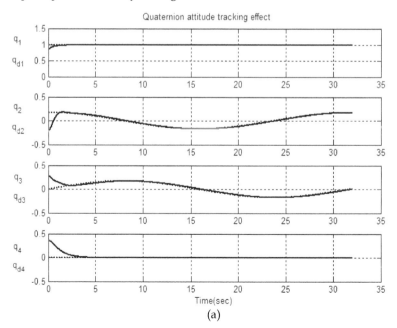

(a)

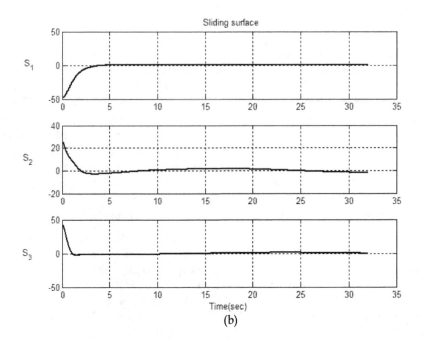

(b)

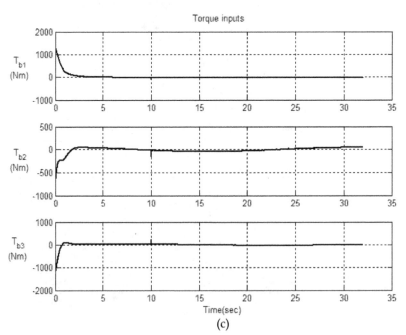

(c)

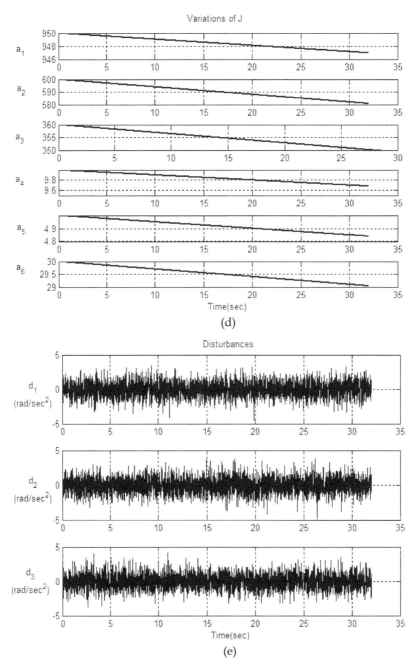

Fig. 5. Simulation results of (a) quaternion attitude tracking effects, (b) the convergence of sliding surface, (c) control torque inputs; (d) variations of J, and (e) real white Gaussian noises of power 1 dBW of the sliding-mode controller for spacecrafts.

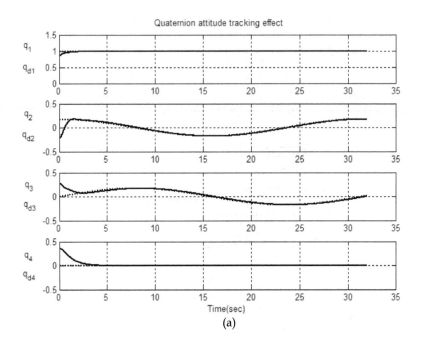

(a)

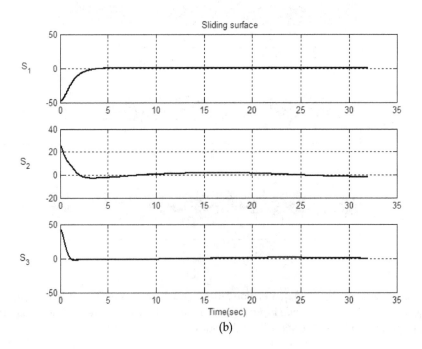

(b)

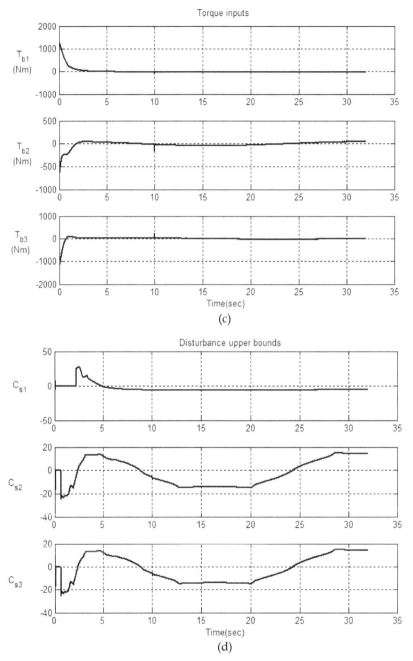

Fig. 6. Simulation results of (a) quaternion attitude tracking effects, (b) the convergence of sliding surface, (c) control torque inputs, and (d) estimated upper bounds of the lumped uncertainty of the fuzzy sliding-mode controller for spacecrafts.

shown in Figs. 6(b) and 6(c), respectively to demonstrate the tracking effect. Slow-varying upper bounds of the lumped uncertainty and disturbances D are shown in Figs. 6(d) and 5(e), respectively.

4.3 Simulation results of adaptive fuzzy sliding-mode attitude controller

The revealing response of the adaptive fuzzy sliding-mode attitude control also presented in quaternion form is given in Fig. 7(a), which also shows the present attitude and the desired one simultaneously. The solid line in each sub-figure denotes the current quaternion of the spacecraft, where the dashed line denotes the desired attitude in Fig. 7(a), we can see the current and desired attitudes are coincident with each other. From Fig. 7(a), the attitude tracking effect can be seen to be fulfilled after about 3 seconds. This is to show feasible of the conclusion from Eq. (30) and to show well results of spacecraft attitude tracking. Here, the robustness and effectiveness can simultaneously be demonstrated by the results shown in Fig. 7 for the attitude control system. The sliding surface and the torque input of the adaptive fuzzy sliding-mode controller with varied desired attitude are shown in Figs. 7(b) and 7(c), respectively to demonstrate the exponential stability and convergence effect. The optimal and estimated upper bounds of the lumped uncertainty and disturbances D are shown in Figs. 7(d) and 5(e), respectively. Here the dashed line in each sub-figure denotes estimated upper bounds of the lumped uncertainty, where the solid line denotes optimal upper bounds of the lumped uncertainty in Fig. 7(d), we can see that optimal and estimated upper bounds are aligned with each other, that is, the tracking effect of the upper-bound

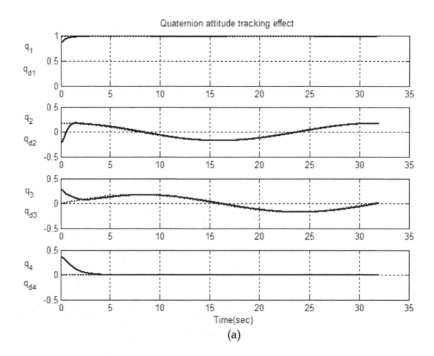

(a)

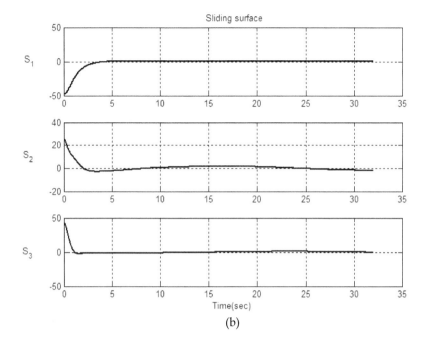

(b)

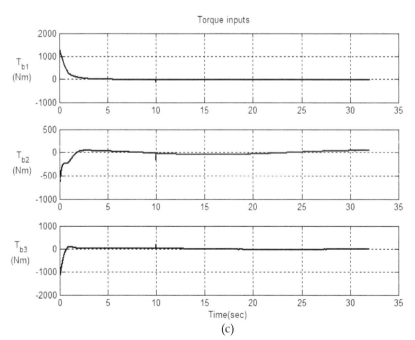

(c)

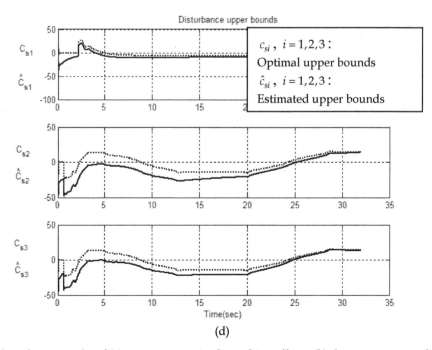

(d)

Fig. 7. Simulation results of (a) quaternion attitude tracking effects, (b) the convergence of sliding surface, (c) control torque inputs, and (d) optimal and estimated upper bounds of the lumped uncertainty of the adaptive sliding-mode controller for spacecrafts.

estimation can be fulfilled finally. Therefore, for upper-bound estimation the persistent exciting of a spacecraft is rich enough for the proposed controller.

5. Conclusions

Since the attitude control and system stability are the key for space technology, we address the nonlinear attitude controller designs of a spacecraft, respectively consisting of the sliding-mode, the fuzzy sliding-mode and the adaptive fuzzy sliding-mode attitude controllers. The fuzzy sliding-mode controller is designed to estimate upper bounds of the lumped uncertainty, and the adaptive fuzzy sliding-mode controller with center adaption of membership functions is designed to estimate optimal upper bounds of the lumped uncertainty, respectively. We prove the exponential stability for the proposed attitude controllers for external disturbances and inertia uncertainties through the aids of the Lyapunov stability analysis and the Barbalat's lemma.

Extensive simulations have been adopted to verify the feasibility for three attitude tracking controllers corresponding to the lumped uncertainty. The system performance and its stability can also be demonstrated by use of the aforementioned theoretical derivations and the realistic simulations.

6. Nomenclature

D Disturbance vector
d_p Pitch angle of propellant
d_y Yaw angle of propellant
F Thrust vector
J Moment of inertia matrix
J_0 Nominal part of J
ΔJ Variation of J
ℓ Distance between nozzle and center of gravity
$L_b = \begin{bmatrix} -\ell & 0 & 0 \end{bmatrix}^T$ Displacement vector
M Moment vector
n Magnitude of movable nozzle thrust
Q Quaternion
S Sliding surface
T Torque
Ω Angular velocity vector

7. Subscripts

b The body coordinate frame
d Desired
e Error
f Fuzzy sliding-mode controller
M Spacecraft
s Sliding-mode controller
T Thrust

8. Appendix 1

8.1 Proof of Lemma 1: Fuzzy sliding-mode attitude controller design

Proof: To achieve the exponential stability and convergence of the spacecraft attitude tracking system designed the fuzzy sliding-mode controller, the dynamic model of a spacecraft, the control torque input, and upper bounds of the lumped uncertainty are respectively defined as Eqs. (1), (6), (22), and (21). And if the Lyapunov function is defined as $V_f = \frac{1}{2}S^T J S$, then the time derivative of the Lyapunov function can be obtained as

$$\dot{V}_f = S^T J \dot{S} + \frac{1}{2}S^T \dot{J} S$$

$$= S^T \left[-\dot{J}\Omega - \Omega \times (J\Omega) + T_b + D - J\dot{\Omega}_d + JP\left(\frac{1}{2}\langle \overline{Q}_e \times \rangle \Omega_e + \frac{1}{2}q_{e4}\Omega_e \right) + \frac{1}{2}\dot{J}S \right]$$

$$= -S^T K_s S - \sum_{i=1}^{3} |s_i| \left[\hat{c}_{si} - \delta_i \, \mathrm{sgn}(s_i) \right]$$

$$\leq -S^T K_s S - \sum_{i=1}^{3} |s_i| \left[\hat{c}_{si} - \delta_i^{\max} \right]$$

$$\leq -\sigma_{\min}(K_s) |S|^2 < 0, \tag{A.1}$$

where the values of $\hat{c}_{s1}$, $\hat{c}_{s2}$, and $\hat{c}_{s3}$ are chosen to be positive big enough and are computed by Eq. (21) to guarantee that the inequality condition shown below can be satisfied

$$\hat{c}_{si}(C_i, W_i) > \delta_i^{\max} \left(Q, \Omega, Q_d, \dot{Q}_d, \ddot{Q}_d \right) \geq |\delta_i|, \tag{A.2}$$

where bounding functions δ_i, $i = 1, 2, 3$ are shown as Eq. (18).

Here we assume that the external disturbance D and the induced 2-norm of $\Delta \dot{J}$ and ΔJ are all bounded, Therefore, the main goal of achieving exponential stability and robustness of the fuzzy sliding-mode attitude controller for the spacecraft attitude tracking system can be satisfied completely.

9. Acknowledgment

The author would like to thank the National Science Council of the Republic of China, Taiwan, for financially supporting this research under Contract No. NSC 100-2221-E-235-004-.

10. References

[1] Peter Nicolas, "Towards a New Inspiring Era of Collaborative Space Exploration," *Humans in Outer Space – Interdisciplinary Odysseys*, vol. 1, chap. 3, pp. 107-118, 2009.
[2] J. C. K. Chou, "Quaternion Kinematic and Dynamic Differential Equations," *IEEE Trans. on Robotics and Automation*, vol. 8, no. 1, pp. 53-64, 1992.
[3] S. Tafazoli and K. Khorasani, "Nonlinear Control and Stability Analysis of Spacecraft Attitude Recovery," *IEEE Transactions on Aerospace and Electronic systems*, vol. 42, no. 3, pp. 825-845, 2006.
[4] A. Mohammad and S. S. Ehsan, "Fuzzy Sliding Mode Controller Design for Spacecraft Tracking in Terms of Quaternion," *Proceedings of the 27th Chinese Control Conference*, pp. 753-757, 2008.
[5] A. Sarlette, R. Sepulchre, and N. E. Leonard, "Autonomous Rigid Body Attitude Synchronization," *Proceedings of the 46th IEEE Conference on Decision and Control*, pp. 2566-2571, 2007.

[6] Q. L. Hu, Z. Wang, and H. Gao, "Sliding Mode and Shaped Input Vibration Control of Flexible Systems," *IEEE Transactions on Aerospace and Electronic systems*, vol. 44, no. 2, pp. 503-519, 2008.

[7] F. K. Yeh, "Sliding-Mode Adaptive Attitude Controller Design for Spacecrafts with Thrusters," *IET Control Theory and Applications*, vol. 4, no. 7, pp. 1254-1264, July 2010.

[8] P. A. Servidia and R. S. Pena, "Practical Stabilization in Attitude Thruster Control," *IEEE Transactions on Aerospace and Electronic systems*, vol. 41, no. 2, pp. 584-598, 2005.

[9] Q. Hu, "Adaptive Output Feedback Sliding-Mode Maneuvering and Vibration Control of Flexible Spacecraft with Input Saturation," *IET Control Theory and Applications*, vol. 2, no. 6, pp. 467-478, 2008.

[10] Y. Xia, Z. Zhu, M. Fu, and S. Wang, "Attitude Tracking of Rigid Spacecraft With Bounded Disturbances," *IEEE Transactions on Industrial Electronics*, vol. 58, no. 2, pp. 647-659, 2011.

[11] K. D. Young, V. I. Utkin, and U. Ozguner, "A Control Engineer's Guide to Sliding Mode Control," *IEEE Transactions on Control Systems Technology*, vol. 7, no. 3, pp. 328-342, 1999.

[12] I. Eker, "Second-Order Sliding Mode Control with Experimental Application," *ISA Transactions*, vol. 49, no. 3, pp. 394-405, 2010.

[13] S. Zimmer, C. Ocampo, and R. Bishop, "Reducing Orbit Covariance for Continuous Thrust Spacecraft Transfers," *IEEE Transactions on Aerospace and Electronics*, vol. 46, no. 2, pp. 771-791, 2010.

[14] P. Janhunen et al., "Electric Solar Wind Sail: Toward Test Missions," *Review of Scientific Instruments*, vol. 81, no. 11, 2010.

[15] J. J. E. Slotine and M. D. Benedetto, "Hamiltonian Adaptive Control of Spacecraft," *IEEE Transaction on Automatic Control*, vol. 35, no. 7, pp.848-852, 1990.

[16] J. J. E. Slotine and W. Li, Applied Nonlinear Control, *Prentice Hall, Englewood Cliffs, New Jersey 07632*, pp. 350-358, 1991.

[17] A. M. Zou, K. D. Kumar, and Z. G. Hou, "Quaternion-Based Adaptive Output Feedback Attitude Control of Spacecraft Using Chebyshev Neural Networks," *IEEE Transaction on Neural Networks*, vol. 21, no. 9, pp. 1457-1471, 2010.

[18] G. S. Huang and H. J. Uang, "Robust Adaptive PID Tracking Control Design for Uncertain Spacecraft Systems: a Fuzzy Approach," *IEEE Transactions on Aerospace and Electronic systems*, vol. 42, no. 4, pp. 1506-1514, 2006.

[19] S. C. Tong, X. L. He, and H. G. Zhang, "A Combined Backstepping and Small-Gain Approach to Robust Adaptive Fuzzy Output Feedback Control," *IEEE Transactions on Fuzzy systems*, vol. 17, no. 5, pp. 1050-1069, 2009.

[20] S. Islam and P. X. Liu, "Robust Adaptive Fuzzy Output Feedback Control System for Robot Manipulators," *IEEE/ASME Transactions on Mechatronics*, vol. 16, no. 2, pp. 188-296, 2011.

[21] J. T. Y. Wen and K. Kreutz-Delgado, "The Attitude Control Problem," *IEEE Trans. on Automatic Control*, vol. 36, no. 10, pp. 1148-1162, 1991.

[22] S. C. Lo and Y. P. Chen, "Smooth Sliding-Mode Control for Spacecraft Attitude Tracking Maneuvers," *Journal of Guidance, Control, and Dynamics,* vol. 18, no. 6, pp. 1345-1349, 1995.

8

Design of Sliding Mode Attitude Control for Communication Spacecraft

Erkan Abdulhamitbilal[1] and Elbrous M. Jafarov[2]
[1]ISTAVIA Engineering, Istanbul
[2]Aeronautics and Astronautics Engineering, Istanbul Technical University, Istanbul
Turkey

1. Introduction

Control problem of a spacecraft is an important topic in automatic control engineering. A body orbiting the Earth in geosynchronous orbit has instabilities in attitude dynamics and disturbances caused by the Earth, the Moon, the Sun and other bodies in space. These effects force the body to lose initial orbit and attitude. Here the control system takes important part of spacecraft missions where it keeps the body in designed orbit and desired attitude. The control system consists of control elements and control algorithms which are developed for the mission by a control engineer [1]. The commonly used control elements for a spacecraft in geosynchronous orbit are thrusters, reaction or momentum wheels, etc.

The sliding mode theory has an attention in the aerospace field. The technique permits the use of a lower order system model for generating control commands. On the other hand, the system is robust to the external disturbances and includes unmodelled dynamics, as well. The theory and methods of sliding mode control design principles are investigated [1]-[3], etc. Variable structure systems with nonlinear control techniques and dead-band on switching function for sliding mode controllers are introduced [4].

A variable structure control design for rigid body spacecraft attitude dynamics with quaternion representation for optimal sliding mode control which consists of three parts: equivalent control, sliding variable, and relay control where simulation results illustrate that the motion along the sliding mode is insensitive to parameter variations and unmodeled effects is given [5]. An automatic controller for active nutation damping in momentum biased stabilized spacecraft is introduced [6], where robust feedback stabilization of roll and yaw angular dynamics are achieved with prescribed qualitative characteristics for a spinning satellite. A smooth sliding mode control which requires well-estimated initial condition for quaternion based spacecraft attitude tracking maneuver is studied [7] where the chattering is eliminated by replacing saturation instead of signum function. A class of uncertain nonlinear systems decoupled by state variable feedback with sliding mode approach for attitude control of an orbiting spacecraft is considered [8] where simulation results show that precise attitude control is accomplished in spite of the uncertainty in the system. As seen from simulations spacecraft is stabilized approximately in 10 seconds. However there is a chattering in control action and thrusters are operating after stabilization of the spacecraft attitude dynamics. A reference book for various spacecraft attitude and orbit dynamics, orbit transfer methods, and different control strategies such as PID and robust control, pulse modulation of thruster control etc. is covered [9]. An attitude control with reaction wheels is evaluated in [10].

There are many papers concerning control of flexible spacecraft. A maneuvering of a flexible spinning spacecraft is treated with variable structure control [11] where system is stabilized between 60-100 seconds for small and large angle maneuvers. An application with one sided dead-band for robust closed-loop control design for a flexible spacecraft slew maneuver using on-off thrusters is studied [12] where analytical simulations and experimental results demonstrate that the proposed switching function provides significant improvement in slew maneuver performance. The size of single-sided dead-band in switching function provides the capability of a tradeoff between maneuver time and fuel expenditure. Rotational maneuver and vibration suppression of an elastic spacecraft is considered [13] where pitch angle trajectories are asymptotically tracked by an adaptive controller. Variable structure control and active suspension of flexible spacecraft during attitude maneuver is studied [14] where positive position feedback technique is used to suspend vibration and variable sliding mode with pulse-width pulse-frequency modulation to eliminate chattering. An adaptive variable structure control of spacecraft dynamics with command input shaping which eliminate residual vibration is studied [15] where PD, conventional and adaptive variable structure output feedback controllers with and without input shaping are compared and simulated. Vibration suspension of flexible spacecraft during attitude maneuvers is considered [16] where PD controller with pulse-with pulse-frequency modulation with positive position feedback is considered for vibration reduction during on-off operation of thrusters. However, as seen from simulation results chattering occurs in control action. A comparison between linear and sliding mode controllers with reaction wheels is studied [17] where only small angle orientations are considered. Designed sliding mode controller stabilizes spacecraft attitude dynamics 30 times faster than output feedback controller with reaction wheels. Station keeping chattering free sliding mode controller design is designed in [18].

A body orbiting the Earth in geosynchronous orbit has instabilities in attitude dynamics and disturbances caused by the Earth, the Moon, the Sun and other bodies in space. These effects force the body to lose initial orbit and attitude. Here the control system takes important part in spacecraft missions where it keeps the body in designed orbit and desired attitude. The commonly used control elements for a spacecraft in geosynchronous orbit are thrusters, reaction, and momentum wheels. Dynamic model of a spacecraft is nonlinear, includes the rigid and flexible mode interaction, and the parameters of the spacecraft are not precisely known. The performance criteria for a spacecraft are fuel expenditure and vibration of flexible structures. The sliding mode technique permits usage of lower order system model for generating control commands, which includes unmodeled dynamics or uncertainties, and stabilizes the plant faster and robustly under bounded disturbance. The chattering at high frequencies is not desired because it may cause vibration. Chattering may be eliminated by replacing saturation instead of signum function. However, in that case non-zero tracking errors exist, which can be made small by taking a tiny region for saturation and also, saturation is limited with hardware capability and reduction of accuracy and robustness as introduced [7] and [8]. On the other hand, chattering may be eliminated by pulse modulation as done [14].

The chapter is organized as follows. Section 1 gives an introduction to sliding mode control of a satellite. Section 2 gives the system description of rigid body nonlinear attitude dynamics. Section 3 evaluate rigid body in circular orbit with internal torquers and introduces equation of motion for a flexible spacecraft. Section 4 includes design of variable structure control systems for nonlinear attitude dynamics of a spacecraft and suspension of vibration of flexible solar arrays. Also design examples and performances comparison are studied. Section 5 concludes the chapter.

2. Rigid body dynamics

Consider a rigid-body with as body-fixed reference frame B with its origin at the center of mass of the rigid body as shown in Figure 1. $\vec{R}_C$ is the position vector of the center of mass from an inertial origin of N, and $\vec{R}$ is the position vector of dm from an inertial origin N.

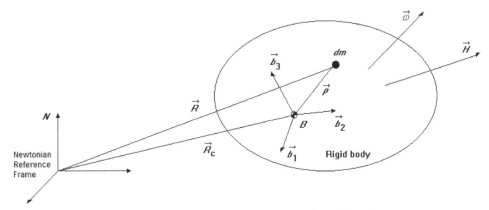

Fig. 1. Body fixed reference frame B at the center of mass of a rigid body.

Let $\vec{\omega} \equiv \vec{\omega}^{B/N}$ be the angular velocity vector of the rigid body in an inertial reference frame N. The angular momentum vector $\vec{H}$ of rigid body about its center of mass can be defined as [9]:

$$\vec{H} = \int \vec{\rho} \times \dot{\vec{R}} \, dm = \int \vec{\rho} \times (\vec{\omega} \times \vec{\rho}) \, dm = I\vec{\omega} \tag{1}$$

The position vector $\vec{\rho}$ of very small mass element dm from the center of mass is defined as

$$\vec{\rho} = \rho_1 \vec{b}_1 + \rho_2 \vec{b}_2 + \rho_3 \vec{b}_3 \tag{2}$$

and finally angular velocity vector $\vec{\omega} \equiv \vec{\omega}^{B/N}$ of a rigid body in an inertial reference frame N can be writen as:

$$\vec{\omega} = \omega_1 \vec{b}_1 + \omega_2 \vec{b}_2 + \omega_3 \vec{b}_3 \tag{3}$$

The angular momentum vector (1) can be rewritten as:

$$\vec{H} = H_1 \vec{b}_1 + H_2 \vec{b}_2 + H_3 \vec{b}_3 \tag{4}$$

where

$$H_1 = I_{11}\omega_1 + I_{12}\omega_2 + I_{13}\omega_3$$
$$H_2 = I_{21}\omega_1 + I_{22}\omega_2 + I_{23}\omega_3 \qquad (5)$$
$$H_3 = I_{31}\omega_1 + I_{32}\omega_2 + I_{33}\omega_3$$

and the matrix form of (5) is:

$$\begin{bmatrix} H_1 \\ H_2 \\ H_3 \end{bmatrix} = \begin{bmatrix} I_{11} & I_{12} & I_{13} \\ I_{21} & I_{22} & I_{23} \\ I_{31} & I_{32} & I_{33} \end{bmatrix} \begin{bmatrix} \omega_1 \\ \omega_2 \\ \omega_3 \end{bmatrix} \qquad (6)$$

The rotational equation of motion of a rigid body in an inertial reference frame N about center of mass is [9]:

$$\int \vec{\rho} \times \ddot{\vec{R}} dm = \vec{M} \qquad (7)$$

Form (1) and (7) we can write a relation between external moment acting on the body about its mass center and angular momentum as [9]-[11]:

$$\vec{M} = \dot{\vec{H}} \qquad (8)$$

The relation above with an angular momentum $\vec{H}$ and external moment $\vec{M}$ is the rotational equation of motion of a rigid body in a circular orbit [9]-[10]:

$$\vec{M} = \dot{\vec{H}} \equiv \left\{ \frac{d\vec{H}}{dt} \right\}_N = \left\{ \frac{d\vec{H}}{dt} \right\}_B + \vec{\omega}^{B/N} \times \vec{H} \qquad (9)$$

Substituting (1) into rotational equation of motion (9) we obtain:

$$\vec{M} = I \cdot \dot{\vec{\omega}} + \vec{\omega} \times I \cdot \vec{\omega} \qquad (10)$$

which is known as Euler's rotational equation of motion.

3. Rigid body in circular orbit

Consider a rigid body orbiting the Earth with a constant radius. A local horizontal and local vertical reference frame A at the center of the mass of an orbiting spacecraft with unit vectors $\vec{a}_1, \vec{a}_2, \vec{a}_3$ as given in Figure 2. $\vec{a}_1$ is along the orbital direction, $\vec{a}_2$ is perpendicular to the orbital plane and $\vec{a}_3$ is always pointing the Earth. The angular velocity of A with respect to N is [10]:

$$\vec{\omega}^{A/N} = -n\vec{a}_2 \qquad (11)$$

where n is the orbital rate defined as

$$n = \sqrt{\mu / R_C^3} \qquad (12)$$

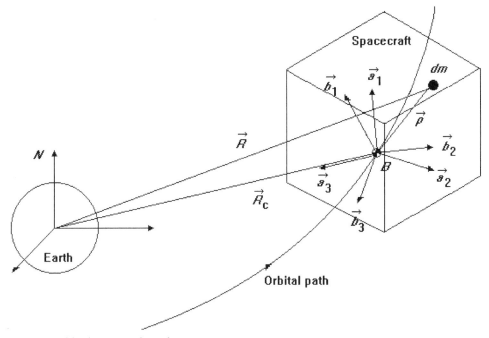

Fig. 2. Rigid body in circular orbit.

Note that, μ is the gravitational constant of the Earth and R_C is the radius of the orbit. The orbital rate for a spacecraft orbiting the Earth in a circular orbit with same angular velocity for one real day can be calculated from relation:

$$n = \frac{2\pi}{23h + 56m + 4.09054s} = 7.2921 \times 10^{-5} s^{-1} \tag{13}$$

The angular velocity of the body-fixed reference frame B with basis vectors $\vec{b}_1, \vec{b}_2, \vec{b}_3$ is given as [9]:

$$\vec{\omega}^{B/N} = \vec{\omega}^{B/A} + \vec{\omega}^{A/N} = \vec{\omega}^{B/A} - n\vec{a}_2 \tag{14}$$

where $\vec{\omega}^{B/A}$ is the angular velocity of B relative to A. To describe orientation of the body-fixed reference frame B with respect to the local vertical local horizontal reference frame A in terms of Euler angles θ_i ($i = 1, 2, 3$), consider the rotational sequence of $C_1(\theta_1) \leftarrow C_2(\theta_2) \leftarrow C_3(\theta_3)$ to B from A. For this sequence relation is:

$$\begin{bmatrix} \vec{b}_1 \\ \vec{b}_2 \\ \vec{b}_3 \end{bmatrix} = \begin{bmatrix} c_2c_3 & c_2s_3 & -s_2 \\ s_1s_2c_3 - s_1s_3 & s_1s_2s_3 - c_1c_3 & s_1c_2 \\ c_1s_2c_3 + s_1s_3 & c_1s_2s_3 - s_1c_3 & c_1c_2 \end{bmatrix} \begin{bmatrix} \vec{a}_1 \\ \vec{a}_2 \\ \vec{a}_3 \end{bmatrix} = \begin{bmatrix} C_{11} & C_{12} & C_{13} \\ C_{21} & C_{22} & C_{23} \\ C_{31} & C_{32} & C_{33} \end{bmatrix} \begin{bmatrix} \vec{a}_1 \\ \vec{a}_2 \\ \vec{a}_3 \end{bmatrix} \tag{15}$$

where $s_i = \sin\theta_i$, $c_i = \cos\theta_i$, $i = 1, 2, 3$.

The gravity gradient torque can be written as [9]:

$$\vec{M}_G = -\mu \int \frac{\vec{\rho} \times \vec{R}_C}{\left|\vec{R}_C + \vec{\rho}\right|^3} dm = 3n^2 \vec{a}_3 \times I \cdot \vec{a}_3 \tag{16}$$

The basis vector on axis 3 of local horizontal and local vertical reference frame A is $\vec{a}_3 = -\vec{R}_C / R_C$. The rotational equation of motion of a rigid body with an angular momentum in a circular orbit can be obtained by substituting $\vec{\omega}^{B/N} \equiv \vec{\omega}$ into (9) as

$$I \cdot \dot{\vec{\omega}} + \vec{\omega} \times I \cdot \vec{\omega} = 3n^2 \vec{a}_3 \times I \cdot \vec{a}_3 \tag{17}$$

$\vec{\omega}$ and $\vec{a}_3$ can be expressed in terms of basis vectors of the body-fixed reference frame B as:

$$\vec{\omega} = \omega_1 \vec{b}_1 + \omega_2 \vec{b}_2 + \omega_3 \vec{b}_3 \tag{18}$$

$$\vec{a}_3 = C_{13} \vec{b}_1 + C_{23} \vec{b}_2 + C_{33} \vec{b}_3 \tag{19}$$

The equation of motion with control torque and disturbance can be written as [5], [7], [9], [17]:

$$I\dot{\omega} + \Omega I \omega = M_G + u_B + d_E(t) \tag{20}$$

where u_B is the body (3×1)-control vector and $d_E(t)$ is an external disturbance (solar radiation, interaction with other bodies in space, etc.). Let us define gravity gradient torque (16) as

$$M_G = 3n^2 \Gamma I C \tag{21}$$

where C is the third column of direction cosine (3×3)-dimensional matrix given in (15), $\Omega = \Omega(\omega)$ and $\Gamma = \Gamma(C)$ are (3×3)-dimensional skew symmetric matrices defined as:

$$C = \begin{bmatrix} C_{13} & C_{23} & C_{33} \end{bmatrix}^T = \begin{bmatrix} -s_2 & s_1 c_2 & c_1 c_2 \end{bmatrix}^T \tag{22}$$

$$\Omega = \begin{bmatrix} 0 & -\omega_3 & \omega_2 \\ \omega_3 & 0 & -\omega_1 \\ -\omega_2 & \omega_1 & 0 \end{bmatrix} \tag{23}$$

$$\Gamma = \begin{bmatrix} 0 & -C_{33} & C_{23} \\ C_{33} & 0 & -C_{13} \\ -C_{23} & C_{13} & 0 \end{bmatrix} \tag{24}$$

Relation between angular velocity ω and attitude angles and their rates for circular orbit can be written as [9]:

$$\begin{bmatrix} \omega_1 \\ \omega_2 \\ \omega_3 \end{bmatrix} = \begin{bmatrix} 1 & 0 & -s_2 \\ 0 & c_1 & s_1 c_2 \\ 0 & -s_1 & c_1 c_2 \end{bmatrix} \begin{bmatrix} \dot{\theta}_1 \\ \dot{\theta}_2 \\ \dot{\theta}_3 \end{bmatrix} - n \begin{bmatrix} c_2 s_3 \\ s_1 s_2 s_3 + c_1 c_3 \\ c_1 s_2 s_3 - s_1 c_3 \end{bmatrix} \cong n_1 \begin{bmatrix} \dot{\theta}_1 \\ \dot{\theta}_2 \\ \dot{\theta}_3 \end{bmatrix} + n_3 \tag{25}$$

or vice versa of (25):

$$\begin{bmatrix} \dot{\theta}_1 \\ \dot{\theta}_2 \\ \dot{\theta}_3 \end{bmatrix} = \frac{1}{c_2} \begin{bmatrix} c_2 & s_1 s_2 & c_1 s_2 \\ 0 & c_1 c_2 & -s_1 c_2 \\ 0 & s_1 & c_1 \end{bmatrix} \begin{bmatrix} \omega_1 \\ \omega_2 \\ \omega_3 \end{bmatrix} + \frac{n}{c_2} \begin{bmatrix} s_3 \\ c_2 c_3 \\ s_2 s_3 \end{bmatrix} \cong n_2 \begin{bmatrix} \omega_1 \\ \omega_2 \\ \omega_3 \end{bmatrix} + n_4 \tag{26}$$

Then the nonlinear equation of motion of three axes stabilized rigid body spacecraft (20) reduces to:

$$In_1 \ddot{\theta} + \left(I \dot{n}_1 - \Omega I n_1 \right) \dot{\theta} - \left(n \left(I \dot{n}_3 - n_3 \right) + 3n^2 \Gamma I C \right) = u_B + d_E(t) \tag{27}$$

where

$$\dot{n}_1 = \begin{bmatrix} 0 & 0 & -c_2 \\ 0 & -s_1 & c_1 c_2 - s_1 s_2 \\ 0 & -c_1 & -c_2 s_1 - c_1 s_2 \end{bmatrix} \tag{28}$$

$$\dot{n}_3 = \begin{bmatrix} c_2 c_3 - s_2 s_3 \\ c_1 s_2 s_3 + c_2 s_1 s_3 + c_3 s_1 s_2 - c_1 s_3 - c_3 s_1 \\ c_1 c_2 s_3 + c_1 c_3 s_2 - s_1 s_2 s_3 + s_1 s_3 - c_1 c_3 \end{bmatrix} \tag{29}$$

Note that, (27) will be used in simulation of attitude dynamics.

3.1 Rigid spacecraft with internal torquers in circular orbit

Here we have considered a three axis stabilized communication satellite with a bias momentum wheel. Some parameters of considered communication satellite are given in Table 1. Internal control torquers for the satellite are reaction wheels mounted on roll and yaw axes, and a momentum wheel is set up on pitch axis which spins along negative direction, see Figure 3. The total angular momentum of spacecraft can be written as [9]:

$$\vec{H} = \left(H_1 + h_1 \right) \vec{b}_1 + \left(H_2 - H_0 + h_2 \right) \vec{b}_2 + \left(H_3 + h_3 \right) \vec{b}_3 \tag{30}$$

where H_1, H_2, H_3 were defined in (5). We can obtain equation of motion for principle axis frame B from rotational equation of motion (9) with considered gravity gradient torque (21), external disturbances and internal torquers as:

$$I \dot{\omega} + \Omega I \omega = M - \left(\dot{h} + \Omega h - \Omega H_0 \right) + d_E(t) \tag{31}$$

Or in term of attitude angles

$$In_1 \ddot{\theta} + \left(I \dot{n}_1 - \Omega I n_1 \right) \dot{\theta} - \left(n \left(I \dot{n}_3 - n_3 \right) + 3n^2 \Gamma I C \right) = -\left(\dot{h} + \Omega h - \Omega H_0 \right) + d_E(t) \tag{32}$$

Properties	Values	Units
Principle moments of inertia, I_{11}, I_{22}, I_{33} ...…............…...	3026 / 440 / 3164	kg.m²
Main body dimensions, x-y-z.…....….........….................…...	1.5 / 1.7 / 2.2	m
Solar arrays (tip-to-tip)....….......................................….......	20	m
Maximum thrust force of thrusters.…….........................…	10	N
Bias momentum.………..............…...............…........….….	91.4	Nms
Array power.…..........…….........…............................…...	1.5	kW
Liquid of bi-propellant thrusters.…...................…...............	N₂O₄/MMH	-

Table 1. Spacecraft Parameters [9].

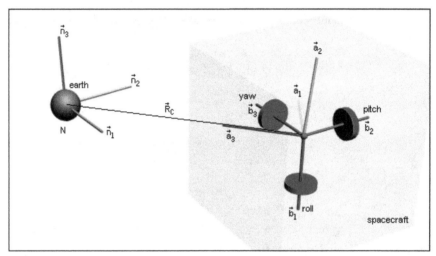

Fig. 3. Rigid spacecraft with internal torquers.

3.2 Communication satellite: rigid body with flexible solar arrays

During on-orbit normal mode operations, both solar arrays always point towards the sun, whereas the main body points towards the Earth. This results in a very slow change of modal frequencies and modal shapes. For control design purposes, however, the spacecraft model will be treated as a time-invariant but nonlinear system with a known range of modal characteristics. The equation of attitude motion of the three axis stabilized with flexible solar array is given [9]-[16]. Extending the rigid body equation (27) with the flexible solar arrays, following set of equations can be obtained for main body:

$$In_1\ddot{\theta} + \left(I\dot{n}_1 - \Omega In_1\right)\dot{\theta} - \left(n\left(I\dot{n}_3 - n_3\right) + 3n^2\Gamma IC\right) + \sqrt{2}\delta\ddot{q} = u_B + d_E(t) \qquad (33)$$

or in terms of ω

$$I\dot{\omega} + \Omega I\omega + \sqrt{2}\delta\ddot{q} = 3n^2\Gamma IC + u_B + d_E(t) \qquad (34)$$

and two solar arrays [9] with control force:

$$\ddot{q} + \sigma^2 q + \sqrt{2}\delta\ddot{\theta} = u_{SA} \qquad (35)$$

where $\delta = diag(\delta_1, \delta_2, \delta_3)$ represents rigid-elastic coupling diagonal matrix of a single solar array (see Table 2), $\sigma = diag(\sigma_1, \sigma_2, \sigma_3)$ are the modal frequencies diagonal matrix, and $q = [q_1 \quad q_2 \quad q_3]^T$ are the modal coordinates, and u_{SA} is the control command produced by solar array drivers. Note that, in this study only the first modes of modal characteristics of flexible solar arrays are taken into account.

Cantilever mode description	Cantilever frequency σ, rad/s	Coupling scalars, $\sqrt{kg \cdot m^2}$		
		Roll, δ_1	Pitch, δ_2	Yaw, δ_3
OP-1	0.885	0	0	35.372
OP-2	6.852	0	0	4.772
OP-3	16.658	0	0	2.347
OP-4	33.326	0	0	0.548
T-1	5.534	0	2.532	0
T-2	17.668	0	0.864	0
T-3	33.805	0	0.381	0
IP-1	1.112	35.865	0	0
IP-2	36.362	2.768	0	0

[1][a]: OP is out-of plane, T is torsion and IP is in-plane

Table 2. Single solar array modes at 6 a.m. [9].

Note that, two solar arrays point towards the Sun. Hence, the solar driving mechanism actuated by Sun Sensors causes the control torque to point solar arrays towards the Sun. In our case we have considered an station keeping controller that can activate the solar array driver mechanism to suspend vibration of flexible solar arrays caused by attitude angle acceleration during maneuvering. The control law will be designed in Section 4.2.

4. Sliding mode control

The sliding mode technique permits usage of lower order system model for generating control commands, which includes unmodeled dynamics or uncertainties, and stabilizes the plant faster and robustly under bounded disturbance. The chattering at high frequencies is not desired because it may cause vibration. Chattering may be eliminated by replacing saturation instead of signum function. However, in that case non-zero tracking errors exist, which can be made small by taking a tiny region for saturation and also, saturation is limited with hardware capability and reduction of accuracy and robustness as introduced [7] and [8]. On the other hand, chattering may be eliminated by pulse modulation as done [16]. In this section we suggest variable structure attitude and station keeping control system design for a communication satellite.

4.1 Sliding mode attitude control system design of a rigid spacecraft

Let s represent a sliding manifold as

$$s = [s_1 ... s_m]^T = \omega + k\theta = 0 \qquad (36)$$

where k is a constant to be selected. The attitude dynamics dominated on the sliding manifold $s = 0$ can be written from [5], [17] as:

$$\dot{\theta} = \begin{bmatrix} -k\theta_3\theta_2 + n\theta_3 - k\theta_1 \\ k\theta_3\theta_1 - k\theta_2 + n \\ -k\theta_2\theta_1 - k\theta_3 \end{bmatrix} = T\theta \tag{37}$$

Assume that the discontinuous control is given by

$$u_B = -I\left(N_1 s + N_2 sign(s)\right) \tag{38}$$

where N_1, N_2 are constant parameters to be selected. To analyze the stability of the system, consider a Lyapunov function candidate

$$V = \frac{1}{2}s^T s \tag{39}$$

Then,

$$\begin{aligned}
\dot{V} = s^T \dot{s} &= s^T \left(\dot{\omega} + k\dot{\theta}\right) \\
&= s^T \left(-I^{-1}\Omega I\omega + 3n^2 I^{-1}\Gamma IC + u + I^{-1}d(t) + k\dot{\theta}\right) \\
&= s^T \left(-I^{-1}\Omega Ik\theta + 3n^2 I^{-1}\Gamma IC + u + I^{-1}d(t) + kT\theta\right)
\end{aligned} \tag{40}$$

Taking the norm of (40) we have

$$\dot{V} \le k\|s\|\left\|I^{-1}\Omega I\right\|\|\theta\| + 3n^2\|s\|\left\|I^{-1}\Gamma IC\right\| + \|s\|\left\|I^{-1}\right\|\|d(t)\| - \|s\|N_1\|s\| - \|s\|N_2 \tag{41}$$

Some matrix and vector norms in (41) satisfy inequalities as below:

$$k\left\|I^{-1}\Omega I\right\| \le L_0 \tag{42}$$

$$3n^2\left\|I^{-1}\Gamma IC\right\| \le M_0 \tag{43}$$

$$\left\|I^{-1}d_E(t)\right\| \le d_0 \tag{44}$$

Therefore, substituting norm values (42)-(44) into (41) we obtain:

$$\begin{aligned}
\dot{V} &\le -\left(N_1\|s\| - L_0\|\theta\|\right)\|s\| - \left(N_2 - M_0 - d_0\right)\|s\| \\
&< 0
\end{aligned} \tag{45}$$

Hence, the sliding mode controller forces the system trajectories toward the sliding manifold asymptotically for $N_2 > d_0 + M_0$ and $N_1 > L_0\|\theta\|/\|s\|$.

4.1.1 Example

Sliding mode control system design is performed firstly with determining proper switching function where the system trajectory is caused to follow the sliding manifold, $s = 0$. Then proposed discontinuous control term is employed to model reaction wheels. Physical sliding surface consists of inputs form the Earth sensor for yaw and pitch attitude angles, from the star tracker for yaw, roll and pitch attitude angles, and from the rate gyro for attitude angle rates. Considered control function (38) and sliding manifold (36) stabilizes the dynamic equation of attitude angles presented via (31) for small attitude angles errors $\theta_1 = 0.3\deg$, $\theta_2 = 0.5\deg$, $\theta_3 = -0.3\deg$ gravity gradient torque and bounded external disturbances as shown in Figure 4.

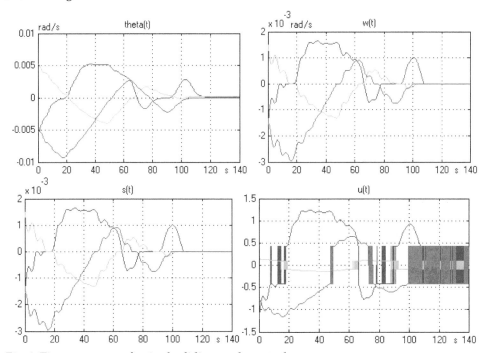

Fig. 4. Time responses of attitude sliding mode control system.

Required parameter k is selected as $k = 0.3$. The disturbance is assumed to be as:

$$d_E(t) = 0.0005 sin(\omega t) \tag{46}$$

The control torque is produced by reaction wheels which have approximately 6000 rpm angular velocity and can produce 1.5 Nm control torque. Therefore, control inputs can be developed for each pitch, roll and yaw axes by sliding mode control approach. Thus attitude sliding mode controller can be obtained for $N_2 > d_0 + M_0$ and $N_1 > L_0 \|\theta\|/\|s\|$ as:

$$u_B = -J(1000s + sign(s)) \tag{47}$$

where J is the inertia matrix of reaction wheels.

As seen from Figure 4, spacecraft attitude errors are eliminated in 120 seconds by sliding mode control approach where some chattering appears in control action. Since control function is applied to electro-motor of reaction wheel, the electro-motors are driven at nominal speeds on chattering phenomena for which sliding motion is conventional [2]. In general, chattering effect can be eliminated by using a saturation element.

4.2 Discontinuous station keeping sliding mode control design of flexible spacecraft

Station keeping sliding mode control algorithm for a three axes stabilized geosynchronous communication satellite is considered in this subsection. The spacecraft is assumed to be controlled externally with small attitude thrust jets. Classical sliding mode technique with chattering free correction and elimination of operation for small attitude angles via dead-band function will be used to model thrust jets variable on-off operation for stabilization of the spacecraft. The performances, modeling and simulation are discussed on a design example by using MATLAB-Simulink programming. The attitude sliding mode controller for geosynchronous satellite with flexible solar arrays will use fuel optimally and adequately as little as possible with proposed control algorithm. Also, attitude sliding mode controller is robust to bounded external disturbances and includes unmodelled dynamics as well.

Two types of simple and easy-to-apply variable structure P+relay controllers different from existing (for example from [5] which includes the linear equivalent and sliding terms plus relay) are proposed for the stabilization of full nonlinear attitude dynamics and vibration control of flexible solar arrays during station keeping maneuvering. Variable structure P+relay control law has only two design parameters. A modified sliding function with a dead-band instead of conventional one is considered to reduce fuel expenditure for small attitude corrections that can be stabilized by internal torquers (reaction or momentum wheels). The size of dead-band provides the capability of a tradeoff between maneuver time and fuel expenditure. The limits of the dead-band of switching function can be determined from maximum available torque produced by reaction wheels. On the other hand, large angle orientation of spacecraft induces structural deformation in the flexible solar arrays. Dynamical models of satellites are nonlinear and include rigid and flexible mode interaction. Therefore, vibration suppression of flexible solar arrays is required. For this case variable structure P+relay algorithm is proposed to eliminate vibration of flexible solar panels.

Desired sliding manifold on which the system equation of motion has good transient performances need to be selected before form a control law. The switching surfaces can be selected [1]-[5], [7], [11], [13]:

$$s = \omega + K\theta \tag{48}$$

where $K = diag(k_1, k_2, k_3)$, in general is a diagonal design matrix to be selected. Particularly, these parameters are selected $k_1 = k_2 = k_3 = k$. Then the sliding manifold is:

$$s = \omega + k\theta \tag{49}$$

After selecting sliding manifold, a variable structure P+relay control algorithm can be formed as follows:

$$u_B = -I\left(N_1 \|\theta\| + N_2\right) sign(s) \tag{50}$$

where N_1 and N_2 are design parameters to be selected. The control law consists of two terms: P term and relay term. The first term is used for the compensation of the model nonlinearities, model and parameter uncertainties. The second term is used to compensate the bounded external disturbances, flexibility effects of solar arrays, and gravity gradient torque. This controller should provide the existence of the sliding mode motion on the selected sliding manifold. So, consider a Lyapunov function candidate:

$$V = \frac{1}{2} s^T s \qquad (51)$$

Now the sliding mode existence condition for the nonlinear satellite equation of motion in large will be investigated. The time derivative of (51) along the state trajectories of dynamics system defined by nonlinear equations (33) or (34) and (35) can be calculated as follows:

$$\dot{V} = s^T \dot{s} = s^T \left(\dot{\omega} + k\dot{\theta} \right) =$$
$$= s^T \left(I^{-1} \left[u_B + d_E(t) - \sqrt{2}\delta\ddot{q} + 3n^2 \Gamma I C \right] - I^{-1}\Omega I n_1 \dot{\theta} + k\dot{\theta} \right) \qquad (52)$$

Taking the norm of (52) where

$$\left\| kT - kI^{-1}\Omega I \right\| \le R_0 \qquad (53)$$

$$\left\| 3n^2 I^{-1}\Gamma I C \right\| \le M_0 \qquad (54)$$

$$\left\| I^{-1} d_E(t) \right\| \le d_0 \qquad (55)$$

we have

$$\dot{V} = s^T \dot{s} \le -\left(N_1 - R_0 \right)\|\theta\|\|s\| - \left(N_2 - d_0 - \delta_0 - M_0 \right)\|s\|$$
$$\le -\left(N_1 - R_0 \right)\|\theta\|\|s\| - \eta\|s\| \qquad (56)$$

where

$$\eta = N_2 - d_0 - d_I \qquad (57)$$

and the internal disturbance is:

$$d_I = \delta_0 - M_0 \qquad (58)$$

For providing negativeness of the $\dot{V}$ it is required that the following sliding mode existence conditions should be satisfied:

$$N_1 - R_0 \ge 0 \quad \text{or} \quad N_1 \ge R_0 \qquad (59)$$

$$\eta \ge 0 \quad \text{or} \quad N_2 \ge d_0 + d_I \qquad (60)$$

Moreover, from (53) and (54) after some evaluations it is easy to design sliding gain constant k as:

$$k \leq \sqrt{R_0/M_0} \tag{61}$$

In result, (56) can be evaluated as:

$$\dot{V} = s^T \dot{s} \leq -\left(N_1 - R_0\right)\|\theta\|\|s\| - \eta\|s\| \leq -\eta\|s\| < 0 \tag{62}$$

Therefore, the sliding manifold $s(t) = 0$ is reached in finite time [2]: $t_{s_1} \leq \|s(0)\|/\eta$.

4.2.1 Modification of switching function

Thrusters apply discontinuous external force for stabilization of the nonlinear attitude dynamics of the spacecraft in finite time with limited thrust force. Control system using attitude thrusters is operated for large attitude angle orientations and its faster stabilization. Note that, thrusters are not required to operate for small attitude correction. Hence, here sliding function s can be modified to two-side dead-band (see Figure 5) to stop thruster operation for small attitude errors:

$$s = \gamma(s) = \begin{cases} s - s_d & s \geq s_d \\ 0 & -s_d < s < s_d \\ s + s_d & s \leq -s_d \end{cases} \tag{63}$$

where $\pm s_d$ is the upper and lower limit of dead-band. Therefore, the control action (50) forces the system to the dead-band limits of sliding manifold $s \pm s_d = 0$ and keeps it in dead zone. As shown [4], dead-zones can have a number of possible effects on control system. Their most common effect is to decrease static output accuracy. They can actually stabilize a system or suppress self-oscillations. For example, if a dead-zone is incorporated into ideal relay, it may lead to the avoidance of the oscillation at the contact point of the relay.

However, the sliding mode existence conditions should be investigated for the following three cases. If $s \geq s_d > 0$, $N \geq R_0$, and $N_2 \geq d_0 + d_I$ then

$$\begin{aligned} \dot{V} &\leq -\|\gamma(s)\|\left(N_1 - R_0\right)\|\theta\| - \|\gamma(s)\|\left(N_2 - d_0 - d_I\right) \\ &\leq -\|s - s_d\|\left(N_1 - R_0\right)\|\theta\| - \|s - s_d\|\left(N_2 - d_0 - d_I\right) \\ &\leq -\eta\|s - s_d\| \end{aligned} \tag{64}$$

If $-s_d < s < 0$ or $s_d > s > 0$, which corresponds to no control action with $\gamma(s) = 0$ we have

$$\dot{V} = \left[R_0\|\theta\| + \left(d_0 + d_I\right)\right]\|\gamma(s)\| \tag{65}$$

If $s \leq s_d < 0$, $N_1 \geq R_0$, and $N_2 \geq d_0 + d_I$ then

$$\begin{aligned} \dot{V} &\leq -\|\gamma(s)\|\left(N_1 - R_0\right)\|\theta\| - \|\gamma(s)\|\left(N_2 - d_0 - d_I\right) \\ &\leq -\|s + s_d\|\left(N_1 - R_0\right)\|\theta\| - \|s + s_d\|\left(N_2 - d_0 - d_I\right) \\ &\leq -\eta\|s + s_d\| \end{aligned} \tag{66}$$

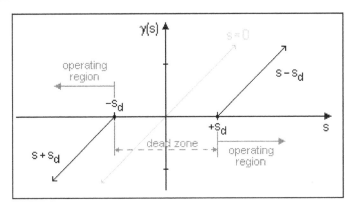

Fig. 5. Dead-band.

Note that, (64) and (66) are operating regime of the attitude sliding mode controller that produces external control torques for stabilization via thrusters. Thus, the controller (50) forces the system to reach dead-band on switching function at finite time. On the other hand, if sliding variable has a value between $-s_d \leq s \leq s_d$ then the sliding mode controller do not take any action. So, the system behavior is determined by attitude dynamics (33) with $u_B = 0$. Therefore, strictly, $\dot{V} \leq 0$ for two operating cases of $\gamma(s)$ (63). This means that system is convergent.

4.2.2 Example

The design of attitude sliding mode controller begins with selection of appropriate sliding manifold. A design parameter k can be selected from (61):

$$k_{max} = \sqrt{\frac{R}{M_0}} = \sqrt{\frac{0.0189}{0.3029}} = 0.25 \tag{67}$$

$$k = 0.25 \leq k_{max} \tag{68}$$

we chose $k = 0.25$ which usually gives better performance for sliding manifold [5]. Inertia matrix can be constructed form Table 1 as:

$$I = \begin{bmatrix} I_{11} & 0 & 0 \\ 0 & I_{22} & 0 \\ 0 & 0 & I_{33} \end{bmatrix} = \begin{bmatrix} 3026 & 0 & 0 \\ 0 & 440 & 0 \\ 0 & 0 & 3164 \end{bmatrix} \tag{69}$$

The design parameters N_1 and N_2 of attitude controller can be determined from conditions (59) and (60). Since control torque is limited by maximum available thrust and geometric configuration of thrusters, attitude controller parameters for a station keeping maneuver must satisfy the following physical condition:

$$I\left(N_1 \|\theta\| + N_2\right) \leq 10 \tag{70}$$

with bounded external disturbance:

$$d_E(t) = 0.5\sin(\omega_d t) \le d_E \tag{71}$$

where ω_d is the frequency of external disturbance, and according to (55) Frobenius norm is $d_E = 0.5$. Consider that a station keeping maneuver at 6 a.m. is required with initial attitude errors as $\theta_1 = 0\,\text{deg}$, $\theta_2 = 10\,\text{deg}$ $(0.1745\,rad)$, $\theta_3 = 0\,\text{deg}$. The variable structure controller parameters can be calculated for considered station keeping maneuver as:

$$N_1 = 0.07 \ge R = 0.0683$$
$$\|\theta\|_F = \left\| \begin{bmatrix} 0 & 0.1745 & 0 \end{bmatrix}^T \right\|_F = 0.1745 \tag{72}$$
$$N_2 = 0.005 \ge \left\| I^{-1}d_E \right\| + \|d_I\| \cong 0.0048$$

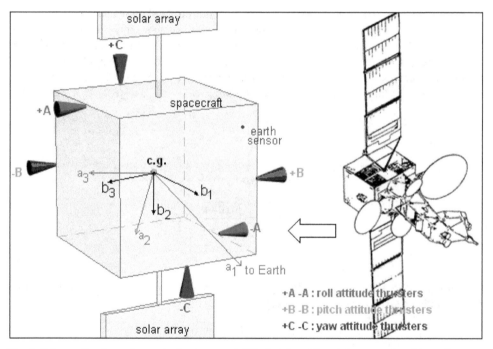

Fig. 6. Schematic view of geosynchronous satellite Intelsat-V and thrusters configuration: +A/-A corresponds to roll attitude thrusters, +B/-B corresponds to pitch attitude thrusters, and +C/-C corresponds to yaw attitude thrusters.

Note that, design parameters N_1, N_2 and $\|\theta\|$ satisfy condition (59) and (60) for considered maneuver. Additionally, the dead-band limit is practically chosen according to thruster performance as $s_d = 0.012$. Therefore, variable structure control algorithm (50) with sliding manifold (49) can be formed as below. Also Figure 7 illustrates behavior of switching function with dead zone

$$u_B = -I\left(0.07\|\theta\| + 0.005\right)sign(\gamma(s))$$

$$\gamma(s) = \begin{cases} s - 0.012 & s \ge 0.012 \\ 0 & -0.012 < s < 0.012 \\ s + 0.012 & s \le -0.012 \end{cases} \qquad (73)$$

$$s = \omega + 0.25\theta$$

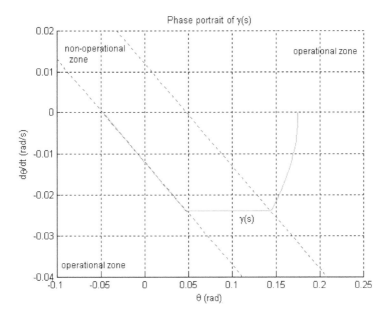

Fig. 7. Phase portrait of switching function of controlled body.

In this section nonlinear spacecraft dynamics (33) with flexible solar arrays (35) of a geosynchronous communication satellite are simulated with variable structure attitude and vibration controllers by Matlab-Simulink with iteration step of 0.1 seconds. Block diagram of satellite control system is shown in Figure 8. The time responses of attitude angles, angular velocities and accelerations; Sliding function and vibration control action and generated by solar array driving mechanism; modal coordinates are given in Figure 9. Note that, control command is illustrated in ($1/s^2$).

As seen from Figure 9 attitude controller stabilizes the nonlinear model of flexible spacecraft approximately in 20 seconds and the sliding manifold is reached in 5 seconds at left side of dead zone. Vibration suppression of flexible solar arrays is achieved about 3-5 seconds and the sliding manifold is reached in 0.8 seconds. The station keeping attitude control performances and vibration suppressions with designed controllers are sufficient for faster stabilization and limited firings.

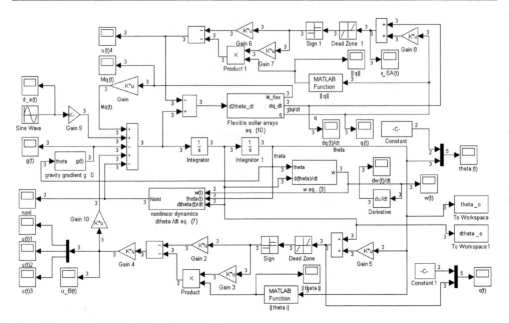

Fig. 8. Block diagram of satellite control system.

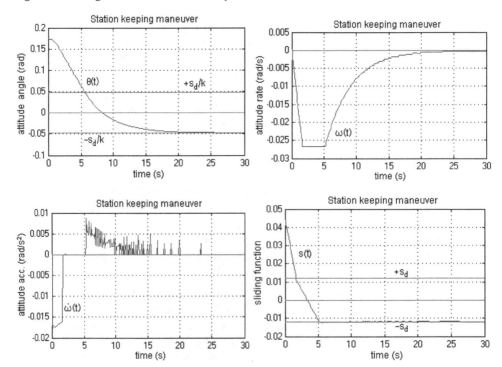

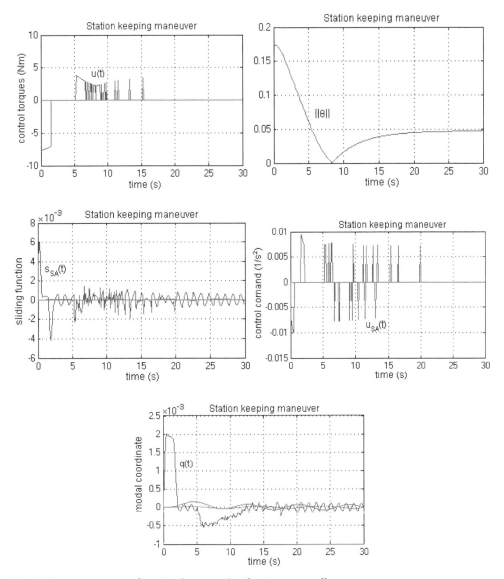

Fig. 9. Time responses of station keeping & vibration controller.

4.3 Comparison

Comparison of simulation results for both designed satellite control systems are done in the table. As seen from first part of Table 3 proposed variable structure controller in Section 4.1 has a large settling time then other considered attitude control systems. Since considered internal actuators has maximum 1.5 Nm torque capability, the settling time is three time shorter than [9] where same satellite (see Table 1) has been considered.

Design	Attitude controller properties			Vibration controller properties	
	controller-type	chattering	settling time	controller-type	settling time
[5]	VSC	no	45 s	-	-
[6]	VSC	-	60 s	-	-
[7]	VSC	no	50 s	-	-
[8]	VSC	yes	10 s	-	-
[9]	PID	-	300 s	-	-
[12]	VSC	yes	15-20 s	-	-
Sec. 4.1	VSC	yes	120 s	-	-
[11]	VSC	little	100 s	VSC	10-15 s
[13]	Adaptive VSC	no	20 s	Linear	15-20 s
[14]	Bang-Bang, nl. s(t)	-	50 s	Pos.Pos.FB	100 s
[15]	Adaptive VSC	no	30 s	Input shaping	10-20 s
[16]	PD+PWPF	yes	20 s	Pos.Pos.FB	10 s
Sec. 4.2	VS P+relay	little	15-20 s	VS P+relay	5 s

Pos.Pos.FB: Positive position feedback, nl: nonlinear

Table 3. Comparison analyses.

Proposed variable structure P+relay attitude controller in Section 4.2 has a little chattering and settling time about 15-20 seconds for a simple station keeping attitude maneuver and VS P+relay vibration controller has also a little chattering (because of introducing dead-band) and settling time about 5 seconds. From second part of Table 3 proposed P+relay controller provides relatively good control performances. Obtained results are preferable for both designs with considered reaction wheels and thrust-jets than that of [6], [7], [8], [11], [14], [15] nevertheless the other results shown in Table 3 also are acceptable for their operational conditions.

5. Conclusions

In this chapter, we have first introduced rigid-body dynamics of orbiting spacecraft. Then we have developed equation of motion of a rigid-body in circular orbit with internal torquers. We have also considered solar arrays and write equation of motion for a flexible spacecraft model. At the second stage we have designed variable structure control system for a rigid body controlled with reaction wheels. Then we have proposed P+relay control technique to design attitude and vibration control systems for a satellite with solar arrays. Finally we had a comparison of simulation results of proposed control techniques with literature.

Modeling and simulation results show that proposed variable structure attitude control system for a rigid body stabilizes nonlinear dynamics successfully and performs satisfactory settling time and control torque with compared results. On the other hand, proposed P+relay attitude and vibration controller for geosynchronous satellite with flexible solar arrays successfully stabilizes the nonlinear model with external disturbances by using minimum fuel for considered initial conditions. We suggest using of P+relay control technique for station keeping maneuvering of flexible satellites.

6. Nomenclature

$\vec{\rho}$: position vector from center of mass of a small mass element

$\vec{R}$: position vector from an inertial origin

dm : a small mass element

n : orbital rate

M_G : gravity gradient torque

$\vec{H}$: angular momentum vector

$\vec{\omega}$: angular velocity vector

$\vec{\theta}$: attitude angle vector

I : inertia matrix

C : direction cosine matrix

Γ : skew symmetric matrix of angular velocity

Ω : skew symmetric matrix of direction matrix

d_E : external disturbance

q : modal coordinates vector

a_1, a_2, a_3 : unit vectors of a local horizontal and a local vertical ref. frame

b_1, b_2, b_3 : unit vectors of body ref. frame

s : sliding manifold

u_B : control vector

V : Lyapunov function candidate

7. References

[1] Utkin, V.I., 1993. "Sliding mode control design principles and applications to electric drives", IEEE Transactions on Industrial Electronics, 40, pp. 23-36.

[2] Utkin, V.I., 1992. Sliding Modes in Control Optimization, Springer-Verlag, Berlin.

[3] Hung. J.Y., 1993. Variable structure control: a survey. IEEE Transactions on Industrial Electronics, 40, pp. 2-22.

[4] Slotine, J-J.E., and Li, W., 1991. Applied Nonlinear Control, Prentice Hall, New Jersey.

[5] Siew-Chong Tan; Lai, Y.M.; Tse, C.K.; Martinez-Salamero, L.; Chi-Kin Wu, 2007. A fast-response sliding-mode controller for boost-type converters with a wide range of operating conditions, IEEE Trans. on Industrial Electronics, vol. 54, no. 6, pp. 3276-3286.

[6] Vadali, S.R., 1996. Variable-structure control of spacecraft large-angle maneuvers, Journal of Guidance, 9, pp. 235-239.

[7] Sira-Ramirez, H., and Dwyer, T.A.W., 1987. Variable structure controller design for a spacecraft nutation damping, IEEE Transactions on Automatic Control, 32, pp. 435-438.

[8] Lo, S-C., Chen, Y-P, 1995. Smooth sliding-mode control for spacecraft attitude tracking maneuvers. Journal of Guidance, Control, and Dynamics, 18, pp.1345-1349.

[9] Singh, S.N., Iyer, A., 1989. Nonlinear decoupling sliding mode control and attitude control of spacecraft. IEEE Transactions on Aerospace and Electronic Systems, 25, pp. 621-633.

[10] Wie, B., 1998. Space Vehicle Dynamics and Control, AIAA, Virginia.

[11] Froelich, R, Papapoff, H., 1959, Reaction wheel attitude control for space vehicles, IRE Transactions on Automatic Control, vol. 4, pp. 139-149

[12] Öz, H., 1993. Variable structure control of flexible spacecraft, Variable Structure Control for Robotics and Aerospace Applications, K.K.D.Young (Edt.), Elsevier, Amsterdam.

[13] Agrawal, B.N., "Bang, H., 1995. Robust closed-loop controller design for spacecraft slew maneuver using thrusters", Journal of Guidance, Control, and Dynamics, 18, pp. 1336-144.

[14] Singh, S.N., De Araujo, A.D., 1999. Adaptive control and stabilization of elastic spacecraft. IEEE Transactions on Aerospace and Electronic Systems, 35, pp. 115-122.

[15] Hu, Q., Ma., G., 2005. Variable structure control and active vibration suppression of flexible spacecraft during attitude maneuver, Aerospace Science and Technology, 9, pp. 307-317.

[16] Hu, Q., Ma., G., 2007. Adaptive variable structure and command shaped vibration control of flexible spacecraft, Journal of Guidance, Control, and Dynamics, 30, pp. 804-815.

[17] Hu, Q., Ma., G., 2005. Vibration suspension of flexible spacecraft during attitude maneuvers, Journal of Guidance, Control, and Dynamics, 28, pp. 377-380.

[18] Abdulhamitbilal, E., and Jafarov, E.M., 2006. "Performances comparison of linear and sliding mode attitude controller for flexible spacecraft with reaction wheels", in Proceeding of 9th Workshop on VSS, Alghero, Italy, June 5-7.

[19] Abdulhamitbilal, E., Jafarov, E.M., 2007, Sliding mode controller design for nonlinear flexible geosynchronous satellite with thrust jets. VSS'08, Proceedings of 10th International Workshop on Variable Structure Systems, Antalya, Turkey, June 8-10.

9

Design and Optimization of HVAC System of Spacecraft

Xiangli Li

*Dalian University of Technology,
China*

1. Introduction

1.1 Background

From Manned spacecraft and space shuttle to the scale of space station, the technology of manned spacecraft has been developing. The astronauts have to work and live in the cabin for much longer time. Therefore, the spacecraft environmental control and life support systems is not only asked to control the cabin environment parameters within a certain range, but also to ensure the cabin environment with high thermal comfort which can meet the physical and psychological needs of astronauts, also improve the efficiency of equipments, structural components in the manned space System. The ventilation, air conditioning problems and the air flow arrangement of the cabin directly affect the environmental parameters controlling and the thermal comfort of the cabin environment. So, it has an important significance to research the ventilation, air quality, thermal environment and comfort of the astronauts in the cabin under the microgravity condition.

There is $10^{-3} \sim 10^{-6}$ -g_0 level of micro-gravity (g_0=9.8 m^2 /s) inside the cabin of spacecraft or the space station. At this point, the phenomena which are common with ground gravity such as natural convection, static pressure differential and sedimentation are greatly reduced. Therefore, forced ventilation is crucially essential to achieve the exchange of matter and energy in cabin under the micro-gravity conditions. With changes of the mission and flight time, improvement of air ventilation system in the manned spacecraft cabin determines the comfort of astronauts. The way of ventilation in such confined spaces like small cabin should give priority to the centralized air supply system.

The environment inside of the space station is similar to a building on the planet. It is quite necessary to solve the design problems of air-conditioning of cabin in order to meet the astronauts' requirement of comfort when they live and work in the space station or the spacecraft, and moreover variety of spacecraft equipments, structural components and the organisms in spacecraft are unable to withstand large temperature fluctuations. In order to ensure equipments working in the normal environment and improve their performance, it is required that the spacecraft thermal control system not only ensures the maintenance of normal temperature, but also provide a constant temperature environment for some equipments. Therefore, temperature and humidity as well as the conditions of ventilation ensure the operating efficiency of equipment, structural components in the spacecraft.

1.2 Particularity of spacecraft cabin air-conditioner design

1.2.1 The cabin is a confined space, where the pressure can be 1 atmospheric pressure (20.95% oxygen) of mixed oxygen and nitrogen like the earth's environment, or 1/3 atmospheric pressure of pure oxygen atmosphere, or 1/2 atmospheric pressure (40% oxygen and 60% nitrogen). With high cabin pressure, thermal capacity and heat transfer capacity, oxygen is provided and regenerated by the ECLSS, and the cabin carbon dioxide produced by human body is also disposed or restored by it.

1.2.2 The heat load mainly comes from the astronauts' metabolic heat (145 W / person), equipment cooling and solar radiation out of the spacecraft or the aerodynamic heat when the spacecraft returns. The bulkhead of manned spacecraft is designed with heat insulation. Personnel thermal load composes about 50%. Moisture load includes human respiration and surface evaporation, which is about 1.83 kg / (person per day). So the cabin heat-moisture ratio F =heat load/moisture load = 6850 kJ / kg (without considering cabin leak). It is necessary to dispose the cabin air with cooling and desiccation.

1.2.3 The recycle and prevention of condensation water in the air. Condensation will cause damage to the equipments, and water exists in the form of droplet under micro-gravity circumstance is also dangerous, which will affect the recycle of precious condensation water. It can be seen from the psychometric chart, the higher the air temperature, the greater the relative humidity and dew point temperature, and vice versa. The most suitable cabin environment is 1 atmospheric pressure (20.95% oxygen, 0.04% carbon dioxide), with temperature of 22 °C~27 °C, relative humidity of 30% to 70%, flow rate 0.2~0.5 m / s, then the dew point temperature is 11 °C~23 °C.

1.2.4 Centralized ventilation helps to balance the cabin temperature and remove the harmful gas by forced convection, which is also helpful for human comfort and equipment use. The temperature is controlled by the volume of the air in the condenser. The humidity is controlled by the dew point temperature. The harmful trace gases and the pressure control will be managed by the ECLSS. The general active temperature control technology utilizes the air through the fan, damper and heat exchangers to achieve the purpose of cooling desiccation, cooperated with fans to ventilate the cabin. Coolant circulation loop accumulates the waste heat and delivers them to the collection equipments like the cooling board, then transfers to the waste heat sink through space radiation radiator.

1.2.5 Because the operating conditions of spacecraft always changes, it requires that the air-conditioning system can meet the multi-state operation mode. Spacecraft's general flight state can be divided into two parts: manned combination flight phase and unmanned flight phase. The design of spacecraft air conditioning system should ensure the requirements of the most adverse conditions and meet the need for checking other operating conditions.

2. The design steps and methods of air-conditioning system in spacecraft

The air-conditioning system of spacecraft can be designed with reference to that of the building air-conditioning system. First, the appropriate air flow and air supply parameters should be determined based on the consideration of heat and moisture load in the cabin. These parameters are not only supposed to meet the requirements of human comfort and ventilation, but also to minimize the amount of air to reduce the size of wind pipe and equipment, also to save the space and reduce aircraft noise within the spacecraft. Hence the optimization of ventilation system parameters is needed to be taken. On this basis, the air flow and piping organization can be designed. Here we use a test chamber to illustrate the design process.

2.1 Principles and processes

The air-conditioning systems of the test cabin can be divided into two parts: instrument zone ventilation system is shown in figure 1 and human activity zone ventilation system is shown in figure 2. The human activity zone is at the middle of test cabin and surrounding area is instrument zone. You can see the arrangement of air-conditioning systems from the figures. The pipe network of instrument zone is consisted of pipe sections and clapboards. This two systems can be combined with some connecting pipe sections.

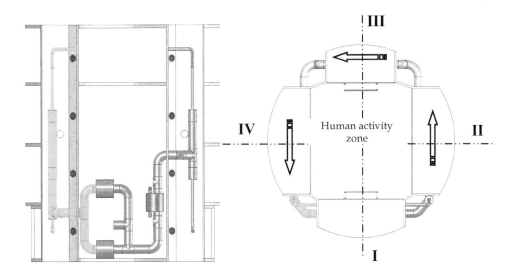

Fig. 1. Model of ventilation system duct layout of instrument zone in test cabin.

According to the different temperature control requirements of human activity area and instrument area, the design of ventilation and air conditioning system should include two independent options based on the two different areas under normal circumstances. Instrument zone generally has no moisture load which is simpler than human activity zone, so the air conditioning system design can refer to the design of human activity zone. The ventilation and air conditioning flow path of human activity zone is shown in Figure 3. The system is composed by condensation dryer, fan, air duct (Pipe network) and some other annexes. The regulation of air temperature and humidity in human activity zone is achieved by regulating the quality of flow into the refrigerant dryers, thereby changing the air supply parameters. System maintains a constant air volume.

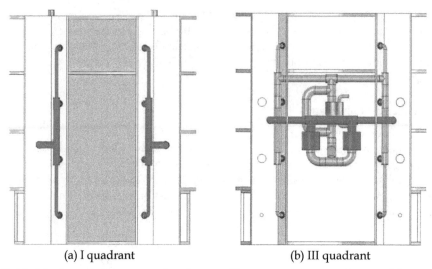

(a) I quadrant   (b) III quadrant

Fig. 2. Model of ventilation system duct layout of human activity zone in test cabin.

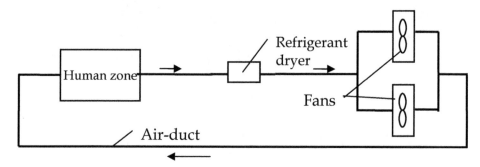

Fig. 3. Schematic diagram of ventilating system in human activity zone.

2.2 Calculation and selection of inlet parameters

In test cabin, the heat load of human activity zone Q =540W, Moisture load W=86.5×3=0.07208g/s. The air temperature control range of human activity area is $21 \pm 4°C$, and the relative humidity is 30%~70%.

The Heat to moisture ratio of the test cabin is:

$$\varepsilon = \frac{Q}{W} = \frac{540}{0.07208} = 7491.67 \tag{1}$$

Under the circumstances that the heat to moisture ratio and design conditions of human activity area are defined, if the air temperature decreases, the air flow of ventilation and air conditioning system increases, and the sense of wind strengthens. A big air temperature difference may affect people's comfort, so it is necessary to determine the appropriate air temperature difference. To compare the supply air temperature difference, assuming the design temperature and relative humidity are t_N =21°C, φ_N =50%, respectively. If the supply air temperature difference is Δt_s, the supply air temperature is 21 °C- Δt_s.

According to the calculation method [1], on the psychometric chart, over the status point N draw heat to moisture ratio line with ε = 7491.67 KJ/Kg, and intersection point with 21°C− Δt_s isotherm line is the air condition point S, and then obtained the air supply volume:

$$G = \frac{Q}{\rho(i_N - i_s)} \tag{2}$$

Where G — air supply volume, m³/s; Q — heat load, 0.54kW; ρ — air density, 1.2kg/ m³; i_N — Air enthalpy of human activity area, 41kJ/kg; i_s — Enthalpy of air supply state point, kJ/kg.

The determination of the air supply state points is shown in Figure 4.

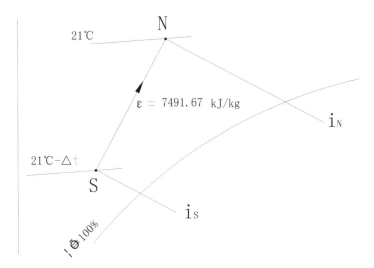

Fig. 4. Establishment of supply air condition-point of human activity zone.

Depending on the different air supply temperature difference, the enthalpy of air supply state point and the corresponding air supply, the calculation results is shown in Table 1.

Δt(°C)	1	2	3	4	5	6	7	8
t_s (°C)	20	19	18	17	16	15	14	13
i_s (kJ/kg)	39.52	37.98	36.43	34.89	33.34	31.81	30.26	28.73
G(m³/min)	17.419	8.729	5.815	4.369	3.493	2.916	2.498	2.188

Table 1. Enthalpy and air supply with different air supply temperature difference.

Calculation results of air supply under different air supply temperature difference are shown in Figure 5.

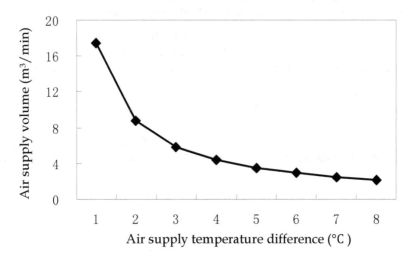

Fig. 5. Air supply volume based on different air supply temperature difference.

It can be seen from Figure 5, the air supply reduces sharply when the air temperature difference begins to increase from 1°C, then the air supply reduces slowly when the air temperature difference reaches to 4°C ~ 5°C. This conclusion is obtained with a precondition that the design temperature and relative humidity are 21 °C and 50%, respectively. Within the allowable range of temperature and humidity of the human activity area, the air supply changes in the same way with the air supply temperature 'variation. Therefore, to the consideration of the comfort and transmission energy consumption of fans, the air supply temperature can be set at 4°C. A larger temperature difference is also fine. So consider the needs of comfort and air flow organization, the general air supply temperature difference can be set at 6 °C or so.

The amount of air supply of human activity area changes with different design parameters after the air supply temperature difference is confirmed. Table 2 shows the calculation results of interior state point's enthalpy, supply air state point's enthalpy and air supply when the indoor temperature is 17 °C, 21 °C and 25 °C, relative humidity is 30%, 50% and 70%, respectively. The calculation results of Air supply under the nine cases are shown in Figure 6.

t_0 °C	17			21			25		
ϕ_N (%)	30%	50%	70%	30%	50%	70%	30%	50%	70%
i_N (kJ/kg)	26.36	32.55	38.78	33.06	41.07	49.15	40.43	50.71	61.13
i_s (kJ/kg)	20.23	26.39	32.57	26.92	34.89	42.93	34.26	44.50	54.83
G(m³/min)	4.405	4.383	4.348	4.397	4.369	4.337	4.379	4.348	4.286

Table 2. Enthalpy and air supply under different design parameters.

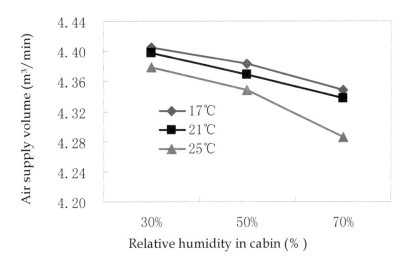

Fig. 6. Air supply under different design parameters.

It can be seen from the calculation results that air supply increases with the reduction of indoor design temperature and design relative humidity. However, the impact of design parameters on the air supply is very small. In the permitted range of temperature and humidity, air supply flow only varies between m³/min~4.405m³/min. In order to ensure reliability, the air supply volume is determined as 4.405 m³/min. A large air flow is beneficial for the uniformity of air distribution.

2.3 The design of air flow and piping organization

Taking into account the characteristics of the spatial shape and equipments' arrangement, the final air flow and piping organization can be describes as follows: double-sided air outlet is deposited on the corners above, while corresponding double-sided air inlet is deposited at the nether corners. Its characteristic is that the working zone is located in the recirculation air flow with an even temperature field. It requires that the outlet is laid close to the top, return air outlet should be located in the same side with supply air outlet. Then the final plan of air organization is determined as centralized air supply in the up corner and air return in the bottom corner. Select the double-outlet louvers with a turbulence coefficient α =0.14 and effective area coefficient is 0.72. Air supply outlet is arranged in the length direction of the human activity zone in test cabin. The calculated process is as follows:

2.3.1 Air outlet type

Taking into account the spatial shape(LxBxH=4mx1.8mx2.0m)of the human activity zone and the features of equipment layout in test cabin, "double-outlet" type is selected, and its turbulence coefficient α =0.14. Air supply outlet is arranged in the length direction of the human activity zone in test cabin, and the range x =B—0.5=1.3m. The minus 0.5m is the no constant temperature district near the cabin wall.

2.3.2 Air supply temperature difference and air supply volume

From the above calculation, air supply temperature difference and air supply are Δt_s=4°C, G=4.75 m³/min, respectively.

2.3.3 Speed of air supply

For the sidewall air supply, the equation (3) gives the calculation method of maximum air supply speed:

$$v_s \leq 0.103 \frac{BHk}{G_s} \tag{3}$$

Where G_s —the air supply volume ,m/s;
         k — coefficient of valid area, k=0.72.

Based on the known parameters and formula (3), B=1.8m, H=2m, the result of v_s is 3.64m/s, $v_s = 3.5$ m/s can be used for the velocity of air supply (To prevent air noise, air supply speed should be within 2 ~ 5m / s, this result meet the requirements.)

2.3.4 The number of air supply outlet

The freedom degree of air supply jet can be calculated by equation (4) as follow:

$$\varpi = \frac{\sqrt{F_n}}{d_0} = 0.89\sqrt{\frac{HBv_s k}{L_s}} \tag{4}$$

Where F_n — cross-area of room space afforded by each outlet, m²;
         d_0 — area equivalent diameter of rectangle outlet, m;
         ϖ — freedom degree of air supply jet

For the human activity zone, the freedom degree of air supply jet is 10

According to the value of $\frac{\Delta t_x}{\Delta t_s}\frac{\sqrt{F_n}}{d_0}$, the zero dimension distance $\bar{x}$ can be obtained by checking the chart of axis temperature difference die-away curve of no equivalent temperature jet flow. Δt_x is the temperature difference between indoor air temperature and axis temperature. In this example, $\Delta t_x = 0.5$°C, Δt_s =4°C

$$\frac{\Delta t_x}{\Delta t_s}\frac{\sqrt{F_n}}{d_0} = 1.25 \tag{5}$$

The $\bar{x}$ is equal to 0.25 when check the curve chart. Then:

$$N = \frac{H \cdot B}{\left(\dfrac{\alpha \cdot x}{\overline{x}}\right)} = \frac{2 \times 1}{\dfrac{0.14 \times 1.3}{0.25}} = 3.77$$

The two-side air supply is used and there are 4 outlets each side. In this calculation, the width of room B is half of practice broad width. So it is 1.0m.

2.3.5 The size of air supply outlet

The area of each air supply outlet is:

$$f = \frac{G_s}{v_s \cdot N \cdot k} = \frac{4.405}{60 \times 3.5 \times 8 \times 0.72} = 0.00364 \text{ m}^2 \tag{6}$$

Area equivalent diameter is determined by equation 7:

$$d_0 = 1.128\sqrt{f} = 1.128 \times \sqrt{0.00364} = 68 \text{ mm} \tag{7}$$

So the sine of double-outlet louvers air supply outlet can be 80mm×80mm The real speed of air supply outlet is $v_s = \dfrac{4.405}{8 \times 60 \times 0.08 \times 0.08} = 1.43$ m/s.

2.3.6 Check the adhesion length

The air conditioning space is available if the adhesion length is larger than the length of cabin. It can be checked by the Archimedes number A_r.

$$A_r = \frac{g d_0 \Delta t_s}{v_0^2 T_n} = \frac{9.8 \times 0.068 \times 4}{3.5^2 \times (273 + 21)}$$

Where the T_n is the absolute temperature of indoor cabin. From the table research by A_r, the adhesion length $x = 3.8$, which is more than range 1.3m. It is noteworthy that the acceleration of gravity is setting for 9.81m/s2. While in the microgravity environment of the aircraft, the Archimedes number A_r will be greatly reduced. So the adhesion lengths significant increases in benefit to the supply air to meet the design requirements.

There are many forms of air supply outlet in air-conditioning systems of the civilian. Here we just calculated the slit-type outlet. Calculate the number of air supply outlet, layout and area of outlet based on air flow organization [2]. Finally, four outlet louvers for each side on the top with the size of 80mm×80mm.The actual wind speed is 1.43m/s.

The ventilation system of test cabin is determined as centralized air supply in the up corner and air return in the bottom corner. The 3D model of human activity area is shown in Figure 7. In addition to the air supply pipe and the return air pipe, condensing dryer, fan (one with a prepared) and air flow control valves and other fixed equipment are placed in the back of quadrant I in test cabin. Return air is dried by condensing dryer and pressured by blower, then passes through the two air ducts of equipment area in quadrant II. The duct layout is shown in Figure 8.

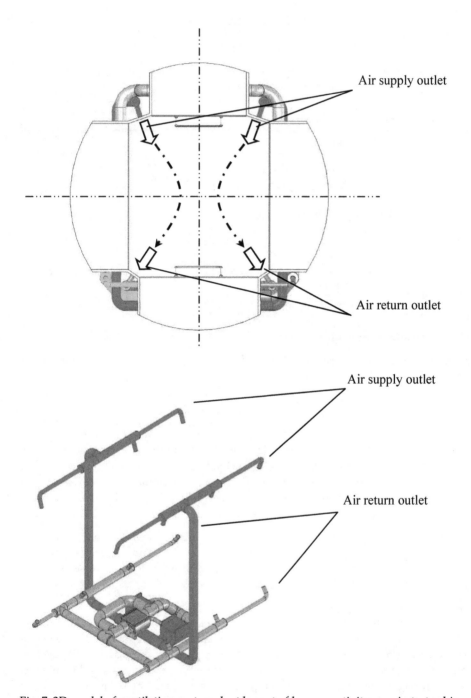

Air supply outlet

Air return outlet

Air supply outlet

Air return outlet

Fig. 7. 3D model of ventilation system duct layout of human activity area in test cabin.

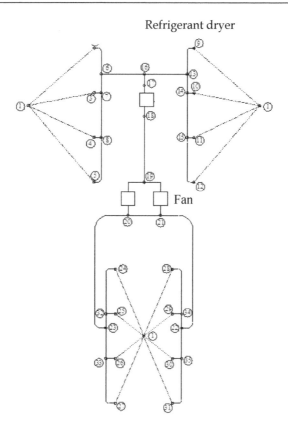

Fig. 8. Ventilation system duct layout diagram of human activity area.

2.4 Hydraulic calculation of the most unfavorable loop

After the pipe layout is done, the pipe diameter and the system resistance should be determined through hydraulic calculations, then determine the fan flow and pressure head, and finish the equipment selection. The flow speed-assumed method can be used in Hydraulic calculation. The recommended value of flow speed is used based on the technical and economic requirements. If the value is relatively large, this can save pipe and space, but the power of device and noise will increase; If the value is relatively small, it will waste the pipe. So, many factors should be taken into account when select pipe diameter. Then calculate the resistance based on the pipe diameter determined by flow speed.

2.4.1 Calculation of the resistance along the way

$$p_f = \lambda \frac{l}{d} \cdot \frac{\rho v^2}{2}$$
(8)

Where l —Pipe length,m;
         d —Pipe diameter,m;

v — Average velocity of cross-section area,m/s;

λ — Resistance coefficient along the way;

ρ — Air density,1.2 kg/m3

The calculation of λ is in accordance with the formula (9), and can be used in the three districts of turbulent.

$$\lambda = 0.11(\frac{K}{d} + \frac{68}{\text{Re}})^{0.25} \tag{9}$$

Where K — Duct roughness, 0.01-0.02 mm for PVC;

Re — Reynolds number, $Re = \dfrac{vd\rho}{\mu}$,where μ is the Dynamic viscosity coefficient

with a value of 1.884×10-5 Pa.s.

2.4.2 Calculation of local resistance

$$p_m = \zeta \cdot \frac{\rho v^2}{2} \tag{10}$$

Where ζ — Local resistance coefficient

For different structural forms of resistance components, the methods of local resistance coefficient are different. Local resistances of all conditions are listed in paper [2].

The design air flow of human activity area in test cabin is 4.405m3/min, the total resistance loss of the most unfavorable loop is 168.13Pa, the resistance loss of the condensation dryer is 50Pa.The fan type is 5-50No2C,and its rated air flow is 4.575m3/min, rated pressure head is 157.7Pa, axis power is 14W.

3. Verification and optimization of hydraulic condition of air-conditioning system

The previous section introduces the preliminary design of the air conditioning system of aircraft. System will form multiple loops when it runs, as there are several function areas (such as human activity areas and equipment areas, etc.) in the aircraft cabin. The systems are independent on the preliminary design stage. When considering system's running conditions, some problems such as whether the previously selected devices (such as air ducts and fans) can meet the requirements of different operating conditions, and whether the selected pipe diameter is reasonable have not been resolved, so the verification and optimization work of hydraulic condition should be carried out.

Two factors need to be considered when carry out the verification and optimization work, one is optimization goal, the other one is simulation of the hydraulic condition. The air flow speed of network is low, so it may consider as steady flow. Basic circuit analysis method or node method [5] can be adopted in hydraulic condition simulation. The frictional resistance coefficient of air duct can be calculated in the explicit format, and the local loss coefficient can be obtained from the manufacturer's manual. Pump head can be approximately expressed by 5-order polynomial. As shown in the figure bellow:

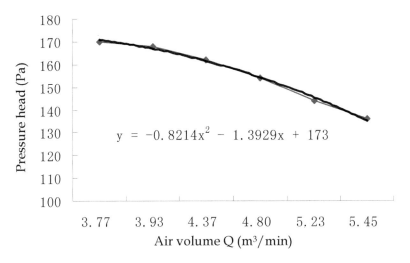

Fig. 9. Fan performance curve of human activity area.

The problems of spacecraft ventilation network is that when the design flow of each user is known, determine the optimization of network loop pipe's adjustment process, and the optimization target is the minimum fan power. This problem can be resolved by penalty function method[6].The method is to add one or more constraint function to the objective function, and the punishment item of the objective function is added based on any punishment against the constraints. The following typical form is:

$$Min L(\mathbf{X},\ U)=f(\mathbf{X})+U\sum_{i=1}^{m}h_i^2(\mathbf{X})+U\sum_{j=1}^{r}\left[\min\left(0,g_j(\mathbf{X})\right)\right]^2 \tag{11}$$

While $L(\mathbf{X},U)$ is the penalty functions; $f(\mathbf{X})$ is the objective function, $\mathbf{X} = (x_1, x_2, \cdots, x_n)^{\mathrm{T}}$;$U$ is the weight factor, or punish parameters; R^n is n dimension Euclid space, $f \cdot h_i \cdot g_j$ are continuous scalar functions on R^n, where h_i is equality constrained conditions, g_j is the inequality constraint conditions.

The optimization target of loop analysis model is the minimum fan power. Its model can be summarized as the formula (12) to equation (14).

$$f(\mathbf{R}_b) = \sum_{i=1}^{b}R_i\left|Q_i^3\right| \tag{12}$$

$$h_i(\mathbf{R}_b) = \mathbf{C}|\mathbf{Q}_b|\mathbf{Q}_b\mathbf{R}_b - \mathbf{Ch}_{\mathrm{F}} = 0 \tag{13}$$

$$g_j(\mathbf{R}_b) = \mathbf{R}_b \geq 0 \tag{14}$$

Where R and Q is the branch resistance and air flow, respectively; b means the number of network branches, i is basic loop number, j for the branch number; C is correlation matrix of the the basic circuits-branch; $\mathbf{R}_b$ is b dimensional column vector, $\mathbf{R}_b = (R_1 \ \ldots \ R_k \ \ldots R_b)^{\mathrm{T}}$;

Fan pressure head $\mathbf{h}_F = (h_{F1} \ \cdots \ h_{Fk} \ \cdots \ h_{Fb})^T$;

The original impedance matrix: $\mathbf{R} = \begin{bmatrix} R_1 & \cdots & 0 & \cdots & 0 \\ \vdots & \vdots & \vdots & \vdots & \vdots \\ 0 & \cdots & R_k & \cdots & 0 \\ \vdots & \vdots & \vdots & \vdots & \vdots \\ 0 & \cdots & 0 & \cdots & R_b \end{bmatrix}$;

$|Q_b|$ and Q_b both are b×b diagonal matrix:

$$|\mathbf{Q}_b| = \begin{bmatrix} |Q_1| & \cdots & 0 & \cdots & 0 \\ \vdots & \vdots & \vdots & \vdots & \vdots \\ 0 & \cdots & |Q_k| & \cdots & 0 \\ \vdots & \vdots & \vdots & \vdots & \vdots \\ 0 & \cdots & 0 & \cdots & |Q_b| \end{bmatrix}, \quad \mathbf{Q}_b = \begin{bmatrix} Q_1 & \cdots & 0 & \cdots & 0 \\ \vdots & \vdots & \vdots & \vdots & \vdots \\ 0 & \cdots & Q_k & \cdots & 0 \\ \vdots & \vdots & \vdots & \vdots & \vdots \\ 0 & \cdots & 0 & \cdots & Q_b \end{bmatrix}.$$

The minimized penalty function which is corresponding to equation (11) is:

$$Min \ L(\mathbf{R}_b, U) = f(\mathbf{R}_b) + U \sum_{i=1}^{m} h_i^2(\mathbf{R}_b) + U \sum_{j=1}^{r} \left[\min\left(0, g_j(\mathbf{R}_b)\right) \right]^2 \tag{15}$$

Because the algorithm requires a gradient, so:

$$\nabla L(\mathbf{R}_b) = \nabla f(\mathbf{R}_b) + U \sum_{i=1}^{m} 2h_i(\mathbf{R}_b)\nabla h_i(\mathbf{R}_b) + U \sum_{j=1}^{r} 2\min\left(0, g_j(\mathbf{R}_b)\right) \cdot \nabla g_j(\mathbf{R}_b) \tag{16}$$

Each constraint item in equation (16) can be removed by the square of the gradient vector and it can be standardized, and then use the gradient's norm of the objective function multiplying each constraint (plus 1 to avoid the gradients close to zero). So the gradient expression of penalty function is given as follows:

$$\nabla L(\mathbf{R}_b) = \nabla f(\mathbf{R}_b) + 2U \sum_{i=1}^{m} \frac{\|\nabla f(\mathbf{R}_b)\| + 1}{\|\nabla h_i(\mathbf{R}_b)\|^2} h_i(\mathbf{R}_b)\nabla h_i(\mathbf{R}_b) +$$
$$+2U \sum_{j=1}^{r} \frac{\|\nabla f(\mathbf{R}_b)\| + 1}{\|\nabla g_j(\mathbf{R}_b)\|^2} \min\left(0, g_j(\mathbf{R}_b)\right) \cdot \nabla g_j(\mathbf{R}_b) \tag{17}$$

In the adjacent area of the solutions, the objective and constraint functions' norm of gradient vector can be considered as constant. As a result, the penalty function which is corresponding to equation (17) can be described as follows:

$$L(\mathbf{R}_b) = f(\mathbf{R}_b) + U' \sum_{i=1}^{m} h_i^2(\mathbf{R}_b) + U'' \sum_{j=1}^{r} \left[\min\left(0, g_j(\mathbf{R}_b)\right) \right]^2 \tag{18}$$

where:

$$U' = U \frac{\|\nabla f(\mathbf{R}_b)\| + 1}{\|\nabla h_i(\mathbf{R}_b)\|^2} \tag{19}$$

$$U'' = U \frac{\|\nabla f(\mathbf{R}_b)\| + 1}{\|\nabla g_j(\mathbf{R}_b)\|^2} \tag{20}$$

In order to avoid too much gradient of penalty function setting, it is necessary to make further adjustments. If the norm of equation (17) is more than the norm of the objective function plus 1, there:

$$\nabla L(\mathbf{R}_b) = U \frac{(\|\nabla f(\mathbf{R}_b)\| + 1)\nabla L(\mathbf{R}_b)}{\|\nabla L(\mathbf{R}_b)\|} \tag{21}$$

So far, each branch's resistance characteristic coefficient R_i can be solved by penalty function method. Above model can be realized through FORTRAN language program.

4. Summary

The design of ventilation and air conditioning system is not big but more complex, and the requirements of reliability is much higher than that of civil air-conditioning, noise and fan energy consumption is also should be strictly controlled. So it is necessary to optimize and adjust the pipeline network after the preliminary design and actual working condition simulation are finished. Before simulation optimization, deviation of some pipe flow is large. The deviation of pipe flow and design flow can be greatly reduced through adjustment of the fan model and part of the pipe diameter. A study shows that, the fan pressure head after optimization is nearly 10% less compared to the total head loss of the most unfavorable loop [7].

5. Symbols

Q	—	heat load	W	—	moisture load
ε	—	heat to moisture ratio	t_N	—	design temperature
ρ	—	density	ϕ_N	—	relative humidity
Δt	—	temperature difference	i	—	enthalpy
G	—	air supply volume	α	—	turbulence coefficient
L	—	length	B	—	width
H	—	high	F	—	area
d	—	diameter	v_0	—	speed of air
$\bar{x}$	—	dimensionless distance	N	—	number of air outlet
A_r	—	the Archimedes number	λ	—	resistance coefficient along the way
p_f	—	resistance along the way	K	—	duct roughness
Re	—	Reynolds number	μ	—	the Dynamic viscosity coefficient
p_m	—	local resistance	ξ	—	local resistance coefficient

6. References

[1] Lu Yajun, Ma Zuiliang, Zou Pinhua. Heating, ventilating and air conditioning [M] Peking: China Architecture & Building Press, 2002

[2] Lu Yaoqing. Practical design manual for heating and air conditioning[M] Peking: China Architecture & Building Press, 2008

[3] Zhao Rongyi, Fan Cunyang, Xue Dianhua. Air condionding[M] Peking: China Architecture & Building Press, 1994

[4] Zheng Zhonghai, Zhang Jili, Liang Zhen. Numerical simulation of upside air-supply bottom-side air-return symmetric ventilation in space station cabin. Journal of Harbin Institute of Technology, 2007, Vol 39, No. 2, p 270-273

[5] Wang Shugang, Sun Duobin. Theory of steady flow pipe network[M] Peking: China coal industry press. 2007

[6] Wan Yaoqing, Liang Gengrong, Chen Zhiqiang. Common program compilation of optimization calculating method. Peking: Worker press, 1983

[7] Li Xiang-li, Wang Shu-Gang, Jiang Jun et al. Hydraulic simulation and optimization of manned spacecraft ventilating system. Dalian Ligong Daxue Xuebao/Journal of Dalian University of Technology, 2010, Vol 50, No. 3, p 399-403

10

Resolving the Difficulties Encountered by JPL Interplanetary Robotic Spacecraft in Flight

Paula S. Morgan

Jet Propulsion Laboratory, California Institute of Technology,
USA

1. Introduction

Although many precautionary measures are taken to preclude failures and malfunctions from occurring in Jet Propulsion Laboratory (JPL) interplanetary robotic spacecraft before launch, unexpected faults and off-nominal conditions do happen in flight. Also, as spacecraft age, electrical and mechanical parts are expected to degrade in performance. Unlike aircraft vehicles, once robotic spacecraft are launched, they cannot be returned to the hangar for repairs. Maintaining the health and functionality of robotic spacecraft, probes, rovers, and their compliment of science instruments is an ongoing challenge which must be met throughout the lifetime of every mission. When unexpected or anomalous events arise, the Spacecraft Operations ground-based Flight Support (SOFS) team of engineers for that particular spacecraft must troubleshoot the problem and implement a solution within the allowable time constraints.

Fig. 1. JPL's Galileo Spacecraft: Mission to Jupiter.

Degradation of spacecraft components can occur from several different sources. Material stresses caused by environmental effects such as solar heating or the cold of deep space and solar radiation bombardment can contribute to malfunctions in subsystem components.

Additionally, autonomously running Flight Software (FSW) sequences and in-flight computer coding upgrades periodically sent to the spacecraft can potentially introduce human-induced faults. Further, as spacecraft design sophistication and complexity increases, failure modes increase in number, and fault diagnosis & resolution becomes a more difficult and time-consuming task for the SOFS team to handle. In order to meet mission constraints, timely solutions must be implemented for handling the task of collecting large volumes of telemetered data from the spacecraft which are compared with archived historical data & spacecraft design information to determine failure causes and implement fault resolution actions. Additionally, interplanetary spacecraft missions that experience large Earth-spacecraft distances (such as exploration missions to the outer planets of our solar system), present an additional challenge since the ever-increasing time delay between commands sent by the SOFS team and return telemetry received by the spacecraft limits the ability to respond to failure occurrences in a timely manner. This Round

Fig. 2. JPL's Cassini Spacecraft: Mission to Saturn.

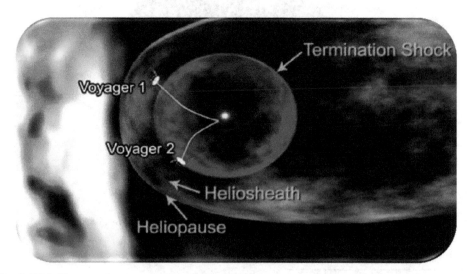

Fig. 3. JPL's Voyager Spacecraft: Interstellar Mission.

Trip Light (and radio) Time (RTLT) delay between ground commanding and spacecraft data delivery back to the SOFS team is especially of concern when critical "one-chance mission events" must take place at a specific time (such as deploying a probe while flying by a planet's moon), or when serious, potentially mission-catastrophic failures occur so quickly that they must be fixed immediately.

To protect robotic spacecraft from these types of hazards and limitations, mission robustness is enhanced by implementing several strategies to provide a spacecraft system with greater integrity and diagnostic capability. This system health management approach is employed by several means: implementing "flight rules" and mission design constraints, applying functional redundancy through FSW, adding redundant hardware, and applying Fault Protection (FP) techniques which consist of automated response routines containing preprogrammed instructions to respond to failure conditions. This FP strategy involves autonomous monitoring of component operation to ensure device health, evaluation of internal and external conditions, and monitoring power allocation to spacecraft devices. In general, most JPL robotic spacecraft require some unique mission specific FP, but the majority of spacecraft configurations contain FP algorithms which protect the command and data processing capabilities, maintaining attitude control of the vehicle, protection against Earth-communication loss with the spacecraft, ensuring that safe external and internal temperature levels are maintained, and recovery from power overloads or power loss. To accommodate the majority of anticipated faults, most spacecraft are equipped with a general-purpose "Safe-Mode response routine" that configures the spacecraft to a reduced power state that is power-positive, thermally stable, in a communicative state, with a known predictable spacecraft configuration so that diagnosis of more complex faults can be addressed by the SOFS Team. Optimization of spacecraft post-fault recovery time is achieved through the development of automated tools and pre-determined "recovery procedures" which contain pre-defined actions for the SOFS team to follow which greatly reduces post-fault recovery time.

This chapter details the challenges and difficulties encountered by several JPL interplanetary spacecraft missions during the course of their mission flight phases and describes the solutions and workarounds implemented by their supporting SOFS ground teams to protect their mission objectives.

2. Background: Health & safety concerns and preventative measures

Once JPL spacecraft are ferried out of Earth's gravity well, usually by multi-stage rocket, it will either enter Earth's orbit or proceed directly out into deep space. Through the use of NASA's Deep Space Network (DSN) radio telescope system, the spacecraft's SOFS team of engineers will stay in contact with the vehicle, providing instructions through "uplinked" commands while the spacecraft's "downlink" telemetry stream of data provides detailed information of all it encounters throughout its mission lifetime. Upon the deployment, configuration, and verification of all its systems following launch, the propulsion system will be utilized to target the spacecraft to the intended destination through its trajectory. JPL's interplanetary spacecraft mission objectives typically consist of orbiting or flying by an object, moon, or planet, or landing the spacecraft or its probe on a target object. The suite of scientific instruments carried onboard the spacecraft will perform many scientific tasks throughout the lifetime of the mission. As spacecraft journey through the vastness of space, many factors will provide a challenge in maintaining spacecraft health and functionality. All of these risk factors must be taken into account when designing JPL spacecraft, even those influences and events which may be unforeseen.

In order for spacecraft systems to function properly, both external and internal temperatures must be monitored, regulated, and controlled during the entire lifetime of the spacecraft's mission. Exposure to the sun's heat is one of the most detrimental external influences on spacecraft operation in the vacuum of space if the vehicle flies in close proximity to this celestial body. The spacecraft's surfaces can superheat when exposed to the sun, while shadowed surfaces can fall to extremely low temperatures. Material stress can result from this thermal expansion-contraction effect, leading to uneven heating. This uneven heating can lead to warpage, breakage of components, or camera distortion. To help alleviate some of these problems, spacecraft are equipped with fault-preventative devices such as optical solar reflectors, mirror tiles, or multi-layer insulation thermal blankets which will reflect the sun's heat and radiation so that the spacecraft is somewhat protected against overheating, while retaining its internal heat to prevent too much cooling. Adverse thermal environmental conditions must be avoided since computers and spacecraft components will cease to work if spacecraft temperatures become too extreme (Qualitative Reasoning Group, 2005). Additional precautions must also be taken to ensure that instruments do not fall out of operating limits, since many devices are designed to operate within a narrow range of temperatures. Also important is the spacecraft's interior environment which must be properly managed as well, since heat build-up can occur from the spacecraft's own systems. One method employed to regulate internal temperatures is circulating the spacecraft's own gas or liquids (fuel) to cool its interior. Equally important is the thermal state of these substances since they must be maintained to ensure that they do not freeze from deep space exposure. This condition would render the propellant unusable so that the spacecraft would not be able to maneuver, eventually becoming misaligned with Earth so that no signals could be sent or received by the spacecraft.

Although precautionary measures are taken to preclude the possibility of human-induced electro-static discharge events (static electricity discharge) within spacecraft components during the manufacturing process, "latent failures" can occur after launch, rendering the device useless or partially useless. Additionally, human error can also be introduced within command sequences which are continuously generated and sent to the spacecraft. These sequences contain instructions for controlling the spacecraft's activities such as tracking Earth, monitoring celestial references for attitude targeting, performing maneuvers to fine-tune the trajectory when required, and carrying out science calibration and operations. These command sequences are all subject to human error which can potentially cause serious faults. One example would be accidentally turning off a radio transmitter or receiver device onboard the spacecraft, thus preventing communication with earth. Another fault could be turning on too many spacecraft instruments and components at the same time so that the spacecraft's power source (solar panels, Radioisotope Thermoelectric Generators (RTGs), fuel cells, etc.) are unable to provide the power required to support all operating systems. This condition is referred to as a "spacecraft-wide under-voltage power-outage" in which loss of power to critical devices can occur, such as the computers which must maintain their power levels to retain computer memory. Automated FP routines are implemented to resolve this type of condition, which is further detailed in an example covered later in this chapter.

Although radio waves travel at the speed of light, making spacecraft-earth transactions almost instantaneous near earth, as the distance between earth and the spacecraft increases, even a signal traveling at the speed of light can take hours. This lag time becomes a high-risk deterrent to fault recovery when spacecraft are sent out great distances like the Galileo, Cassini, and Voyager missions. Under some anomalous conditions, it is impossible for spacecraft to respond to ground commands quickly enough to preclude a catastrophic failure from occurring. An example would be the failure of a latch valve to close properly in the propulsion maneuvering system after re-pressurization of the spacecraft's fuel tanks has commenced. This type of fault can cause the tank pressure to rise substantially in a very short amount of time. If this condition were to occur on the Cassini spacecraft (Mission-to-Saturn) where the RTLT is approximately 3 hours, the pressure level could potentially reach a catastrophic point before the pressure measurement data could even reach earth to indicate that the fault condition has occurred, since Cassini's telemetry stream takes well over an hour to reach SOFS personnel from its Saturn-Titan orbit position. This "lag time" problem especially becomes a concern for spacecraft missions that contain one-time opportunities such as planet/moon encounters. For these events, the timing is crucial since only one opportunity exists to meet the objective and there may be no second chance. These unique events must proceed without the threat of fault interference in order for the spacecraft's mission to be successful (Morgan, 2011).

Another concern for spacecraft systems is Electromagnetic Compatibility (EMC) between components. When designing spacecraft subsystems, Electromagnetic Interference (EMI) effects must be minimized so that the spacecraft's systems function properly within their intended operational environment, without adversely affecting or being adversely effected by other spacecraft components. Spacecraft subsystems can become ineffective or malfunction if neighboring devices are not designed to minimize their EMI effects when operating simultaneously. To ensure component compatibility, EMI assessment and testing are required pre-launch to avoid undesirable electromagnetic fields, conducted voltages, and currents. As an example, the Cassini mission implemented a study to preclude EMI effects from other subsystem devices on the Duel Technique Magnetometer (MAG) science instrument. The MAG device consists of two instruments which have been mounted along an 11-meter boom apparatus to minimize spacecraft component EMI effects. During the early project phase, several engineering components and science instruments were identified to be potential magnetic interference sources (e.g. Traveling Wave Tube Amplifiers (TWTA), Propulsion Module Subsystem (PMS) multiple latch valves, Power & Pyro Subsystem (PPS) latch relays, etc.). Pre-launch preliminary assessments indicated that the permanent magnets contained in these subsystem devices had the potential to impact the upcoming MAG science experiments. A Magnetics Control Review Board (MCRB) was established to address EMC issues to ensure that magnetic cleanliness was maintained between devices (Narvaez, 2002). Participants representing these subsystem devices discussed precautionary measures such as shielding methods, implementation of magnetic compensation, wiring layouts to minimize loop areas, and replacing magnetic materials with non-magnetic materials. This effort led to the establishment of requirements and guidelines to assist hardware designers in developing EMC strategies which would produce minimal magnetic field output. The MCRB committee stressed implementing these fixes as early as possible in the design phase to allow for flexibility in the available solutions. Amongst the fixes implemented on behalf of the MAG instrument EMI reduction effort were: 1) both TWTAs were packed side-by-side within their housing so that their respective magnetic field polarities would be configured in opposing directions, 2) the PPS subsystem arranged all magnetic latch relays to occur in pairs, with their respective magnetic poles opposite to each other (provides self-cancelation; for odd number relays, a small compensation magnet was installed to neutralize the field), 3) a theoretical model was produced for the four RTGs which provided optimum compensation for the selected arrangement of clocking angles. For those subsystems which could not reduce their EMI effects or replace high magnetic materials with non-magnetic materials, magnetic compensation was implemented. The most significant magnetic compensation was installed into the PMS latch valve components. In this case, each latch valve was measured and magnetically compensated with magnets which contained the same dipole moment (opposing). Following these EMI reduction applications, each one of Cassini's components was tested in order to verify its respective magnetic cleanliness for the overall system, prior to its final installation on the spacecraft.

In addition to the above challenges, many spacecraft designs have become more complex throughout the last several years. As a result, fault diagnosis and resolution becomes a more difficult and time-consuming task to undertake since fault causes can lead to a plethora of possibilities for these very complicated systems. This poses a substantial challenge for the SOFS Team whose task it is to collect large volumes of telemetry data needed to diagnose faults and propose resolution actions. This can be an arduous, time consuming manual process, sometimes requiring hundreds of data products from the spacecraft's telemetry stream to be compared to archived historical data, as well as design information in order to evaluate the problem to propose a solution. To aid the fault diagnosis and solution process, automated FP routines are typically implemented into the spacecraft's FSW to deal with the majority of possible failure conditions; this FP is designed to protect for any Single Point Failure (SPF) conditions that might arise (unless proven extremely unlikely; waiver issued), with the following priorities in mind:

1. Protect critical spacecraft functionality
2. Protect spacecraft performance and consumables
3. Minimize disruptions to normal sequence operations
4. Simplify SOFS recovery response

These FP groundrules are typically implemented with the following principle in mind, following any anomaly: Ensure the spacecraft's commandability remains intact as well as the maintenance of its systems; to remain in a stable, safe state for a pre-determined period of time following any anomaly (e.g. for the Cassini spacecraft, this period is two weeks, by which time the SOFS team should be able to recover the spacecraft and restart its onboard sequence). Sections 3.2.1 & 3.2.2 detail two examples of FP implementation from the Cassini mission.

3. FP groundrules for JPL spacecraft

Each JPL spacecraft is unique in its configuration and mission objectives and the task of implementing autonomous FP must be considered carefully according to its configuration, expected environment, component design, and its operational tasks, although some FP is approached in a generic manner. In general, autonomous fault protection should only be implemented on-board the spacecraft for those fault conditions where a ground response is not feasible or practical, or if fault resolution action is required within a pre-defined period of time of detecting the failure. Otherwise, the ground system should have adequate time to respond to the fault and should be responsible for the fault recovery. In both cases, the ground is responsible for failure diagnosis and re-configuration of the spacecraft to nominal operations after the fault. Some spacecraft designs may be quite simple (e.g. lack propulsion and attitude control subsystems entirely, such as an atmospheric probe), and some spacecraft are quite complex, but many spacecraft share common systems which require a similar approach in FP design (Morgan, 2005).

3.1 Fault protection typically implemented into JPL spacecraft

Some spacecraft have design configurations simple enough to warrant only minimal fault protection which is meant to address any type of fault condition that might occur, yet other spacecraft designs are so complex and sophisticated, with long mission durations, that they must maintain a system which may present numerous error possibilities. Most spacecraft typically rely on a "general-purpose, Safe-Mode" fault response which typically configures the spacecraft to a lower power state by turning off all nonessential spacecraft loads, commanding a thermally safe attitude, providing a safe state for the hardware, establishing an uplink and a downlink, reconfiguring to a low-gain antenna, and terminating the command sequence currently executing on the spacecraft. This type of response is used to configure the spacecraft into safe and predictable state so that the SOFS team has enough time to evaluate the fault causes and determine a solution.

FP typically implemented into JPL spacecraft designs also includes an automated response to address "loss of spacecraft signal" faults that affect the SOFS team's ability to communicate with the spacecraft. Failure to receive the spacecraft's uplink signal can be caused by a number of problems which include ground antenna failures, environmental interferences, spacecraft hardware failures, as well as an erroneous spacecraft attitude (pointing error), radio frequency interferences, or an error introduced in an uplinked sequence (e.g. radio transmitter device accidentally turned off). If the spacecraft has experienced these types of failures and is no longer able to receive commands from the ground, a FP response can be implemented to help re-establish the uplink. This type of FP is referred to as a "Command Loss Response" (from the perspective of the spacecraft, that it is no longer receiving ground commands) which is typically an "endless-loop" response (see Section 3.2.1).

Another FP algorithm typically installed into spacecraft is for recovery from a system-wide loss of power. This is referred to as "Under-Voltage" recovery, and can be caused by a number of fault conditions depending on the spacecraft design (i.e. oversubscribing the power available, a short in the power system, or a communications bus overload). Should a system-wide power loss occur, not even the Safe-Mode response will execute since the main computer will also lose power thus causing loss of the mission. Therefore, FP must be implemented to detect the power level drop so that the system may automatically shed its non-essential loads from the communications bus, isolate the defective device, and re-establish essential hardware. The quick actions of this response allow critical spacecraft memories to be maintained throughout the Under-Voltage event (see Section 3.2.2).

FP monitors detect anomalous conditions using predefined "trigger values" which are referred to as "thresholds" or "redlines," that represent the value at which an anomalous condition is present. The monitor design may also include logic which detects for, and ignores data from failed sensors. "Consecutive occurrence counters" are also used in some FP monitors; these are referred to as "persistence filters" and are implemented for a variety of reasons: to ensure that transient occurrences do not trigger a response, to satisfy hardware turn-on constraints, or to allow other FP monitors to detect faults first. SOFS personnel can also enable or disable the spacecraft's monitors and responses during the mission as appropriate. This is accomplished through a FSW flag which may be manipulated by the team. For the most part, the FP is designed assuming that these flags will be enabled throughout the mission; however, some exceptions to this strategy exist:

- The response is only appropriate when the associated device is powered on & operating
- The response is required only for specific mission events
- The response is not appropriate for a particular event
- The response is not compatible with the currently operating sequence

Figures (4a) through (4d) depict four JPL spacecraft designs with quite different mission objectives, which employ typical and mission unique FP (Ball Corp., 2001; JPL, 1997, 2005).

CloudSat: Earth Orbiting Satellite
Standard FP: 3 Safe-Mode Responses
         5 Under-voltage Responses
         Memory Scrubber & Bus FP
Unique FP: Significant computer & thermal FP

Stardust Spacecraft: Inner Solar System; Comet Explorer
Standard FP: 1 Safe-Mode Response
         1 Under-voltage Response
         1 Command Loss Response
         Memory Scrubber & Bus FP
Unique FP: Some computer & thermal FP

Fig. (4a). CloudSat Spacecraft FP Allocation.   Fig. (4b). Stardust Spacecraft FP Allocation.

Cassini Spacecraft: Outer Solar System Explorer
Standard FP: 1 Safe-Mode Response
         1 Post-Safe Mode Response
         1 Under-voltage Response
         1 Command Loss Response
         Memory Scrubber & Bus FP
Unique FP: Significant command & data processing computer FP, radio unit FP, thermal FP, fuel tank pressure FP, attitude articulation and control computer FP

Voyager Spacecraft: *Interstellar Explorer*
Standard FP: 1 Under-voltage Response
         1 Command Loss Response
Unique FP: Computer FP, attitude articulation and control computer FP, and radio unit FP

Fig. (4c). Cassini Spacecraft FP Allocation.

Fig. (4d). Voyager Spacecraft FP Allocation.

3.2 FP examples from the Cassini-Huygens mission-to-Saturn spacecraft

The Cassini–Huygens spacecraft is a joint NASA/ESA/ASI mission to the Saturnian system sent to study the planet and its many natural satellites. The craft was launched from Cape Canaveral on October 15, 1997 following nearly two decades of development. It is comprised of a Saturn orbiter (shown in Figure 2) and an atmospheric probe/lander to investigate the moon Titan. The Cassini spacecraft has also returned data on a wide variety of tasks including assessment of the heliosphere, planet Jupiter, and has conducted relativity tests. During the early part of its seven-year cruise phase, Cassini's trajectory was fine-tuned by performing "gravity-assist flyby" maneuvers which utilized the inner planets of the solar system. Two of these gravity assist flybys were implemented around Venus (April 26, 1998 & June 21, 1999), one around Earth (August 18, 1999), and one around Jupiter (December 30, 2000) as shown in Figure 5. With the use of this VVEJGA (Venus-Venus-Earth-Jupiter Gravity Assist) trajectory, it took 6.7 years for the Cassini spacecraft to arrive at Saturn in July 2004.

During the 6.7 year cruise phase, several Trajectory Control Maneuvers (TCM) were performed using Cassini's Main Engine (ME) and Reaction Control System (RCS) jets to guide the Spacecraft to its intended destination. Once near the Saturnian system, the "Saturn Orbit Insertion (SOI)" burn maneuver was implemented to slow the craft down so that it could be captured into Saturn's orbit. This marked the beginning of its four-year Orbital Tour phase around Saturn's complex system of moons which is shown in Figure 6. The probe was separated on Christmas Eve 2004, landing on the Titan moon in January 2005. The current end-of-mission plan is for a controlled 2017 Saturn impact (Smith & TPS, 2009).

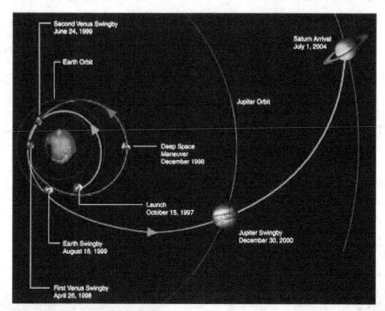

Fig. 5. Cassini's Inner Planet Flyby Schedule.

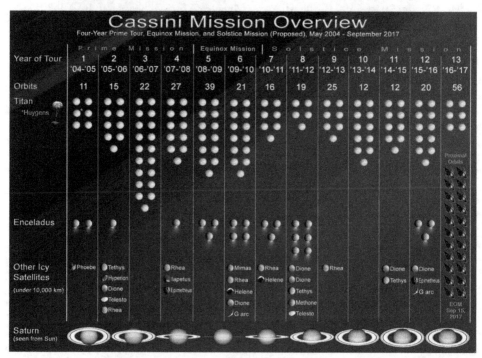

Fig. 6. Cassini's Prime Mission, Extended Mission (XM), and Extended-Extended (XXM) Mission.

3.2.1 Cassini's command loss algorithm

Figure 7 illustrates how the Cassini Spacecraft's Command Loss FP Algorithm addresses faults that can cause ground-spacecraft communications loss; this condition is referred to as "loss of spacecraft commandability." A special "countdown timer" has been implemented into the onboard CDS FSW to keep track of the last time an uplink command was received from the ground operators. This timer decrements continuously (at one second intervals) and is reset back to its "default value" (several days for Cassini) each time an uplink command is received by the spacecraft. The extended absence of uplink commands will eventually lead to the monitor's request for the response, since the timer will eventually decrement to "0". Under these conditions, the assumption is that the spacecraft has experienced a failure where it can no longer receive commands.

Fig. 7. Cassini's Command Loss Response Chain for One CDS Cycle (Endless Loop Response).

Cassini contains redundant units for the Command & Data Computer (CDS), Radio Frequency (RFS) devices, (Deep Space Transponders, TWTAs, Telemetry Control Units (TCU)), as well as three antennas (one High Gain Antenna (HGA) and two Low Gain Antennas (LGA)). The Command Loss Response is divided up into "Command Groups" with "Command Pauses" installed after each group of commands has been executed. These pauses allow several hours (the equivalent of at least two RTLT periods) for the SOFS team to attempt re-acquisition of the spacecraft using the newly response-commanded spacecraft configuration. As shown in the figure, the first Command Group will select the auxiliary oscillator and execute the Safe-Mode response which turns off non-essential loads, commands the spacecraft's High Gain Antenna to the Sun, and places the spacecraft in a known uplink & downlink state. A 15 hour wait period has been installed after this first Command Group to allow sufficient time for the SOFS team to re-establish the uplink, if possible, before hardware swaps begin. If this attempt is unsuccessful, the response will proceed with the next course of actions in Command Group #2 which is to start the series of RFS hardware unit swaps. Five to seven hour wait periods are installed between each subsequent Command Group to allow the SOFS team adequate time to send commands to the spacecraft to re-establish the uplink on the new commanded configuration. At the end of the response chain (approx. 5 days 20 hrs), a swap to the redundant CDS is initiated and the response will activate on the other computer's FSW (the response runs endlessly until an uplink command is received by the ground). The goal of Command Loss FP is to perform hardware swaps and/or re-command the S/C attitude until the ground acquisition is restored. Once the spacecraft successfully receives a command from the ground and the uplink has been re-established, the response will terminate and reset its countdown timer, thus leaving the spacecraft on the last successfully commanded configuration.

3.2.2 Cassini's under-voltage trip algorithm

Cassini's "Under-voltage Trip" monitor and response are shown in Figure 8, "Cassini Spacecraft's Under Voltage FP Actions for Shorted RTG" in which a RTG power unit (one of three on this spacecraft), has shorted. In this example, the Power Subsystem FP senses a power drop below the predefined threshold for the duration of the persistence filter. The

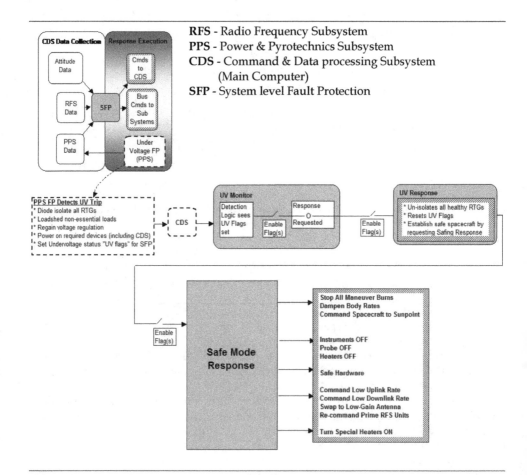

Fig. 8. Cassini's Under Voltage Fault Protection Actions for a Shorted RTG.

first action taken by the Power Subsystem FP is to diode-isolate all three RTGs, turn off (loadshed) all spacecraft non-essential loads, regain the voltage regulation to 30 watts, and then turn on all essential hardware. It also sets three "UV Status Flags" (one for each RTG) to notify System-level FP (SFP) that an Under-Voltage trip event has occurred. Once the CDS becomes operational, it will deliver these UV Status Flags to SFP. SFP's Under Voltage monitor will examine the state of each RTG and if enabled, will request the Under Voltage response. The SFP response un-isolates any correctly operating RTG, unsets its corresponding UV Status Flag, and establishes a predictable, safe spacecraft state by executing the Safe-Mode response.

3.3 Cassini safing response activations to date

On the Cassini project, is the responsibility of the SOFS ground team to support spacecraft activities via the established Mission Plan, to follow established constraints, flight rules, agreed upon waivers, and requirements documentation in order to support the following activities:

- Real-time & near real-time monitoring of subsystem performance
- Ensure subsystem health and safety of all subsystems is maintained
- Develop onboard Spacecraft Sequences (8 – 10 wk segments for Cassini)
- Develop & support ME & RCS maneuvers
- Support science activities (data collection)

as well as designing and uplinking required FSW updates which are needed to meet ongoing mission goals and upcoming events. SOFS tasks also include FSW parameter upgrades, producing Engineering Change Requests (ECR) to implement FSW changes, validation & verification testing, command & uplink strategy of activities, and development of "instructional procedures" for initiating coordinated spacecraft activities between SOFS team members. On Cassini, anomaly resolution is initiated through "recovery procedures" when FP activates, which consists of steps to verify the state of the spacecraft through its telemetry stream, determine the FP that activated, and coordination of recovery steps between team experts (representing RFS, PPS, CDS, SFP etc.) in order to determine the fault cause and resolve the problem as well as reactivating the spacecraft's onboard sequence. To date, 6 activations of the Safing Response (and 'parent' FP routines) have been triggered:

1. 1998 Mar24 – Fault Cause: When the redundant Stellar Reference Unit (SRU) was turned on, it was misaligned with its counterpart SRU
   Diagnosis: This was an unforeseen incident which cannot be modeled in Cassini test facility
   FP Activated: Spacecraft Safing Response
2. 1999 Jan11 – Fault Cause: An overly sensitive attitude control target parameter was implemented in FSW
   Diagnosis: Only flight experience can reveal this problem
   FP Activated: Spacecraft Safing Response
3. 2001 May10 – Fault Cause: The onboard sequence was missing a telemetry mode in the redundant CDS unit which caused the counterpart CDS to Reset
   Diagnosis: Operator error

FP Activated: Internal CDS FP + Spacecraft Safing Response

4. 2003 May12 – Fault Cause: Missing attitude control pointing vector in onboard sequence

    Diagnosis: Operator error

    FP Activated: Spacecraft Safing Response

5. 2007 Sept11 – Fault Cause: Cosmic ray hit the spacecraft (referred to as Single Event Upset (SEU)) which caused the TWTA to turn off

    Diagnosis: Due to environmental effects

    FP Activated: 3 response cycles of TWTA FP + Spacecraft Safing Response (TWTA swap to redundant unit)

6. 2010 Nov02 – Fault Cause: Cosmic ray hit an uplinked command which caused the CDS to swap to its redundant unit

    Diagnosis: Due to environmental effects

    FP Activated: Internal CDS FP + Spacecraft Safing Response

As stated previously, the Safing Response (and internal FP routines) provide a safe spacecraft state with low uplink & downlink rates for the SOFS team to diagnose the fault condition, recover the spacecraft systems, and reactive the onboard sequence. Two weeks was originally allocated for this recovery period. But once the Cassini spacecraft reached the Saturnian system and began its mission Tour phase in 2004, three Orbital Trim Maneuvers (OTMs) were now required for each loop around Saturn-Titan, making the two-week-turnaround period infeasible since a lengthy spacecraft recovery period could cause Cassini to fall off the Tour trajectory. A High Gain Antenna Swap (HAS) Response was therefore designed into the FSW to help the SOFS team improve recovery time. This HAS response executes one hour after the Safing Response activation to increase the downlink rate from 5bps (bits per second) => 1896bps, and the uplink rate from 7.8125bps => 250bps. Figure 9 shows the HAS Response following the Safing Response activation.

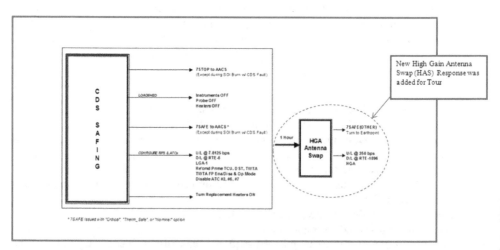

Fig. 9. New HGA Response was added in 2003; Configured to Run 1 hr. after Safing Response Executes.

4. Examples of unforeseen, unplanned for spacecraft problems & SOFS team solutions

Not all spacecraft faults will activate the Safing Response and terminate the onboard command sequence. Some faults are benign enough to allow the sequence to remain in progress since the FP can fix the conditions without intervention from the SOFS team. However, some fault conditions are unforeseen prelaunch, presenting themselves as a new challenge for the SOFS to resolve during the actual flight phase. This section lists a few examples of unexpected faults that have occurred on several JPL spacecraft, without the benefit of preventative FSW, FP, or redundancy to fix the problem. In spite of this fact, all SOFS teams realized that in any spacecraft mission there is always the possibility that new problems can arise due to unknown environmental effects, human errors, or component/science instrument aging.

4.1 Unexpected events for the Cassini-Huygens spacecraft encountered during flight

4.1.1 Environmental errors

4.1.1.1 Solid state power switch SEUs

The Cassini spacecraft consists of 192 SSPS switches which are susceptible to SEUs, caused by galactic rays within the flight environment. One or more photon hits can occur on the voltage comparator resulting in a false indication that the current load is anomalously high. When this condition occurs, the SSPS switch transitions from either an "on" or "off" state to "tripped." The result of this condition can be benign to serious, depending on which switch is tripped, and if it is in use at the time. In May 2005, a SSPS trip event on the spacecraft's ultra stable oscillator caused the SOFS ground team to lose communication with the spacecraft for a short period of time. In September 2007, the TWTA device tripped which activated a FP response, thus causing a Power-On-Reset of the RFS system, and hardware swaps to the redundant Telemetry Control Unit and TWTA device; the Safing Response was also activated (see Section 3.3). Although nothing can be done to reduce or inhibit the occurrence of SEU induced SSPS trips (which are unpredictable and occur sporadically), the SOFS team designed a new algorithm in CDS FSW to respond to these upset events. This new "SSPS Fault Protection" algorithm cycles through one SSPS per second (of 192 switches) and responds to the tripped condition if three consecutive passes through the monitor logic determines that a tripped switch condition is present. A series of predetermined actions have been coded into FSW to respond to the "tripped" condition for each switch, depending on the appropriate action for that load. An example is shown below for the CAssini Plasma

SSPS Switch Number	SSPS Switch (Load)	Log Event?	Cmd Switch "Off"?	Cmd Switch "On"?	Cmd an Alt. Switch?	Alt. Switch SSPS No.	Cmd'ed Switch State
28	CAPS_Elec_LC	Y	Y	N	Y	31	ON
31	CAPS_RHtr_LC (Alternate Switch)						

Table 1. SSPS FP for CAssini Plasma Spectrometer (CAPS) Instrument.

Spectrometer (CAPS) Instrument where if its electrical load current is tripped, the FP will log the event, command the switch "off", and then command its CAPS Replacement Heater Load Current (CAPS_RHtr_LC) "on" to protect the thermal integrity of the device:

As of this the date of this writing, there have been 33 SSPS trip events (25 during the prime mission).

4.1.1.2 Main Engine Assembly (MEA) cover degradation

Cassini's ME assembly requires a cover which must be deployed (closed) when the engines require protection from micrometeoroid and on-orbit dust impacts which often surrounds Saturn and its moons. Shortly before the Deep Space Maneuver (DSM) burn when the MEA cover was stowed (opened), the cover assembly did not open as far as was observed in ground tests. The cover opened 14 degrees less than expected, but the SOFS team demonstrated that this opening angle was adequate to allow for successful main engine burns to commence (on either nozzle). The cause of this degradation in performance of the MEA cover was attributed to the increased stiffness in the cover material (kapton & beta cloth) due to exposure to the space environment which was experienced during flight within the inner solar system, although a period of disuse also contributed to this increased stiffness. These environmental effects cannot be adequately modeled in ground tests. The SOFS team's ongoing response to this unexpected behavior of the cover actuation was to monitor its behavior closely (along with device experts) with results to date demonstrating that the opening angle has remained acceptable through several dozen cycles, with no further signs of degradation observed as depicted above (Figure 10). As of the date of this writing, 66 in-flight cycles have now been performed (Millard & Somawardhana, 2009).

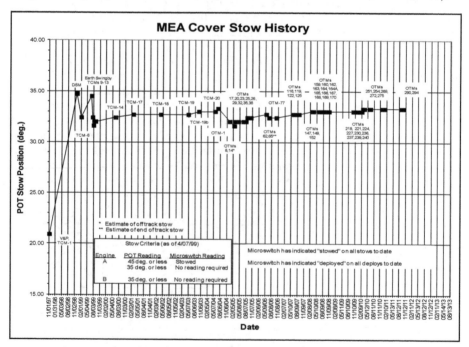

Fig. 10. Main Engine Assembly (MEA) Cover Stow (Open) History ~ Position vs. Mission Time.

4.1.2 Human induced errors

4.1.2.1 Probe transmission design error during Titan moon encounter

The European Space Agency's (ESA) Huygens Probe was piggybacked aboard the Cassini orbiter to capture data from Titan's atmosphere and measure wind effects and surface features once deployed onto this moon. Return of the probe's data was a key element to the success of the joint Cassini-Huygens mission. Since the Huygens Probe had minimal onboard data storage capability, data was to be transmitted to the Cassini orbiter immediately during the Entry, Descent, and Landing (EDL) phase of the probe mission so that the orbiter itself provided the bulk data storage that was needed. To prepare for the probe deployment and relay of its Titan data back to earth, which was to commence in January 2005, end-to-end in-flight tests of the Probe Relay link were performed in February 2000. This was necessary in order to characterize the behavior of the combined Cassini-Huygens system, where the real probe signal was simulated in-flight from the DSN to the spacecraft. During these tests, the signal and data detection thresholds of the receiver were of particular interest. Results confirmed that there was sufficient margin to maintain the carrier and subcarrier lock for the duration of the probe mission, but the digital circuitry which decodes the data from the subcarrier did not have sufficient bandwidth to properly process the data from the subcarrier once it was Doppler shifted by the nominal 5.6 km/s velocity difference between the orbiter and the probe. Since the digital circuit design did not adequately account for the probe data's full Doppler shift, the affect of this anomaly was that it would lead to an unacceptable loss of data during the probe descent to Titan phase. This lead to the formulation of the Huygens Recovery Task Force (HRTF) team, a joint effort between ESA/NASA group of experts to troubleshoot the problem in January 2001. Their efforts led to a three-part solution which allowed recovery of the Titan data.

Firstly, the mission profile was redesigned to provide the Huygens Probe with a trajectory which allowed a low Doppler shift in the probe-Cassini orbiter radio link. This impacted the early part of the Saturn Tour phase resulting in a higher Cassini orbiter flyby altitude of Titan, at 60,000 km, which required redesigning the first two revolutions around Saturn into three revolutions, and then resuming the original planned tour (at a moderate ΔV cost). Second, the Probe Support Avionics assembly was to be commanded to the base frequency (called BITE Mode – a test mode that holds the lockup frequency at a level equivalent to -1m/s relative velocity) by the Cassini orbiter, instead of utilizing the signal at the expected Doppler frequency. This mode of operation was commanded at 12sec intervals (issued by FP; a reserve FP algorithm slot in FSW was utilized to aid in this solution), to ensure that BITE Mode was maintained. Thirdly, the probe's transmitters were pre-heated before probe descent into Titan's atmosphere to optimize the transmit frequency.

The Huygens Probe mission was very successful, with the exception of one (of two) data transmission channel to the orbiter which was not received and recorded (human error). Since all instrument data was duplicated between the two data channel streams, most data was collected with the exception of the Doppler Wind Experiment which relied upon receipt of both channels (Allestad & Standley, 2006.).

4.1.2.2 Imaging Science Subsystem (ISS) haze anomaly

The ISS instrument is a remote sensing device that captures most images in visible light, as well as some infrared and ultraviolet images. By radio telemetry the ISS has returned hundreds of thousands of images of Saturn, its rings, and its moons. The ISS device consists of a Wide-Angle Camera (WAC) to photograph large areas, and a Narrow-Angle Camera (NAC) for areas of fine detail. Each of these cameras utilizes a sensitive Charge-Coupled Device (CCD) as its electromagnetic wave detector, with each CCD having a 1,024 square array of pixels. Both WAC & NAC cameras are configured with spectral filters that rotate on a wheel in order to view different bands within the electromagnetic spectrum ranging from 0.2 to 1.1 μm.

In 2001 (five months after the Jupiter flyby), a distinct haze was observed around Saturn images that were captured by the NAC which had not been seen in previous images. Further analysis of these images indicated that this anomaly was caused by contamination of extremely small particles which resided either upon the filter assembly or the CCD window. The investigation pointed to a decontamination cycle that was performed on May 25, 2001, thirteen months after the previous decontamination cycle which occurred prior to the Jupiter flyby. This indicated that there had been a longer than usual time period for contamination to build up. Additionally, this decontamination cycle had started from a temperature of -90 deg C, whereas all previous cycles had started at 0 deg C. This meant that the Periodic Instrument Maintenance (PIM) had a temperature swing of 120 degrees instead of 30 degrees. A series of decontamination cycles commenced, ranging from seven to fifty-seven days in length. In July 2002, after the final cycle, the haze was no longer present in the images. A new flight rule was instated to prohibit the use of the Level 1 and Level 2 heaters at the same time which prevented heating to 30 deg C and experiencing a large temperature swing such as this event which cause the anomaly (Haemmerle & Gerhard, Undated).

4.1.3 Occurrence of waived failure in flight: Leaking prime PMS regulator

One month after Cassini launched (Nov. 1997), a waived, potentially mission catastrophic Single Point Failure (SPF) occurred in flight. FP design typically dictates that no credible SPF shall prevent attainment of mission objectives or result in a significantly degraded mission, with the exception of the class of faults exempted by waiver due to low probability of occurrence. In this case, a pre-launch waived failure of the Prime Regulator within the PMS failed to properly close. In fact, the regulator exhibited a significant leak rate when the fuel & oxidizer tanks were pressurized for the first time during the Trajectory Correction Maneuver #1 (TCM-1). The leak rate was determined to be 1700 cc/min compared to the expected 1.70 cc/min "worst case leak rate" which was observed in testing. It was determined that the first pyro valve firing prior to TCM-1 event was the cause of this high leak rate, due to a stuck particle in the regulator (from pyro firing debris). The subsequent 90 min DSM burn (initiated at launch +14 months) exhibited an even higher leak rate at an increase of 6.6 times larger than TCM-1. This behavior suggested that an even larger particle had become trapped in the regulator. With this anomaly in place (which was not correctable), all non-critical ME burns to commence during the mission were affected, as well as the critical Saturn Orbit Insertion (SOI) burn maneuver coming up in July 2004.

Fortunately, the Prime Regulator leak problem was discovered several years before the SOI burn was to commence, thereby allowing sufficient time to evaluate the history behind this problem and discover the cause of the anomaly (an important 'lessons learned' for future spacecraft in their development phases) and determine a fix to the mission design. Cassini's pre-launch Regulator design was based upon Galileo's Teflon "soft-seat" configuration which had demonstrated very good performance in flight, exhibiting excellent leakage behavior. However, cold-flow tests indicated that this type of soft-seat design was likely to experience a blocked flow passage due to seat extrusion (potentially a mission catastrophic failure). Galileo's test data was unavailable to evaluate this problem, so that Cassini's soft-seat was replaced with a "hard-seat" to avoid susceptibility to this failed-block condition, with a slight performance difference: the specified leak rate is increased by a factor of "10" with this hard-seat design:

Soft Seat:	1.0×10^{-3} scc/s	3.6 scc/hr
Hard Seat:	1.0×10^{-2} scc/s	36 scc/hr

Table 2. Established Leak Specifications for Soft Seat & Hard Seat Regulator Designs.

Enhancements were incorporated into Cassini's PMS design due to this increased risk in leak rate. A redundant, backup regulator was installed, as well as two new 'Over Pressure' (OP) FP algorithms, which were designed to detect any tank over-pressurization within the fuel and oxidizer tanks, which was to be use for all non-critical ME burns (non-critical mission phases; i.e. not used during the critical SOI Burn event). The pre-launch mission design called for the PMS system to be characterized 30 days prior to the SOI Burn maneuver, so that the OP FP could be disabled. Leak mitigation measures were also added to the PMS plumbing: Two high-pressure helium latch valves (LV10 & LV11), a pyro-isolation ladder upstream of the regulators (PV10-PV15), plus several filters as depicted in Figure 11 (Barber, 2002; Leeds et al., 1996):

These design changes led to a heightened confidence which drove Cassini's mission design and led to the implementation of two waivers for the critical SOI Burn; so that the OP FP algorithms were NOT required during the SOI maneuver:

Waiver #1: Any "Under Pressure" condition is negligible
Waiver #2: Any "Over Pressure" condition is extremely improbable

Fixing the Problem: Cassini's original flight strategy was to lock-up the prime regulator one month after launch, command the LV-10 helium latch valve open to feed helium into the tanks, and then leave LV-10 open for the remainder of the mission. With the leaky Prime Regulator in place, *any* pressurized ME burn had the potential to increase the leak rate. Since the Backup Regulator was also subject to a particulate-induced leak, swapping devices was deemed impractical unless the leak rate increased substantially on the Prime Regulator. Therefore, it was decided that LV10 must be opened just before to any ME pressurization activity, and must be closed as soon as the desired pressure levels were reached. Hence all ME burns had to be initiated via uplinked autonomous command sequences to ensure that the proper timing was maintained. This solution was not applicable to the SOI Burn which was critical to the Cassini mission in that the spacecraft must be decelerated sufficiently in order to be captured into Saturn's orbit (Morgan, 2010).

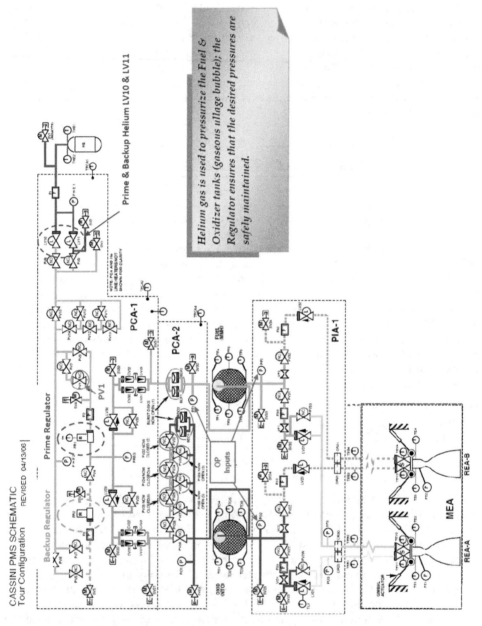

Fig. 11. Cassini's Propulsion System Schematic (ME Only).

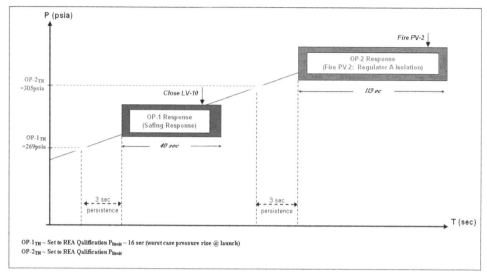

Fig. 12. OP-1 & OP-2 Monitor and Response Configuration.

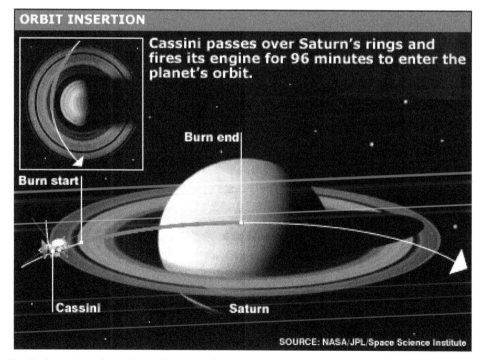

Fig. 13. Insertion of the Cassini Spacecraft into the Saturnian System.

Since the PMS system could no longer be characterized and pressurized 30 days prior to the SOI Burn, the solution was to open LV-10 70sec before SOI Burn would commence, and close LV-10 when the desired tank pressure levels were reached. Identification of new failure modes associated with these changes in SOI Burn strategy were also necessary (e.g. helium LV-10 could become stuck closed, thus requiring an automated swap to the redundant helium LV-11 via FP; the Prime Regulator could fail wide-open or completely closed, thus requiring a swap to the redundant Backup Regulator), and these studies were conducted during the cruise phase of the mission before reaching Saturn.

New/augmented FP changes were incorporated in FSW and uplinked prior to the SOI Burn event, as well as performing characterization studies of LV-10 leakage performance to ensure proper behavior (leak rate within spec). The SOI Burn commenced in July 2004 and was very successful with no faults present; regulator performance was also very good (no increase in leak rate or significant rise in tank pressure level).

4.2 Galileo Mission-to-Jupiter spacecraft unexpected events

4.2.1 Missed launch due to STS-51L Shuttle challenger explosion

The Mission-to-Jupiter Galileo spacecraft was finally launched via Space Shuttle (STS-34) on October 18, 1989 after 11 years of development effort and 6 major mission redesigns. Once completed, Galileo was scheduled to launch onboard Shuttle Atlantis, STS-61G in 1986. The Centaur-G liquid hydrogen-fueled booster stage was to be utilized for a direct trajectory to Planet Jupiter. However, the mission was delayed by the interruption in launches that occurred following the STS-51L Shuttle Challenger disaster. Implemented were new safety protocols as a result of the tragedy which prohibited the use of the Centaur-G stage on Space Shuttle flights, forcing Galileo to use a lower-powered Inertial Upper Stage solid-fuel booster. During the down-time between 1986 and 1988 while the Space Shuttle Investigation was underway, the Galileo team evaluated alternative measures, since the low-powered

Fig. 14. Galileo's Launch.

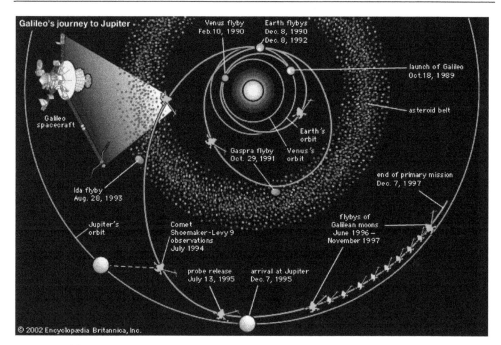

Fig. 15. Galileo's Mission Trajectory.

booster option presented a crisis in that the energy required to achieve a direct trajectory to Jupiter would no longer be possible. The mission was re-profiled to use several gravitational slingshot maneuvers of the spacecraft by the solar system's inner planets, so that a Venus-Earth-Earth Gravity Assist (VEEGA) strategy was designed and implemented in order to provide the additional velocity required to reach its destination. Galileo flew by Venus on February 10, 1990 gaining 8,030 km/hr; flew by Earth twice, the first time on December 8, 1990, then a second flyby of Earth on December 8, 1992, adding 3.7 km/sec to its cumulative speed. In 1994, Galileo was perfectly positioned to observe the fragments of Comet Shoemaker-Levy 9 crash into Jupiter. Galileo released its probe on July 13, 1995, and became the first man-made satellite on December 8, 1995 to enter Jupiter in a 198-day parking orbit.

By the clever use of gravity assists from the inner planets Venus and Earth, a viable mission was possible, although required a much longer flight time to Jupiter. This extended journey required several design modifications which included adding several sun shields to protect the vehicle when flying by Venus. To ensure its systems would survive, Galileo also added operations modifications which included a delay in the deployment of the High Gain Antenna (HGA) until the spacecraft was past the first Earth flyby event.

4.2.2 High gain antenna deployment failure

Galileo's HGA consisted of a metalized mesh fortified by a set of ribs (i.e. similar to an inverted umbrella), held to the support tower by a series of pins and retaining rods. These retaining rods were release shortly after launch but the HGA was maintained in a closed configuration that was thermally protected from the sun until the spacecraft was > 1 AU away (after the first Earth flyby). The SOFS Team communicated with Galileo through its two Low Gain Antennas (LGA). When commanded to deploy on April 11, 1991, the HGA only partially deployed, leaving the HGA mission in jeopardy. An investigation team was organized to rectify the problem where numerous attempts were made to fully deploy the antenna over the next two years, while investigating the alternative of using the LGA to support the Jovian operations segment of the mission. All attempts to fully deploy the HGA were unsuccessful, leaving the HGA antenna nearly useless.

In order to redesign the Galileo mission for LGA use only, the telecommunications link architecture was redesigned. The current architecture only supported 10 bps at Jupiter which was less than 1/10,000th of the 134 kilobits per second (Kbps) required. Since modifications to the spacecraft's hardware to boost the transmit power was not possible, receiving capability of Earth's ground stations and developing a more efficient data and telecommunications architecture was the primary focus of the needed upgrades. Arraying the DSN antennas increased the rate by a factor of 2.5, and modifications to the receivers and telecommunications link parameters, improving encoding and onboard data compression further increased the downlink from 10 bps to 4.5Kbps. Since these improvements were insufficient to bring down all science data objectives, the SOFS team negotiated with the science team to prioritize science goals, develop new science plans, and periodically update spacecraft FSW to increase data efficiency. Also, as a backup to the downlinked data, the onboard Data Memory Subsystem (DMS) tape recorder was utilized during selected high activity periods (Nilsen & Jansma, 2011).

Fig. 16. Galileo's Undeployed HGA.

5. Contending with mission difficulties

5.1 Mars exploration rover wheel failures

The Mars Exploration Rover (MER) mission is an ongoing, scientific undertaking involving two golf cart-sized robotic rovers. This mission is part of NASA's Mars Exploration Program, which includes two previous Viking program landers (1976) and the Mars Pathfinder probe (1997). The six-wheeled MER robotic vehicles, Spirit and Opportunity, landed in 2004 to explore the Martian surface and its geology. The mission's primary objective is to search for and characterize a wide range of rocks and soils that hold clues to past water activity on Mars. Originally a three-month mission, the MER mission was extended to present day. To date, much evidence has been collected to indicate that Mars was once a wetter and warmer place than has previously been determined.

Fig. 17. Mars Exploration Rover.

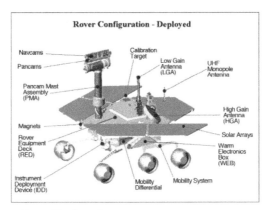

Fig. 18. Rover Configuration - Deployed.

During the mission, one of Spirit's six wheels stopped working. Its right-front wheel became a concern once before, when it began drawing unusually high current five months after the January 2004 landing. The SOFS team decided to drive Spirit backwards, which redistributed its lubricant and actually returned the wheel to normal operation. However, during the 779th Martian day, the motor that rotates that same wheel ceased working. One possibility considered by the SOFS team was that the motor's brushes or contacts that deliver power to the rotating part of the motor had lost contact. As a result of dragging Spirit's right front wheel, it cut a furrow in the Martian soil, revealing the layer beneath the surface, and in doing so, unearthed a material which significantly changed our thinking about Mars. Spirit found the evidence for a hydrothermal system, not only proving the existence of liquid water on Mars, but that there were energy sources coincident with that of liquid water, revealing the potential for support of an ecosystem.

Rover Spirit became trapped in soft sand in 2009, and eventually ceased communicating with Earth in March 2010. Nearly seven years after launch, Rover Opportunity is still healthy, although the SOFS team has been driving this vehicle backward for the last two years in order to spread wear more evenly within its gear mechanisms (Callas, 2006).

5.2 Voyager interstellar mission RTLT

The Voyager Spacecraft Program consists of two scientific probes; Voyager 1 and Voyager 2. Both were launched in 1977 to take advantage of the favorable planetary alignment of the outer planets. Although officially designated to study just Jupiter and Saturn, the probes were able to continue their mission into the outer solar system, and as of June 2011, have exited the solar system and currently reside within the Heliosheath (region between the Termination Shock and the Heliopause). Voyager 1 is currently the farthest human-made object from Earth; as of July 2011:

Fig. 19. Voyager Spacecraft.

	Distance from Sun	Round-trip Light Time
Voyager 1	17.59 billion km (117.6 AU)	32h 23m 55s
Voyager 2	14.33 billion km (95.8 AU)	26h 19m 38s

Table 3. Voyager Distance / RTLT from the Sun.

On December 10, 2007, instruments onboard Voyager 2 sent data back to Earth indicating that the solar system is asymmetrical; the Heliopause remains an unknown distance ahead.

The number of real-time commands must be kept to a minimum for spacecraft missions that must endure long RTLTs. For the Voyager spacecraft, the SOFS team performs commanding primarily by uplinked sequences containing several instructions. Since scheduled DSN antenna time coverage (~9hrs) is too short to wait for real-time verification of these commands, the team utilizes an alternative method to verify command success: the command sequence is transmitted by the DSN antenna (ground station) to that point in space where the spacecraft *will be* a OWLT (One Way Light Time) hence. For the Voyager 1 in July 2011, this OWLT was approximately 16.2hrs, respectively; downlink telemetry is then transmitted from the spacecraft and received by the DSN antenna from the point in space where the spacecraft *was* OWLT ago (Medina, Sedlacko, & Angrum, 2010).

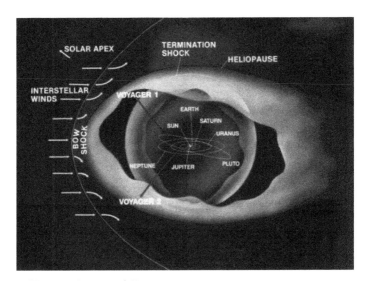

Fig. 20. Current Voyager Spacecraft Position.

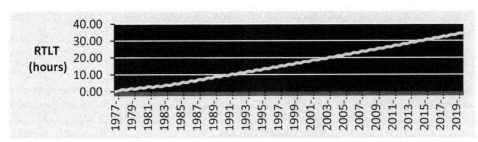

Fig. 21. Voyager 1 RTLT vs. Time.

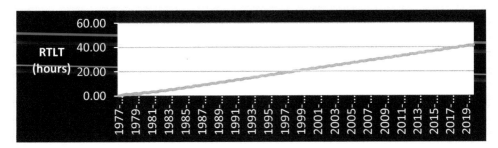

Fig. 22. Voyager 2 RTLT vs. Time.

A total time allotment of 32hrs, 23min, and 55sec plus sequence execution time, FSW processing time, and command execution time is required in order to verify each command sequence (i.e. more than one DSN antenna pass coverage is required to uplink and verify every command/command sequence; see Figure 23).

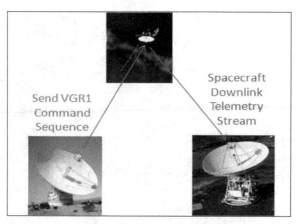

Fig. 23. Voyager DSN Strategy.

SOFS teams managing missions with long RTLTs such as the Voyager spacecraft must minimize real-time commanding. When commands are sent, typically the team must verify these commands the following day (or two). To date, both Voyager spacecraft have adequate electrical power and propellant margin to maintain systems and attitude control until around 2025, at which time, science data return and spacecraft operations will cease.

6. Lessons learned

Although numerous precautionary measures are implemented into JPL robotic spacecraft missions to preclude faults and prevent failures, many unforeseen problems can occur throughout its journey. "Lessons learned" documentation captured from previously flown spacecraft can be of great help when designing future missions. For the most part, autonomous FP algorithms are based upon past flight experience, but new mission destinations can present challenges never before encountered by spacecraft.

Overall, for spacecraft to function properly without significant risk or degradation to the mission and its objectives, autonomous FP must be implemented to ensure that detection and resolution of fault occurrences are dealt with properly so that the spacecraft may preserve its overall health and provide a system with adequate diagnostic capabilities. This effort requires that subsystems are characterized accurately. The approval of Cassini's pre-launch PMS regulator waivers is a good example of a mistake in ruling out the possibility of malfunction based upon surmised flight experience, without supporting test data (from Galileo) for adequate evaluation. Unfortunately, the enhancements to FSW and hardware boosted confidence in the upgraded design changes and drove Cassini's FP design strategy as well as its mission profile, as its most critical maneuver relied solely upon the successful initiation of the 30-day Pre-SOI Burn Characterization Task. Yet even under these circumstances, FP modifications and additions were successfully designed and uplinked to the spacecraft to preserve the mission, it's three tour phases, and safeguard its science data collection objectives since designers provided for the possibility of extra FP slots in FSW.

In any case, the experience gained though Cassini's leaky regulator problem, and lessons learned in many other JPL spacecraft missions, has demonstrated that these types of unexpected failures can be resolved though re-evaluation and implementation of new FSW/FP in-flight; an endeavor which is possible during spacecraft missions if enough time is available. Flight experience has also taught us that the development of post-FP response recovery procedures which contain pre-defined actions for the SOFS team to follow greatly reduces post-fault recovery time and accuracy in diagnosing faults. New strategies such as the "planet-flyby gravitational slingshot" concept developed for the Galileo mission provide innovative ideas which may be utilized on upcoming spacecraft designs; in this case, for boosting the heavy, two-story sized Cassini vehicle into deep space, thereby reducing propellant requirements by as substantial margin.

7. Acknowledgments

This research was carried out at the Jet Propulsion Laboratory, California Institute of Technology, under a contract with the National Aeronautics and Space Administration.

8. References

Allestad, D. & Standley, S. (2006.) *Cassini Orbiter Operations Lessons Learned for the Huygens Probe Mission,* Dated 2006.

Ball Aerospace & Technologies Corp. (2001). *CLOUDSAT: Automated Fault Protection Implementation, System Engineering Report,* Dated 2001.

Barber, T. (2002). *Initial Cassini Propulsion System in-Flight Characterization,* AIAA 2002-4152. Dated 2002.

Callas, J. (2006). *Mars Rovers Get New Manger During Challenging Period,* NASA News (NASA Home > Mission Sections > Mars > Missions > Mer), Dated 2006

Haemmerle , V. & Gerhard , J. (Undated). *Cassini Camera Contamination Anomaly: Experiences and Lessons Learned,* AIAA Paper.

Ingham, M., Mishkin, A., Rasmussen, R., & Starbird, T. (2004). *Application of State Analysis and Goal-Based Operations to a MER Mission Scenario,* Jet Propulsion Laboratory/California Institute of Technology. Dated 2004.

Jet Propulsion Laboratory/California Institute of Technology & Lockheed-Martin. (1997). *STARDUST Spacecraft Fault Protection Review.* Dated 1997.

Jet Propulsion Laboratory/California Institute of Technology. (2005). Technology Section: *Mars Reconnaissance Orbiter, Ulysses, Topex/Poseidon, Mars Global Surveyor, Multi-angle Imaging SpectroRadiometer, Deep Impact, Galileo, Stardust, Voyager I, Voyager II.* Home page. http://www.jpl.nasa.gov/missions/

Leeds, M., Eberhardt N., & Berry, R. (1996). *Development of the Cassini Spacecraft Propulsion System,* AIAA Paper 96-2864. Dated 1996.

Medina, E., Sedlacko, D. & Angrum, A. (2010). NASA new article, *Voyager's Thirty Year Plan,* Jet Propulsion Laboratory Dated October 18, 2010.

Millard, J. & Somawardhana, R. (2009). *Cassini Main Engine Assembly Cover Flight Management and Performance,* Jet Propulsion Laboratory/California Institute of Technology. Dated 2009.

Morgan, P. (2011) *Robotic Spacecraft Health Management,* Textbook: 'System Health Management: with Aerospace Applications'; Chapter 34, Dated 2011.

Morgan, P. (2010). *Cassini Spacecraft's In-Flight Fault Protection Redesign for Unexpected Regulator Malfunction,* Jet Propulsion Laboratory/California Institute of Technology, proceedings from IEEE Conference Big Sky, Montana. Dated March 2010.

Morgan, P. (2005). *Fault Protection Techniques in JPL Spacecraft,* proceedings from First International Forum on Integrated System Health Engineering and Management in Aerospace (ISHEM), Napa, California, Dated November 2005.

Narvaez, P. (2002). *The Magnetostatic Cleanliness Program for the Cassini Spacecraft,* Jet Propulsion Laboratory/California Institute of Technology. Dated 2002.

Nilsen, E. & Jansma , T. (2011). NASA new article, *"Galileo's Rocky Road to Jupiter,"* NASA News (NASA Home > Office > OCE > APPEL > Ask > Issues > 42), Dated May 23, 2011.

Smith, J. & The Planetary Society. (2009). *"Designing the Cassini Tour,"* Home Page: http://www.planetary.org/blog/article/00001979/ Dated 2009.

11

Autonomous Terrain Classification for Planetary Rover

Koki Fujita
Kyushu University
Japan

1. Introduction

In order to improve autonomous mobility of planetary rover, many works have recently focused on non-geometric features of surrounding terrain such as color, texture, and wheel-soil interaction mechanics (Dima et al., 2004; Halatci et al., 2007; Helmick et al., 2009; Ishigami et al., 2007). To tackle with the issue, most of them propose to utilize on-board sensors such as multi-spectral imagers, CCD cameras, laser range sensor, and accelerometer.

This study aims at classifying textures and physical properties of homogeneously-distributed terrain which are originated from the sizes of soil particles as well as mechanical interaction properties between rover body and soil.

As for the imaging sensors involved in the past works, while they discretly utilize image data, this work proposes to utilize whole of the motion image sequence taking terrain surface from rover on-board camera. Unlike the conventional techniques to classify terrain surfaces based on single or stereo camera images, the proposed method improves discrimination ability for visual salience and has possibility to remotely estimate properties of dynamic interaction between rover body (wheels) and terrain surface, such as relative velocity, slippage, and sinkage.

Given constant linear motion of camera, and homogeneous and isotropic properties of terrain texture, motion image sequence can be reduced to a set of parameters of the Dynamic Texture model (Saisan et al., 2001). The estimated parameters contain unique properties not only with visual salience in terrain surface but also with dynamics in camera (or vehicle) motion and terra-mechanics associated with surrounding terrain.

Aiming at validating the concept to classify terrain image sequences based on the Dynamic Texture model, this work shows experimental results for different types of soils and translational motions of camera by using a testbed. Results of a cross validation test and a receiver operating characteristic (ROC) analysis shows feasibility of the proposed method, and issues to be improved in future work.

2. Overview of the proposed method

In this work, a terrain classification method is proposed as an online estimation scheme installed for planetary rover. The schematic view of the proposed method is shown in Fig. 1.

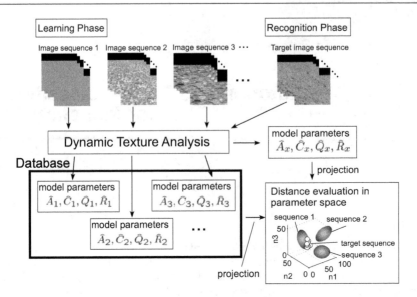

Fig. 1. Overview of terrain classification scheme utilizing Dynamic Texture.

As shown in the figure, the shceme is divided into two phases called "Learning Phase" and "Recognition Phase". Each phase is briefly described as follows.

Learning Phase:

1. Acquire video sequences for various types of terrains (e.g. fine regolith, sand, gravel, etc.) taken from view points in the vehicle's steady-state motion.
2. Estimate the parameters of the Dynamic Texture model.
3. Construct a database of the estimated parameter sets for all the different types of terrain sequences.

Recognition Phase:

1. Acquire a target image sequence.
2. Estimate the parameters of the Dynamic Texture model.
3. Compute the distances between the dynamical system model for the target sequence and the ones registered in the database.
4. Classifying the target image sequence as the one closest to the terrain types in the database.

3. Dynamic texture model

Given constant linear motion of camera mounted on the vehicle and homogeneous and isotropic properties of the terrain texture, the motion image sequence captured from the camera can be reduced to a set of parameters in a linear dynamical system model as follows:

$$\begin{cases} x(k+1) = Ax(k) + v(k), \ v(k) \sim \mathcal{N}(0,Q); \ x(0) = x_0, \\ y(k) = Cx(k) + w(k), \qquad w(k) \sim \mathcal{N}(0,R), \end{cases} \tag{1}$$

where $k = 0, 1, 2, \cdots$ is the discrete time instant, $y(k) \in \mathbb{R}^m$ is a vector of measured pixel brightness values in the k-th image frame, m equals the number of pixels in an image frame, $x(k) \in \mathbb{R}^n$ is an n-dimensional state vector, and $v(k) \in \mathbb{R}^n$ and $w(k) \in \mathbb{R}^m$ are white Gaussian random vectors. As seen in the equations, the above dynamical system is characterized by the parameter matrices $A \in \mathbb{R}^{n \times n}$, $C \in \mathbb{R}^{m \times n}$, $Q \in \mathbb{R}^{n \times n}$, and $R \in \mathbb{R}^{m \times m}$.

Whereas these parameters can be estimated using a system identification theory such as N4SID (Overschee & Moor, 1994), the computational load tends to be large for video sequences which contain substantial data. Previous work on Dynamic Textures (Saisan et al., 2001) proposed to apply a suboptimal estimation algorithm utilizing a principal component analysis (PCA-ID) in order to decrease the dimensionality of the state-space model. However, with this PCA-ID algorithm, not only the accuracy issue on the estimated dynamical system model still remains, but also computational load could be serious for relatively large size of the image frames due to the algorithm of PCA.

In this work, instead of the conventional PCA-ID algorithm, a new algorithm based on the components of 2-dimensional discrete cosine transform (2D-DCT) and a system identification algorithm, N4SID. The proposed method has an advantage in that optimal solution for the dynamical model is obtained within less computational time.

The proposed method contains two steps as follows:

STEP1: Original $M \times N$ pixel data from the terrain image sequence, $f_{i,j}$ ($i = 1, 2, \cdots, M$, $j = 1, 2, \cdots, N$) are transformed into $F_{k,l}$ ($k = 1, 2, \cdots, M$, $l = 1, 2, \cdots, N$) such that

$$F_{k,l} = C_k C_l \sum_{i=1}^{M} \sum_{j=1}^{N} f_{i,j} \cos\left(\frac{(2i-1)k\pi}{2M}\right) \cos\left(\frac{(2j-1)l\pi}{2N}\right)$$

(2)

$$\text{where} \quad C_{k \text{ or } l} = \begin{cases} 1/\sqrt{2}, & \text{if } k \text{ or } l = 1 \\ 1, & \text{else} \end{cases}$$

Since $F_{k,l}$ is obtained by a linear transformation from the original image data, their principal properties should be preserved in the output components for the lower dimensional spatial frequencies. Thus, among m ($= M \times N$) components of 2D-DCT output for the original image, only m_c ($= M_c \times N_c$, $m_c < m$) ones are applied to the N4SID algorithm. If $y_c(k)$ is defined as $[F_{1,1}(k), F_{1,2}(k), \cdots F_{M_c, N_c}(k)]^T \in \mathbb{R}^{m_c}$, the dynamical system model corresponding to Eq. (1) is described such that

$$\begin{cases} x_c(k+1) = A_c x_c(k) + v_c(k), \quad v_c(k) \sim \mathcal{N}(0, Q_c); \ x_c(0) = x_{c0}, \\ y_c(k) = C_c x_c(k) + w_c(k), \quad w_c(k) \sim \mathcal{N}(0, R_c), \end{cases}$$

(3)

where the subscript c denotes the vectors or the matrices for the low-dimensional 2-D DCT components.

STEP2: N4SID algorithm (Overschee & Moor, 1994) is applied to $y_c(k)$ ($k = 1, 2, \cdots, K$) in **STEP1**, and the linear dynamical system paraters such as A_c, C_c are computed for given order of the system n.

4. Recognition of dynamic texture model

Since the linear dynamical system models as shown in Eq. (1) are characterized by the parameter matrices A and C, they can be identified with column space of the extended observability matrix:

$$\mathcal{O}_\infty(M) = \left[C^T \ (CA)^T \ (CA^2)^T \ \cdots \right]^T. \tag{4}$$

For a large enough number n, the above extended observability matrix is approximated by the finite observability matrix:

$$\mathcal{O}_n(M) = \left[C^T \ (CA)^T \ (CA^2)^T \ \cdots (CA^{n-1})^T \right]^T. \tag{5}$$

In order to recognize different Dynamic Texture models, the follwoing three typical metrics can be introduced for measuring distances between the dynamical models in parameter space.

1. **Euclidean distance:** For the observability matrix of Eq. (5), a distance metric for models M_1 and M_2 can be defined as a simple but appropriate one to represent a difference in dynamical property as follows:

$$d_E(M_1, M_2) = \sqrt{\sum_{i=1}^{n}(\sigma_i(M_1) - \sigma_i(M_2))^2}, \tag{6}$$

   where $\sigma_i(M_1)$ and $\sigma_i(M_2)$ are the i-th order singular values of $\mathcal{O}_n(M_1)$ and $\mathcal{O}_n(M_2)$, respectively.

2. **Martin's distance:** Martin's distance (Martin, 2000) is a distance metric derived for a linear dynamical system model, ARMA model, which is equivalent to Eq. (1). It can also be applied to distinguish different Dynamic Texture models. If p principal angles $\theta_k \in [0, \pi/2]$ between the ranges of the matrices A and B are recursively defined for $k = 1, 2, \cdots, n$ as

$$\cos\theta_1 = \max_{x \in \mathbb{R}^p, \, y \in \mathbb{R}^q} \frac{|x^T A^T B y|}{||Ax||_2 ||By||_2} = \frac{|x_1^T A^T B y_1|}{||Ax_1||_2 ||By_1||_2},$$

$$\cos\theta_k = \max_{x \in \mathbb{R}^p, \, y \in \mathbb{R}^q} \frac{|x^T A^T B y|}{||Ax||_2 ||By||_2} = \frac{|x_k^T A^T B y_k|}{||Ax_k||_2 ||By_k||_2} \quad \text{for } k = 2, \cdots, q, \tag{7}$$

   subject to $x_i^T A^T A x = 0$ and $y_i^T B^T B y = 0$ for $i = 1, 2, \cdots, k-1$, the Martin's distance is derived as follows (De Cock & De Moor, 2000):

$$d_M(M_1, \ M_2) = \sqrt{\ln \prod_{i=1}^{n} \frac{1}{\cos^2 \theta_i}}. \tag{8}$$

3. **Kernel density function (KDF) on Stiefel manifold:** While the above two metrics are derived to directly measure the distance between two subspaces spanned by the column vectors of the observability matrices, distance metrics on special manifold such as Stiefel and Grassmann manifolds, on which the parameters of the dynamical system model lie have also been proposed (Turaga et al., 2008). In the previous work, a metric using a kernel density function based on a Procrustes representation for the distance metric is introduced.

The Stiefel manifold $V_{n,m}$ is a space whose points are n-frames in $\mathbb{R}^m$, and each point on the manifold can be represented as $m \times n$ matrix X such that $X^T X = I_n$, where I_n is $n \times n$ identity matrix. By singular value decomposition of $\mathcal{O}_n(M)$ such that

$$\mathcal{O}_n(M) = U\Sigma V^*, \tag{9}$$

the orthonormal matrix $U \in \mathbb{R}^{m \times n}$ $(U^T U = I_n)$ is regarded as a point on the Stiefel manifold retaining the column space property of the original observabiliry matrix. Although the Stiefel manifold is endowed with a Riemannian structure and a geodesic computation for distances between points on the manifold is possible, the other distance metric called a "Procrutes distance" is introduced to estimate a class conditional probability density in an ambient Euclidean space.

The Procrustes distance is defined for two matrices X_1 and X_2 on $V_{k,m}$ as follows:

$$d_V^2(X_1, X_2) = \min_{R>0} tr(X_1 - X_2 R)^T (X_1 - X_2 R)$$
$$= \min_{R>0} tr(R^T R - 2X_1^T X_2 R + I_k). \tag{10}$$

If R varies over the space $\mathbb{R}^{k \times k}$, the Procrustes distance is minimized at $R = X_1^T X_2$, so that $d_V^2(X_1, X_2)$ is equal to $tr(I_k - X_2^T X_1 X_1^T X_2)$. Also, the class conditional density for this Procrustes distance metric can be estimated by using the following function (Chikuse, 2003):

$$\hat{f}(X; P_s) = \frac{1}{n} C(P_s) \sum_{i=1}^{n} K(P_s^{-1/2}(I_n - X_{2,i}^T X_1 X_1^T X_{2,i}) P_s^{-1/2}), \tag{11}$$

where $X_{2,i}$ $(i = 1, \cdots, n)$ are the sample matrices on the Stiefel manifold from the same class of the model. $K(A)$ is the kernel function for a matrix A, P_s is $n \times n$ positive definite matrix as a smoothing parameter, and $C(P_s)$ is the normalizing factor selected so that the estimated kernel density integrate to unity.

In this paper, X_1 and $X_{2,i}$ are the matrices on the Stiefel manifold constructed by the models M_1 and M_2, respectively, and these matrices correspond to U derived from the singular value decomposition of the observability matrix $\mathcal{O}_n(M)$. As a kernel function to compute $\hat{f}(X; P_s)$, the exponential kernel $K(A) = \exp(-tr(A))$ is treated. Since the output of $\hat{f}(X; P_s)$ ranges between 0 and 1 and increases inversely with the distance between two models, the following function is defined as an actual distance metric:

$$d_K(M_1, M_2) = 1.0 - \hat{f}(X; P_s) \tag{12}$$

5. Recognition test for real image sequences

In order to validate the effectiveness of the proposed methods, an experiment was conducted by using a testbed as shown in Fig. 2. Real image sequences for four types of the terrain textures (magnesium lime, fine and coarse sand, and gravel) were obtained using a CCD camera (SONY XCD-V60CR). On this testbed, translational motions are given to the camera fixed on a wheeled structure. The wheel is driven by constant torque from a brushless DC motor, which gives averagely constant velocity to the CCD camera on flat surface. The experimental environment is shown in Table 1.

Real image sequences as shown in Fig. 3 were applied to the proposed methods. Each terrain sequence depicts different soil particles identically-distributed in the image frames. In order to

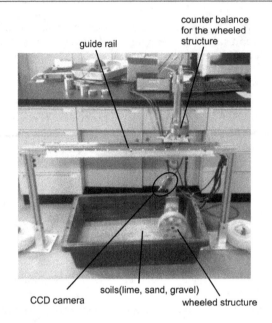

Fig. 2. Testbed for acquiring terrain image sequences.

Specification of CCD camera	Focal length: 8 mm Field of view: 34.0×25.6 deg CCD image resolution (original): 640×480 pixels Frame rate: 30 fps
Height of the camera	330 mm
Wheel diameter	181.7 mm
Velocity of the camera (mean value)	V_1: 17.4 mm/sec, V_2: 35.0 mm/sec, V_3: 53.5 mm/sec

Table 1. Experimental environment

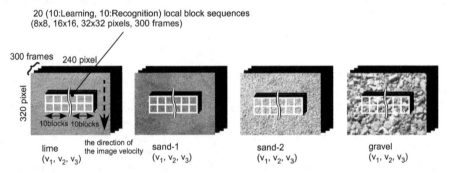

Fig. 3. Real image sequences applied to the proposed methods.

		Terrain type			
		lime (A)	sand-1 (B)	sand-2 (C)	gravel (D)
Image	v_1 (a)	Aa	Ba	Ca	Da
velocity	v_2 (b)	Ab	Bb	Cb	Db
	v_3 (c)	Ac	Bc	Cc	Dc

Table 2. Table of combination

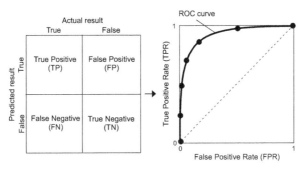

Fig. 4. A receiver operating characteristic (ROC) analysis.

see discriminative ability not only for terrain textures but also for rover translational motion, three constant torques were given to the DC motor for each type of terrain texture, which generated different velocities, V_1 to V_3 (as shown in Table 1). These velocities of the camera resulted in the different image velocity fields named as v_1, v_2, and v_3. The combination for all the experimental parameters with the terrain textures and the image velocities is shown in Table 2. As shown in Fig. 3, 20 local block images of 8×8 (CASE 1), 16×16 (CASE 2), and 32×32 (CASE 3) pixels were cropped from the original images, and the block image sequences consisting of 300 frames (for about 10sec) were applied to the proposed scheme.

In this study, considering sufficient accuracy for estimating the Dynamic Texture model by using N4SID or PCA-ID algorithm, the dimension of the finite observability matrix, n in Eq. (5) was fixed at 10. Also, the number of the 2D-DCT components in Eq. (3) was fixed such that $M_c = N_c = 8$ (i.e. $m_c = M_c \times N_c = 64$). As for the smoothing parameter in the KDF on the Stiefel manifold defined as Eq. (11), it was set such that $P_s = 100$ after trying to apply several values.

Recognition rate was evaluated through 2-fold cross validation test, that is, while half of the block image sequences were applied for the Learning Phase, the rest of the target sequences were for the Recognition Phase. The same process was repeated after exchanging the block image sequences for each phase. Note here that the block image sequences for the both phases were selected so that they never overlap with each other in the spatiotemporal domain as shown in Fig. 3.

The terrain image sequences were recognized using threshold values of each distance metric, which were coincident with the maximum distances among the same image sequences in the Learning Phase. Aiming at seeing sensitivity to the threshold values, a receiver operating characteristic (ROC) (Witten et al., 2011) analysis was conducted at the same time.

If the relation between predicted result and actual result for a discrimination threshold is shown as a cross tabulation in Fig.4, two types of the evaluation metrics, true positive rate (TPR) and false positive rate (FPR) are derived as follows:

$$\text{TPR} = N_{TP}/(N_{TP} + N_{FN}), \quad \text{FPR} = N_{FP}/(N_{FP} + N_{TN}), \tag{13}$$

where N_{TP}, N_{FN}, N_{FP}, and N_{TN} mean the numbers of the true positive, the false negative, the false positive, and the true negative, respectively. For these two operating characteristics (TPR and FPR) computed from various threshold values, ROC curve is plotted as shown in the right-hand side of Fig. 4.

For any classification problem, the false positive rate increases with the true positive rate. Since optimal classifier should obtain enough high true positive rate relative to the corresponding false positive rate, curve in ROC space is desirable to be skewed to upper left corner as shown in Fig. 4.

Results of the cross validation test are shown in the following figures and tables. Figs. 5 to 10 show correlation maps in which each grayscale block image implies the correlation between the learned sequences (aligned in columns) and the target sequences (aligned in rows) for the twelve parameter combinations. The grayscale level is computed from the mean values of the distances among twenty block image sequences, such that the darker image shows the shorter distance. To increase visibility of the block images, the grayscale levels are normalized for each map. The results of the PCA-ID algorithm are also shown to compare the proposed 2D-DCT+N4SID algorithm with the conventional one.

As shown in these figures, all the distance metrics show clear correlations for the same terrain textures labeled as A, B, C, and D. They are appeared as the darker block images along the diagonal line from the upper left corner to the lower right corner on each map. While in most of the correlation maps, the differences between the correlation strengths for different terrain textures are not necessarily clear except for the Euclidean distance computed from the proposed 2D-DCT+N4SID algorithm, which only shows the gradual changes of the grayscale levels for different terrain textures.

Tables 3 to 8 show results of the true positive rates defined in Eq. (13). In this study, two classes of recognized features named "dynamic texture class" and "static texture class" are focused on. While the static texture class is categorized only according to terrain types (i.e. A, B, C and D in Table 2), the dynamic texture class is categorized according to image velocity as well as to terrain type (i.e. Aa, Ab, $\cdots$, Dc in Table 2). In the tables, the results obtained for these two categories are shown. The results of the recognition rates for the conventional PCA-ID algorithm are also shown for comparison.

As shown in the tables, for the Euclidiean and the Martin's distances, the true positive rates are relatively high over 86.5% for the 2D-DCT+N4SID algorithm or 89.9% for the PCA-ID algorithm for the both feature classes. On the other hand, the KDF on the Stiefel manifold shows lower rates especially for the dynamic texture class, which results in at most 23.3% for the 2D-DCT+N4SID algorithm or 70.8% for the PCA-ID algorithm, and significantly decreasing with the increasing block image sizes. As for the test results, the effect of the block image size is only seen in those for the KDF on the Stiefel manifold. One of the issues of the KDF on the Stiefel manifold is considered that this metric needs enough sample matrices on the Stiefel manifold from the same image sequences to compute the kernel density function, which may not be satisfied for relatively large size of the image sequences. Acoording

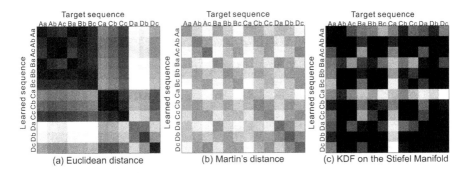

Fig. 5. Correlation map for CASE 1 (The proposed 2D-DCT+N4SID algorithm, 8×8 pixel-block seqeunces, the 1st test).

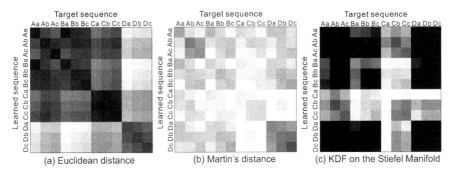

Fig. 6. Correlation map for CASE 2 (The proposed 2D-DCT+N4SID algorithm, 16×16 pixel-block seqeunces, the 1st test).

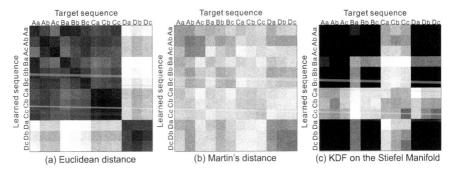

Fig. 7. Correlation map for CASE 3 (The proposed 2D-DCT+N4SID algorithm, 32×32 pixel-block seqeunces, the 1st test).

to the results of the true positive rates, although it seems that the conventional PCA-ID algorithm achieves better performance than the proposed algorithm does, a different view can be obtained from the following results of the ROC analysis.

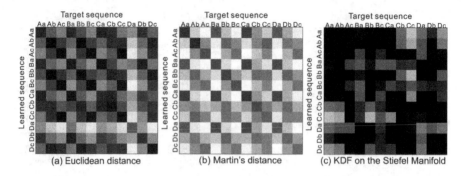

Fig. 8. Correlation map for CASE 1 (The conventional PCA-ID algorithm, 8×8 pixel-block seqeunces, the 1st test).

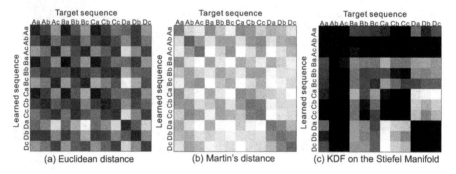

Fig. 9. Correlation map for CASE 2 (The conventional PCA-ID algorithm, 16×16 pixel-block seqeunces, the 1st test).

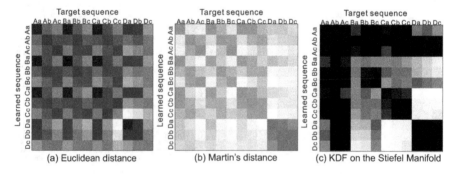

Fig. 10. Correlation map for CASE 3 (The conventional PCA-ID algorithm, 32×32 pixel-block seqeunces, the 1st test).

Some ROC curves for the same experimental results are shown in Figs. 11 to 16. They are plotted for 30 threshold values equally sampled between the maximum and the minimum values for the learned sequences. For all the results, while the plots close to the lower left

Dynamic texture class	1st	2nd	mean	Static texture class	1st	2nd	mean
Euclidean dist.	92.0%	83.5%	87.8%	Euclidean dist.	94.5%	98.7%	96.6%
Martin's dist.	94.3%	98.5%	96.4%	Martin's dist.	99.8%	99.6%	99.7%
KDF on the Stiefel manifold	19.2%	27.5%	23.3%	KDF on the Stiefel manifold	90.3%	93.3%	91.8%

Table 3. True positive rates for CASE 1 (The proposed 2D-DCT+N4SID algorithm, 8×8 pixel-block sequences).

Dynamic texture class	1st	2nd	mean	Static texture class	1st	2nd	mean
Euclidean dist.	90.5%	82.4%	86.5%	Euclidean dist.	98.9%	97.3%	98.1%
Martin's dist.	96.4%	95.8%	96.1%	Martin's dist.	99.9%	99.7%	99.8%
KDF on the Stiefel manifold	5.0%	9.2%	7.1%	KDF on the Stiefel manifold	83.3%	87.5%	85.4%

Table 4. True positive rates for CASE 2 (The proposed 2D-DCT+N4SID algorithm, 16×16 pixel-block sequences).

Dynamic texture class	1st	2nd	mean	Static texture class	1st	2nd	mean
Euclidean dist.	85.8%	88.8%	87.3%	Euclidean dist.	97.8%	100.0%	98.9%
Martin's dist.	96.9%	98.8%	97.8%	Martin's dist.	99.6%	99.9%	99.7%
KDF on the Stiefel manifold	5.8%	10.0%	7.9%	KDF on the Stiefel manifold	85.8%	84.7%	85.3%

Table 5. True positive rates for CASE 3 (The proposed 2D-DCT+N4SID algorithm, 32×32 pixel-block sequences).

corner, $(\text{FPR}, \text{TPR}) = (0,0)$ show the ones for the minimum threshold values, the plots on upper right portions show the ones for the maximum threshold values.

Although all the ROC plots start from the lower left corners, they don't necessarily reach to the upper right corners. Most of the plots for the KDF on the Stiefel manifold end in the middle of the ROC spaces, and especially for the dynamic texture class, they end up very low recognition rates, which is similarly seen in the previous tables of the true positive rate. On the other hand, the proposed 2D-DCT+N4SID algorithm depicts more desirable curves for the Euclidean distance, while the highest positive rates don't necessarily exceed those for the conventional PCA-ID algorithm.

Comparing between the two feature classes, the dynamic texture class and the static texture class, the recognition rates for the latter class show higher rates for a certain threshold values of each distance metric as shown in Tables 3 to 8. However, their recognition performances

Dynamic texture class				Static texture class			
	1st	2nd	mean		1st	2nd	mean
Euclidean dist.	88.3%	91.5%	89.9%	Euclidean dist.	95.9%	99.7%	97.8%
Martin's dist.	95.9%	96.8%	96.3%	Martin's dist.	99.7%	99.8%	99.8%
KDF on the Stiefel manifold	60.8%	80.8%	70.8%	KDF on the Stiefel manifold	100.0%	95.3%	97.6%

Table 6. True positive rates for CASE 1 (The conventional PCA-ID algorithm, 8×8 pixel-block sequences).

Dynamic texture class				Static texture class			
	1st	2nd	mean		1st	2nd	mean
Euclidean dist.	93.0%	96.1%	94.5%	Euclidean dist.	99.3%	99.7%	99.5%
Martin's dist.	94.0%	93.1%	93.5%	Martin's dist.	99.9%	99.8%	99.9%
KDF on the Stiefel manifold	8.3%	4.2%	6.3%	KDF on the Stiefel manifold	93.6%	91.1%	92.4%

Table 7. True positive rates for CASE 2 (The conventional PCA-ID algorithm, 16×16 pixel-block sequences).

Dynamic texture class				Static texture class			
	1st	2nd	mean		1st	2nd	mean
Euclidean dist.	91.1%	89.8%	90.4%	Euclidean dist.	99.5%	99.6%	99.5%
Martin's dist.	87.3%	94.9%	91.1%	Martin's dist.	99.1%	99.4%	99.3%
KDF on the Stiefel manifold	0.0%	0.8%	0.4%	KDF on the Stiefel manifold	80.0%	83.6%	81.8%

Table 8. True positive rates for CASE 3 (The conventional PCA-ID algorithm, 32×32 pixel-block sequences).

in the ROC space are not necessarily better than for the dynamic texture class as depicted in Figs. 11 to 16. From these results, the three distance metrics introduced in this study are perceived as the ones intimately involved with the dynamical properties not only with the static visual salience.

The merit of the proposed algorithm is also seen in the computational time for the Dynamic Texture model learning. Table 9 shows the computational time for each process in the recognition test using a PC (CPU: Intel Core i7-640LM, RAM: 8MB). This result clearly shows that the conventional PCA-based algorithm becomes ineffective with the increasing size of the block image, and the computational time for the proposed algorithm does not depend on the size of the block images so much as the conventional algorithm does.

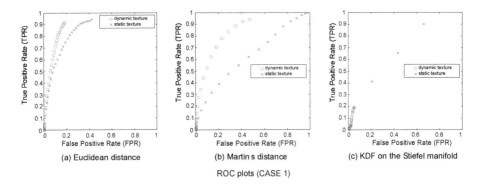

Fig. 11. ROC plots for CASE 1 (The proposed 2D-DCT+N4SID algorithm, 8×8 pixel-block sequences, the 1st test).

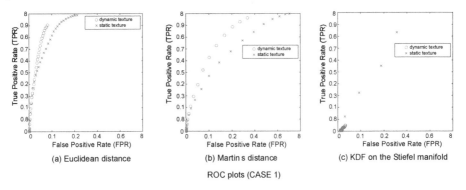

Fig. 12. ROC plots for CASE 2 (The proposed 2D-DCT+N4SID algorithm, 16×16 pixel-block sequences, the 1st test).

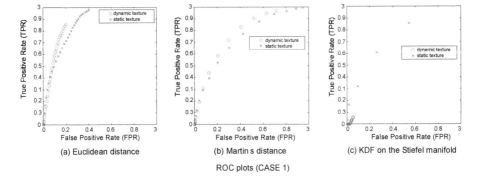

Fig. 13. ROC plots for CASE 3 (The proposed 2D-DCT+N4SID algorithm, 32×32 pixel-block sequences, the 1st test).

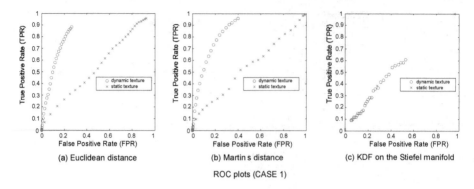

(a) Euclidean distance   (b) Martin s distance   (c) KDF on the Stiefel manifold

ROC plots (CASE 1)

Fig. 14. ROC plots for CASE 1 (The conventional PCA-ID algorithm, 8×8 pixel-block sequences, the 1st test).

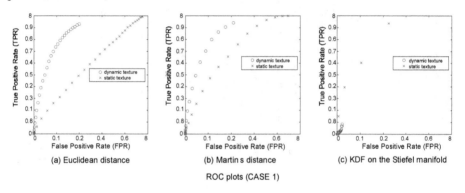

(a) Euclidean distance   (b) Martin s distance   (c) KDF on the Stiefel manifold

ROC plots (CASE 1)

Fig. 15. ROC plots for CASE 2 (The conventional PCA-ID algorithm, 16×16 pixel-block sequences, the 1st test).

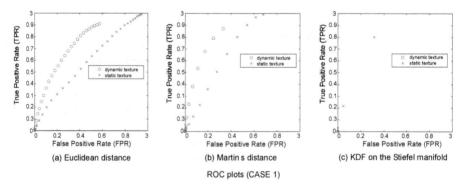

(a) Euclidean distance   (b) Martin s distance   (c) KDF on the Stiefel manifold

ROC plots (CASE 1)

Fig. 16. ROC plots for CASE 3 (The conventional PCA-ID algorithm, 32×32 pixel-block sequences, the 1st test).

Computational time (2D-DCT+N4SID)	Computational time (PCA-ID)
CASE 1 (8 × 8 pixels): 35.2 sec	CASE 1 (8 × 8 pixels): 30.0 sec
CASE 2 (16 × 16 pixels): 36.2 sec	CASE 2 (16 × 16 pixels): 131.8 sec
CASE 3 (32 × 32 pixels): 38.1 sec	CASE 3 (32 × 32 pixels): 1184.2 sec

Table 9. The computational times for the recognition test.

It should be discussed in future work which metric is more appropriate to discriminate more various types of the terrain texures and dynamical properties caused by rover motion, considering the validity of the model estimation algorithms.

6. Conclusion

This paper proposes a novel terrain classification method for planetary rover utilizing Dynamic Texture. The recognition rates computed from several distance measures for the estimated Dynamic Texture models were evaluated through the experiments using a testbed. According to the experimental results, some distance metrics show relatively high true positive rates to discriminate not only terrain textures but also rover translational motion. Also, one of the metrics computed from the proposed model estimation algorithm shows more desirable characteristic in the ROC space.

7. Future work

In future works, ditance metric suitable to distinguish various types of terrain textures as well as dynamical properties of rover such as translational velocity, slippage, and sinkage is going to be discussed in detail. At the same time, the validity of the model estimation algorithms based on a linear dynamical system model is further evaluated.

8. Acknowledgement

The author would like to thank Dr. N. Ichimura in National Institute of Advanced Industrial Science and Technology (AIST) in Japan for his kind interest and valuable comments. The author also thanks the Japan Society for the Promotion of Science for supporting this study as a part of Grants-in-Aid for Scientific Research (C) (No. 22500176).

9. References

Chikuse, Y. (2003). *Statistics on special manifolds, Lecture Notes in Statistics*, Springer, New York, NY.

De Cock, K. & De Moor, B. (2000). Subspace angles between linear stochastic models, *Proceedings of the 39th IEEE Conference on Decision and Control (CDC2000), Vol.2*, pp. 1561–1566.

Dima, C., Vandapel, D. & Hebert, M. (2004). Classifier fusion for outdoor obstacle detection, *Proceedings of the IEEE International Conference on Robotics and Automation (ICRA)*, pp. 665–671.

Halatci, I., Brooks, C. & Iagnemma, K. (2007). Terrain classification and classifier fusion for planetary exploration rovers, *Proceedings of the IEEE Aerospace Conference*, pp. 1–11.

Helmick, D., Angelova, A. & Matthies, L. (2009). Terrain adaptive navigation for planetary rovers, *Journal of Field Robotics* Vol. 26(No. 4): 391–410.

Ishigami, G., Nagatani, K. & Yoshida, K. (2007). Path planning for planetary exploration rovers and its evaluation based on wheel slip dynamics, *Proceedings of the IEEE International Conference on Robotics and Automation (ICRA)*, pp. 2361–2366.

Martin, R. (2000). A metric for ARMA processes, *IEEE Transactions on Signal Processing* Vol. 48(No. 4): 1164–1170.

Overschee, P. & Moor, B. (1994). N4SID: Subspace algorithms for the identification of combined deterministic-stochastic systems, *Automatica*, Vol.30(Issue 1): 75–93.

Saisan, P., Doretto, G., Wu, Y. & Soatto, S. (2001). Dynamic texture recognition, *Proceedings of the IEEE Conference on Computer Vision and Pattern Recognition (CVPR'01)*, Vol.2, pp. 58–63.

Turaga, P., Veeraraghavan, A. & Chellappa, R. (2008). Statistical analysis on Stiefel and Grassmann Manifolds with applications in Computer Vision, *Proceedings of the IEEE Conference on Computer Vision and Pattern Recognition (CVPR'08)*, pp. 1–8.

Witten, I., Frank, E. & Hall, M. (2011). *Data Mining: Practical Machine Learning Tools and Techniques 3rd ed.*, Morgan Kaufmann, Burlington, MA.

Permissions

The contributors of this book come from diverse backgrounds, making this book a truly international effort. This book will bring forth new frontiers with its revolutionizing research information and detailed analysis of the nascent developments around the world.

We would like to thank Dr. Rushi Ghadawala, for lending his expertise to make the book truly unique. He has played a crucial role in the development of this book. Without his invaluable contribution this book wouldn't have been possible. He has made vital efforts to compile up to date information on the varied aspects of this subject to make this book a valuable addition to the collection of many professionals and students.

This book was conceptualized with the vision of imparting up-to-date information and advanced data in this field. To ensure the same, a matchless editorial board was set up. Every individual on the board went through rigorous rounds of assessment to prove their worth. After which they invested a large part of their time researching and compiling the most relevant data for our readers. Conferences and sessions were held from time to time between the editorial board and the contributing authors to present the data in the most comprehensible form. The editorial team has worked tirelessly to provide valuable and valid information to help people across the globe.

Every chapter published in this book has been scrutinized by our experts. Their significance has been extensively debated. The topics covered herein carry significant findings which will fuel the growth of the discipline. They may even be implemented as practical applications or may be referred to as a beginning point for another development. Chapters in this book were first published by InTech; hereby published with permission under the Creative Commons Attribution License or equivalent.

The editorial board has been involved in producing this book since its inception. They have spent rigorous hours researching and exploring the diverse topics which have resulted in the successful publishing of this book. They have passed on their knowledge of decades through this book. To expedite this challenging task, the publisher supported the team at every step. A small team of assistant editors was also appointed to further simplify the editing procedure and attain best results for the readers.

Our editorial team has been hand-picked from every corner of the world. Their multi-ethnicity adds dynamic inputs to the discussions which result in innovative outcomes. These outcomes are then further discussed with the researchers and contributors who give their valuable feedback and opinion regarding the same. The feedback is then collaborated with the researches and they are edited in a comprehensive manner to aid the understanding of the subject.

Apart from the editorial board, the designing team has also invested a significant amount of their time in understanding the subject and creating the most relevant covers. They scrutinized every image to scout for the most suitable representation of the subject and create an appropriate cover for the book.

The publishing team has been involved in this book since its early stages. They were actively engaged in every process, be it collecting the data, connecting with the contributors or procuring relevant information. The team has been an ardent support to the editorial, designing and production team. Their endless efforts to recruit the best for this project, has resulted in the accomplishment of this book. They are a veteran in the field of academics and their pool of knowledge is as vast as their experience in printing. Their expertise and guidance has proved useful at every step. Their uncompromising quality standards have made this book an exceptional effort. Their encouragement from time to time has been an inspiration for everyone.

The publisher and the editorial board hope that this book will prove to be a valuable piece of knowledge for researchers, students, practitioners and scholars across the globe.

List of Contributors

T. A. Sands
Columbia University, USA

Daniel Condurache
"Gheorghe Asachi" Technical University of IASI, Romania

Cheng Ling Kuo
Department of Physics, National Cheng Kung University, Taiwan

Pan Xiaogang, Wang Jiongqi and Zhou Haiyin
National University of Defence Technology, China

Min Hu and Hong Yao
Academy of Equipment, Beijing, China

Guoqiang Zeng
College of Aerospace and Material Engineering, National University of Defense Technology, Changsha, China

Leonardo M. Reyneri, Danilo Roascio, Claudio Passerone, Stefano Iannone, Juan Carlos de los Rios, Giorgio Capovilla and Jairo Alberto Hurtado
Department of Electronics and Telecommunications, Politecnico di Torino, Torino, Italy

Antonio Martínez-Álvarez
Department of Computer Technology, University of Alicante, Alicante, Spain

Fu-Kuang Yeh
Department of Computer Science and Information Engineering, Chung Chou University of Science and Technology, Changhua, Taiwan, R.O.C

Erkan Abdulhamitbilal
ISTAVIA Engineering, Istanbul, Turkey

Elbrous M. Jafarov
Aeronautics and Astronautics Engineering, Istanbul Technical University, Istanbul, Turkey

Xiangli Li
Dalian University of Technology, China

Paula S. Morgan
Jet Propulsion Laboratory, California Institute of Technology, USA

Koki Fujita
Kyushu University, Japan

Printed in the USA
CPSIA information can be obtained
at www.ICGtesting.com
JSHW011448221024
72173JS00004B/989

9 781632 401359